Chemical Product Design

Second edition

The chemical industry is changing, going beyond commodity chemicals to a palette of higher value added products. This ground-breaking book, now revised and expanded, documents this change and shows how to meet the challenges implied. Presenting a four-step design process – needs, ideas, selection, manufacture – the authors supply readers with a simple design template that can be applied to a wide variety of products. Four new chapters on commodities, devices, molecules, and microstructures show how this template can be applied to products including oxygen for emphysema patients, pharmaceuticals like taxol, dietary supplements like lutein, and beverages that are more satisfying. For different groups of products the authors supply both strategies for design and summaries of relevant science. Economic analysis is expanded, emphasizing the importance of speed-to-market, selling ideas to investors, and an expectation of limited time in the market. Extra examples, homework problems, and a solutions manual are available.

E. L. Cussler is Distinguished Institute Professor at the University of Minnesota. The author of the text *Diffusion*, he has received the Colburn and Lewis awards from the American Institute of Chemical Engineers and is a member of the National Academy of Engineering.

G. D. Moggridge is a Senior Lecturer at the University of Cambridge. He has taught chemical product design since 1998, receiving the Entec Medal and Frank Morton Prize from the Institution of Chemical Engineers and a Pilkington Teaching Prize.

Chemical Product Design

Second edition

E. L. Cussler
University of Minnesota

G. D. Moggridge
University of Cambridge

CAMBRIDGE
UNIVERSITY PRESS

CAMBRIDGE UNIVERSITY PRESS
Cambridge, New York, Melbourne, Madrid, Cape Town,
Singapore, São Paulo, Delhi, Tokyo, Mexico City

Cambridge University Press
The Edinburgh Building, Cambridge CB2 8RU, UK

Published in the United States of America by Cambridge University Press, New York

www.cambridge.org
Information on this title: www.cambridge.org/9780521168229

First published 2001
Second edition 2011

Printed in the United Kingdom at the University Press, Cambridge

A catalogue record for this publication is available from the British Library

ISBN 978-0-521-16822-9 Paperback

Additional resources for this publication at www.cambridge.org/9780521168229

10062456 X

For Betsy and Liana, who tolerated numbered wine glasses.

Revised
Sunscreen
Specifications

Barbie

A PERFECT COFFEE CUP

SCRUBBING
NITROGEN FROM
NATURAL GAS

Lotus Effect

pigments for farmed salmon

Silver
Bullets for
Zebra
Mussels

The
sensory
experience
of
drink
in
the
throat

Home Oxygen Supply

The Case of

muffler design

pacemakers

for pets

High Level
Radioactive
Waste

Dolls

Osmotic Pumps

Pure Water
for
Bangladesh

Prozac
Synthesis

ATHLETIC SHOES

chewing gum
flavor

Methanol
from
Biomass

snack food
crispness

Removal of SO_2 at the
DRAX Power Station

the Swedish Meatballs

A BREATHABLE
BOTTLE CAP

Fuel Cell Catalysts

Taxol
Synthesis

Liquid Bandages

Contents

List of Symbols

a	surface area per volume
A	area
b	constant
B	bottoms in distillation (Section 6.4)
c	total concentration
c_i	concentration of species "i", in either moles per volume or mass per volume
$\tilde{C}_p,\ \hat{C}_p$	molar and specific heat capacities respectively at constant pressure
$\tilde{C}_v,\ \hat{C}_v$	molar and specific heat capacities respectively at constant volume
d	diameter or other characteristic length
D	diffusion coefficient
D	distillate (Section 6.4)
D	decimal reduction time (Section 9.3)
E	activation energy
f	friction factor
F	feed
g	acceleration due to gravity
G	molar flux of gas
G	Gibbs free energy
G	crystal growth rate (Section 8.4)
Gr	Graetz number ($d^2 v/Dl$)
$h,\ h_i$	heat transfer coefficients
H	partition coefficient
H	enthalpy
$\tilde{H},\ \hat{H}$	molar and specific enthalpies
HTU	height of transfer unit
IRR	internal rate of return
j_i	diffusion flux of solute "i", moles or mass per area per time
J_i	total flux of solute "i", moles or mass per time
$k,\ k_D$	mass transfer coefficient
$k,\ k_B$	Boltzmann's constant
$k,\ k_R$	reaction rate constant
$k,\ k_T$	thermal conductivity

K	equilibrium constant
K	overall mass transfer coefficient
K_{OW}, K_{OA}	partition coefficients between octanol and water or air (Section 2.4)
Kn	Knudsen number (λ/d)
l, L	length
l'	length of unused bed (Section 8.4)
L	molar flux of liquid
m	partition coefficient relating mole fractions in gas and liquid
m	molecular mass
M	total mass
$\tilde{M}_i$	molecular weight of species "i"
n_i	average number concentration per volume (Section 9.4)
N	number of ideal stages
$\tilde{N}$	Avogadro's number
N_i	number of particles in volume V
NPV	net present value
NTU	number of transfer units
p	pressure
$p(N_i)$	probability that volume V contains N_i particles (Section 9.4)
P	power
Pe	Péclet number (dv/D)
q	energy flux
q_i	concentration of solute "i" per mass of adsorbent (Section 8.4)
Q	heat
Q	volumetric flow
r	radius
r	correlation coefficient (Section 9.2)
r, r_i	rate of chemical reaction
R	gas constant
R_D	reflux ratio (Section 6.4)
Re	Reynolds number ($dv\rho/\mu$)
RIPP	acronym for separations of fermentation products (Section 8.2)
ROI	return on investment
S	entropy
Sc	Schmidt number (v/D)
St	Stanton number (k_D/v)
t	time
T	temperature
U	overall heat transfer coefficient
$\hat{U}$	specific internal energy
v	velocity
V	volume
W	work
We	Weber number ($\rho v^2 l/\gamma$)
x, x_i	mole fraction in liquid
X	fraction conversion in chemical reaction
y, y_i	mole fraction in vapor
z	position

α	thermal diffusivity
α	relative volatility (Section 6.4)
γ	surface tension
γ_i	activity coefficient of species "i"
δ	thickness of thin layer, especially a boundary layer
δ_i	solubility parameter of species "i"
ε	void fraction
η	efficiency
θ	fraction of unused adsorption bed (Section 8.4)
κ	reciprocal of Debye length
λ	mean free path
μ	viscosity
μ_i	chemical potential of species "i"
ν	kinematic viscosity
ν	stoichiometric coefficient
ρ	density
σ	collision diameter
σ	concentration fluctuation (Section 9.4)
τ	characteristic time
τ	tortuosity
ϕ	volume fraction
φ	electrochemical potential
ω	angular velocity
ω	regular solution parameter
ω_i	weighting factor for attribute "i"

Preface

Since its inception around a century ago, the chemical industry has focused on the manufacture of commodities. A commodity chemical, produced at over 1000 tons per year, is sold into a world market where the products are differentiated only by price. Benzene, polypropylene, and titanium dioxide are examples.

This industry had its Golden Age from 1940 to 1980, with growth equivalent to that of the modern software industry. Commodities of course continue to be made – the world needs toluene, ammonia and methanol just as it always has. However commodities are made by a dwindling number of ultra-efficient companies, which employ relatively few people. Sometimes, these companies are private, allowing them more easily to ride out the trade cycles typical of commodity businesses. Increasingly, the companies are associated with national oil companies and so have captive petroleum-based feedstocks, the most common raw materials for these commodity products.

More recently, as market growth has slowed, chemical companies without these captive feedstocks have moved towards higher value added products. These products are distinct from commodities in three ways: quantity, value, and structure. They are produced in small quantities, often less than 10 tons per year. The archetype is the active ingredients of a drug, where a few kilograms can command millions of dollars. These higher value added products are made of ingredients which cost a tenth or less of their selling price. These products gain their value from a molecular or micro structure which gives them better performance.

A company's advantage in making these products does not come from having a cheap, reliable feedstock. The advantage comes from a better product. The company does not sustain its position by always becoming a more efficient producer. The company keeps its advantage by continuous innovation, by patent protection, and by trade secrets. It needs better chemistry and engineering, all the time.

The movement of chemical companies towards higher value added products is reflected in the employment of new graduates, most of whom start work not on commodities, but on specialty chemical products. However, most new graduates are trained largely, if not exclusively, to serve the commodity chemical industry. To train these graduates more effectively, traditional education of chemical professionals must expand. The expanded education must not focus on the traditional question, "How should we make this commodity product?" The expanded education should focus on what is increasingly the more relevant question, "What high value added products are we going to make?"

This book describes chemical product design, and so attempts to fill part of the educational gap between the commodity chemical industry and the new, more fragmented, high value added chemical product industry. The book is divided into two parts. First, we present a template for chemical product design, which starts to answer the question about what we should make. Second, we apply the template to different types of chemical products.

Thus, the first part of the book requires chemists and engineers to go beyond their traditional role of how they will make a chemical commodity chosen by others. This part assumes that these chemists and engineers will be working as part of a project team. Such teams will include those representing marketing, research, manufacturing, and sales. This first part of the book expands material in the first edition. While the book aims to help engineers think about these other aspects of the chemical enterprise, it may also introduce non-engineers to ideas and constraints of engineering. Understanding this introduction requires knowledge of calculus and chemistry.

The second part of the book, which is new to this second edition, is specific to particular parts of the chemical industry. For example, Chapter 6 reviews commodities, Chapter 8 centers on active ingredients, like pharmaceuticals, and Chapter 9 includes personal-care products. Those already working in one product area may find some chapters are more useful than others. Those who seek an overview should be able to gain from all parts.

One market for the book is those trained in commodity chemicals but now involved with other types of chemicals with higher added value. The book also is suitable as a text for university courses. We and others have used the material in this book in a required course, originally taught along the same lines as chemical process design. Such courses involved lectures, use of process simulators, and one large report written by teams of students. All found teaching and learning the material challenging. However, although our courses have always been highly rated, we believe that the courses were not wholly satisfactory. While we could point to a few students whose work was so strong that it attracted venture capital, we always knew that the students' average experience was uneven.

As a result, we now use the book's content in two different ways, which we find more effective than imitating courses in chemical process design. First, we teach a separate course of about forty classes restricted to product design. The classes are split between lectures and smaller recitations, where the specific problems are discussed. The students do not write one large report, but six to ten shorter reports. They are encouraged to develop ideas as teams, but they are expected to write each report individually. In the second type of new course, we teach about twelve classes as a supplement to chemical process design. Again, about half of these are lectures; again, each student writes not one but three or four individual essays. Both of these strategies work better for us.

In addition to our teaching in universities, we have presented this material as short courses to several companies. We have found that different people have trouble with different parts of product design. Inexperienced students are wonderful at generating new product ideas, but they have trouble making estimates which let them quickly select among possible alternative products. Experienced chemists and engineers have no trouble making quick estimates and sensible selections, but they are less effective at suggesting new ideas. Despite these differences, both groups have benefited from and enjoyed their efforts to get better at chemical product design using the methods described in this book. We challenge you, either as a professor or as a student: while this material is hard to teach and hard to learn, it will often be among the most satisfying parts of your education. Have fun with your designs.

We are indebted to many who helped us write this book. We benefited from the encouragement of Professor John Bridgwater, who arranged our collaboration at the University of Cambridge. We were strongly influenced by the excellent book, *Product Design*, by Ulrich and Eppinger, which showed us how this subject could be effectively taught in mechanical engineering. Finally, we would like to thank our students, who have been generously tolerant as we shaped a few slogans into an educational experience.

E. L. Cussler and G. D. Moggridge
Cambridge, UK

"I have been anticipating the launch of this book since Dr. Moggridge told me back in 2009 that he had been working on the second edition. This is the long-awaited update of the book which was the first of its kind in the chemical engineering literature. The book had served as the foundation of my chemical product design course. Cussler and Moggridge's clear writing style and abundance of real-world examples in the book make it a must-have for any faculty interested in this fast emerging field of product design in chemical engineering."

Sin-Moh Cheah, Singapore Polytechnic

"Nowadays, Chemical Process Industries mostly produce products with target end use properties, and not simply molecules. This textbook offers the opportunity, for students, chemists, engineers or professors, to discover the framework, methodology and building concepts of the emerging 'chemical product design' discipline. The 2nd edition proposes an increased number of relevant and novel examples, treated thanks to a unique common approach. This reference textbook should be strongly recommended to anyone who wants to take into account the evolution of the chemical industry, and its incidence on teaching applied chemistry or chemical engineering."

Eric Favre, ENSIC, Nancy, France

"After their pioneering first edition on 'Chemical Product Design', Cussler and Moggridge have produced a second edition in which they have adopted a more systematic approach to this topic, which should make it easier to teach at the undergraduate level in chemical engineering courses, and more readable by industrial practitioners. The authors describe a four-step design strategy, involving a template composed of needs, ideas, selection, and manufacture that can be applied in principle to any product, from commodities to novel devices to molecular products to microstructures. The authors have also expanded the economic analysis to emphasize the importance of speed-to-market. Any instructor teaching a course on product design will find this book to be a very useful textbook. Industrial practitioners should find this book to be an excellent reference for promoting innovation in their organizations."

Ignacio E. Grossmann, Carnegie Mellon University

"There is a difficult transition from asking 'How do we make this commodity more cheaply?' to asking 'What should we be making in the first place?', but Cussler & Moggridge clearly guide the way. This is the seminal textbook on Chemical Product Design, demonstrating how chemical engineering fundamentals can be effectively applied to product design. The new and expanded material in the second edition greatly improves the text, illustrating how to apply their design template with industrially relevant problems. I highly recommend this book to everyone in the field of chemical product design and development."

Michael Hill, Columbia University

"Back in 2001 Cussler and Moggridge pioneered Chemical Product Design and introduced its first ever textbook. Now, one decade later, they do it again, through a substantially revised new edition that covers a broader range of topics (chapters for different product categories, coverage of business idea evaluation/presentation tools, defining future trends and establishing sets of final commandments for chemical product development), and will for sure become the book of reference in this field for the forthcoming years, a mandatory presence in many many shelves, including mine!"

Pedro Saraiva, University of Coimbra

"This second edition of a pioneering and well-received book has been carefully updated and enlarged by the addition of four new exemplifying chapters. Appearing 10 years ago as the first text on chemical product design, the book has been an important source of inspiration for chemists and chemical engineers. The systematic treatment of this diverse discipline and the many practical examples make the book very useful as an introductory text for both a university course and for practicing engineers."

Søren Kiil, DTU-Chemical Engineering

"The book builds on the well established four-step design process presented in the ground-breaking first edition. New additions emphasize the differences in design and manufacturing characteristics of molecules, micro-structured products and devices as opposed to chemical commodities. Students and practitioners may find this an invaluable introduction into the methodology of chemical product design and the use of engineering principles to support the screening and selection of product options."

Ton Broekhuis, University of Groningen

"Revisions to the second half of this excellent text by Cussler and Moggridge have made the book even more relevant and valuable to the challenge of contemporary chemical product design"

Keith Alexander, University of California, Berkeley

An Introduction to Chemical Product Design

This chapter explains what this book is about and why its subject is important. This is a book about the design of chemical products. In our definition of chemical products, we include four categories. The first, commodity products, is familiar. Second, there are molecular products, which provide a specific benefit. Pharmaceuticals and pesticides are obvious examples. Third, there are products whose microstructure, rather than molecular structure, creates value. Paint and ice cream are examples. The fourth category, chemical products, comprises devices which effect chemical change. An example is the blood oxygenator used in open-heart surgery.

The nature of chemical product design is described in Section 1.1. It emphasizes decisions made before those of chemical process design, a more familiar topic. Chemical product design is a response to major changes in the chemical industry which have occurred in recent decades. These changes, described in Sections 1.2 and 1.3, involve a split in the industry between manufacturers of commodity chemicals and developers of specialty chemicals and other chemical products. The former are best served by process design, and the latter by product design.

The fourth section of this chapter outlines the product design procedure that we will use in the remainder of the book. This procedure is a simplification of those already used in business development. Such a simplification clarifies the basic sequence of ideas involved. Moreover, the simple procedure allows us to consider in considerable detail the technical questions implied in specific products. This technical approach is suitable for those with formal training in engineering and chemistry, and may also be challenging for those whose training is largely in business. Chapters 2–5 give more detail on how to apply this simple design template to chemical products.

Further detail of the different categories of chemical product – commodities, molecular products, microstructured products, and devices – are given in the final section of this chapter. The distinction between these categories is rationalized on the basis of the length scale which is key to their performance and manufacture. We argue that the four-step template described in Section 1.4 can be applied usefully to each of these categories of product, but that the emphasis is different in each case. These differences of emphasis are the subject of the second half of the book: in Chapters 6–9 we discuss the key steps, and the most important technical tools, for the design of each category of product.

1.1   What Is Chemical Product Design?

Imagine four chemically based products: an amine for scrubbing acid gases, a pollution-preventing ink, an electrode separator for high-performance batteries, and a ventilator for a well-insulated house.

These four products may seem to have nothing in common. The amine is chemically well defined: a single chemical species capable of selectively reacting with sulfur oxides. The ink is a chemical mixture, which includes both a pigment and a polydisperse polymer "resin." The electrode separator should provide a safeguard against explosion if the battery accidentally shorts out. The ventilator both provides fresh air and recovers the energy carefully conserved by insulating the house in the first place.

What these products do have in common is the procedure by which they may be designed. In each case, we begin by defining what we need. Next, we think of ideas to meet this need. We then select the best of these ideas. Finally, we decide what the product should look like and how it should be manufactured.

We define chemical product design as this entire procedure. At the start of the procedure, when we are deciding what the product should do, we expect major input from both marketing and research, as well as from science and engineering. By the end of the procedure, when we are focused on the manufacturing process, we expect a reduced role for marketing, and a major effort from engineering. However, we believe that the entire effort is best viewed as a whole, carried out by integrated teams drawn from marketing, research, and engineering.

We can see how product design develops by considering three of the products already mentioned in somewhat more detail. For example, for the pollution-preventing ink, our original need may be to reduce emissions of volatile solvents in the ink by 90%. Our ideas to meet this need include reformulating the polymer resin in the ink in two different ways. First, by using a polydisperse resin of broad molecular weight distribution, we can control the ink rheology and so eliminate the need for volatile solvents in the ink itself. Thus, there will be no emissions during printing. Second, by adding pendant carboxylic acid groups to the resin, we can make the resin not only an effective component of the ink but also an emulsifying agent in dilute base. If we wash the presses with dilute base, we can clean them without volatile solvents and without solvent-soaked shop rags. The manufacture of the new ink will be very similar to that used for the existing ink.

Consider the amine for scrubbing acid gases. Current acid-gas treatment often uses aqueous solutions of amines, such as monoethanol amine. After these solutions absorb acid gases like carbon dioxide and sulfur oxides, they are regenerated by heating. Though this heating gives an efficient regeneration, it can be expensive. The need is for amines that can be more easily regenerated. Our idea is to effect the regeneration with changes in pressure. We would absorb the acid gases at high pressure and regenerate the amines at low pressure, where the acid gases just bubble out of solution. In order to achieve this end we have little idea how to proceed, so we are forced to synthesize small amounts of a large number of sterically hindered amines. We will test all candidates to find the best ones. We will then manufacture the winners. Like many high value added chemicals, these will be custom syntheses, made in batch in equipment used for a wide variety of products. This obviates the need for intensive process design in many of the chemical products which we are considering.

A third example of product design is house ventilation. Well-insulated houses are energy efficient, costing little to heat, but they may exchange air at less than one tenth of the recommended rate. To get more fresh air, we can open a window, but this sacrifices our efforts at good insulation. The need is for a fresh air exchanger that captures the heat and humidity of our snug house, but exhausts stale air, with smells and carbon dioxide. Our idea is for an exchanger for both heat and water vapor for this energy-efficient house. We can manufacture this in the same way as other low-cost, cross-flow heat exchangers. In this example, our product is a device – not a chemical – that increases health and comfort in the house.

The designs of the ink, the acid-gas absorbant, and the home ventilator are examples of the subject of this book. This subject is different from chemical process design. In process design, we normally begin by knowing what the product is, and what it is for. Most commonly, it is a commodity chemical of carefully defined purity; ethylene and benzene are good examples. This material will be sold into the existing market for such a commodity, so we know the price we can expect. The focus of our process design will be efficient manufacture. We will usually use a continuous process, depending on optimized, dedicated equipment, which has been thoroughly energy integrated. This type of careful process design is essential in order to compete successfully in the commodity chemical business.

The chemical products discussed in this book are different. Their promise stems less from their efficient manufacture, and more from their special functions. They will usually be made in batch, in generic equipment; or will themselves be small pieces of equipment. Process efficiency may be less important than speed to reach the market place. Energy integration may be of secondary value. Indeed, most product design may occur before manufacture is even an issue.

We believe that product design merits increased emphasis because of major changes in the chemical industry. We do not argue that the chemical engineer's concern with process design should disappear. However, we do assert that the topics we study should reflect the chemical industry of today. How this has developed is outlined in the next section.

1.2   Why Chemical Product Design is Important

Chemical product design has become more important because of major changes in the chemical industry. To understand these changes, we will review the history

Table 1.2–1 *Growth of textile fibers in 10^6 lb/year*
From 1950 to 1970, synthetic fibers grew about 20% per year. Since then, their growth is 5% per year. (Source: Spitz; US Department of Commerce).

	1948	1969	1989
Cotton, wool	4353	4285	4794
Synthetics	92	3480	8612

of the industry, using as an example the development of synthetic textile fibers. We also need to examine how these changes have affected employment.

1.2–1 CHANGES IN THE CHEMICAL INDUSTRY

From 1950 to 1970 the chemical industry produced ever increasing amounts of synthetic textile fibers, as shown in Table 1.2–1. Over these decades, while the production of natural fibers was about constant, the production of synthetics grew 20% per year. This growth is comparable to that of the software industry today; Du Pont can be seen as the Microsoft of the 1950s. This was a golden age for chemicals.

However, from 1970 to 1990, synthetic textile fibers grew by less than 5% per year, about the same as the growth of the world population. From 1970 to 1990, the industry stayed profitable by using larger and larger facilities. Bigger profits came from consolidating production into bigger plants, designed for greater efficiency in making one particular product. Interest in computer-optimized design was a consequence of this consolidation. Such optimization meant small producers were forced out. For example, the number of companies making vinyl chloride in the USA shrank from twelve in 1964 to only six in 1972 (Spitz, 1988).

More recently, the industry has required new strategies to stay profitable. These strategies often centered on restructuring, which was three times more likely to affect engineers than the general working population. Whether called "restructuring," "downsizing," "rightsizing," or "rationalization," the strategy meant many mid-career engineers were suddenly looking for new jobs. The Engineering Workforce Commission in the USA now feels that engineers will average seven different jobs per career, a dramatic change from two per career in the recent past (National Science Board, 2003). Middle management, that traditional goal of bright but not brilliant students, is no longer a safe haven. While starting salaries remain high, the envy of other technical professions, these salaries have not increased faster than average wage inflation in 30 years. In this environment, professional organizations like the American Institute of Chemical Engineers now provide more help in job transitions and financial planning. Such organizations can no longer behave only as nineteenth-century-style learned societies.

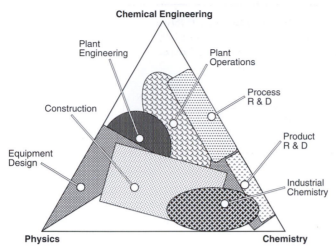

Figure 1.2–1 *Skills learned by chemical engineers* Traditional skills are a blend of physics, chemistry, and engineering. These are sufficient for chemical product design.

Having exhausted optimization and restructuring as ways to stay profitable, chemical companies now have three remaining options. First, they can leave the chemical business. This option seems reasonable to a surprising number, including some petrochemical businesses. Second, chemical companies can focus exclusively on commodities. This seems a preferred strategy for some private companies, who may be better able to handle the ebb and flow of the profits from a commodity business. It implies a ruthless minimization of research and a concentration on in-house efficiency.

The third strategy open to these chemical companies is to focus their growth on specialty chemicals or high-performance materials. Such chemicals, produced in much smaller volumes than commodities, typically have much higher added value as well. This higher added value means that more research and higher profits are possible. Not surprisingly, many chemical companies are turning their focus to specialty chemicals or high-performance materials.

Interestingly, this new focus has not changed the skills that companies demand from chemists and chemical engineers, though it has changed the jobs that they do. The various subjects which chemical engineers learn can be positioned on the triangular diagram in Figure 1.2–1. The three corners of this plot represent training in physical sciences, in the chemical sciences, and in chemical engineering subjects. Different jobs use these three elements in different proportions, as shown in the figure. There is no surprise in this: plant engineering will demand a greater knowledge of mechanics and a smaller background in chemistry than those involved in research and development. Figure 1.2–1 also suggests national averages. British chemical engineers have more chemical engineering and less chemistry than their counterparts in the USA. Please do not take this diagram too literally; use it instead as a catalyst for thought. We maintain that the basic skills

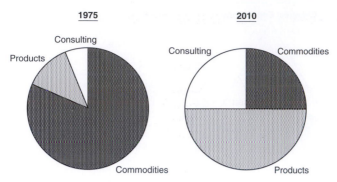

Figure 1.2–2 *Changes in employment* The dominance of commodity chemicals has been eclipsed by efforts on other types of products.

needed by chemical engineers have always been diverse and have not altered dramatically.

1.2–2 CHANGES IN EMPLOYMENT

While changes in the chemical industry may not have changed the skills needed, the focus of chemical companies on specialties and new materials has had a major impact on the jobs which chemists and chemical engineers do. To examine this impact, we compare the jobs taken by recent graduates with those taken by graduates twenty five years ago. Our data for this are fragmentary, taken from records of graduates from the universities of Cambridge and Minnesota. They are probably biased towards large corporations, about whom our university placement offices have better records.

The available data suggest major changes, as shown in Figure 1.2–2. In 1975, three quarters of chemical engineering graduates went to work in the commodity chemicals business. The small number who did not were split between work on products, either design or development; and work in other areas, which for convenience we have labelled "consulting." This category includes those working directly for consulting firms as well as those carrying out specific tasks like environmental-impact statements.

More recently, the distribution of jobs has become completely different. The largest group of chemical engineering graduates, in Minnesota's case more than half, now work primarily on products. This includes students who work on materials, on coatings, on adhesives, and on specialty chemicals. The number who work in commodity chemicals has dropped so that it is now less than a quarter of new graduates. The number who work in consulting has risen dramatically, as commodity chemical businesses outsource many of the functions which they used to do in house. For example, in one case, a commodity chemical company has taken its process engineering group from 1500 persons to fewer than 50. This is not a business cycle; this is a change in the way in which that company expects

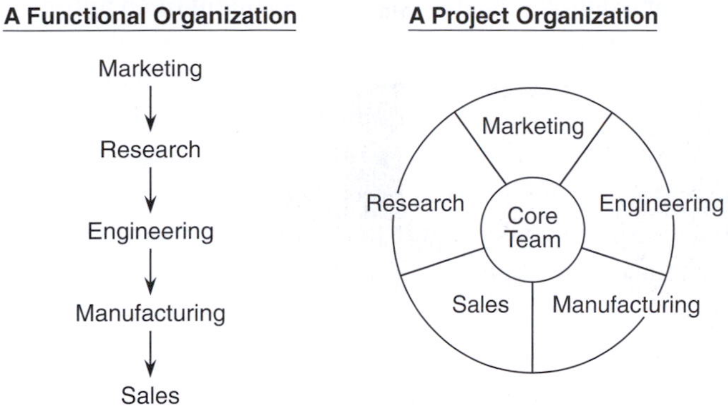

A Functional Organization   **A Project Organization**

Figure 1.3–1 *Two limiting types of corporate organization* The project organization is believed to give greater speed and synergy but greater management complexity.

to do business: they will buy the process engineering they need from consultants. This is why the number of people involved in consulting has risen.

The emergence of products as a focus for chemical engineers implies changes in what chemical engineers do. In the past, we chemical engineers could limit our thinking to reaction engineering and unit operations, waiting for the marketing division to tell us what chemicals needed to be made, and in what amounts. Such intellectual isolation is no longer possible. Now we can expect to be involved in teams, a consequence of a new corporate culture.

1.3 Changes in Corporate Culture

At the same time, there have been changes in corporate culture, in the ways in which companies do business. These are consequences of the changes in the chemical industry discussed in the previous section, and they are at least as important, because they alter the ways chemical engineers work. Two major changes are especially important: the way in which corporations organize their product design, and the ways in which corporate strategy affects jobs. Each is discussed below.

1.3–1 CORPORATE ORGANIZATION

The organization of product development is most easily discussed by comparison of two limiting cases: organization by function, and organization by project. These are shown schematically in Figure 1.3–1. Both can be effective.

In a functional organization, different divisions have different responsibilities: marketing, research and development, engineering, legal affairs, and so on. Product development proceeds by each division doing its job, and then passing its results on to the next division. The result is like chemical reactions in series, as suggested by Figure 1.3–1. This organization is especially associated with large, established industrial companies which have major capital investments in manufacturing. For example, the marketing department of an automobile company could discover that consumers want better climate control, i.e. better heating and air-conditioning. Marketing would report their results to research, who would develop the electronic controls required for this goal. Engineering would extend the research results so that the new controls could be manufactured cheaply and efficiently. Throughout the process of product design, the development is sequential: marketing talks largely to research, only rarely to engineering. Such a functional organization can be effective, but it is almost always slow.

A common alternative is a project organization. In a project organization, a core team is formed from the different divisions. The team will normally include representatives from marketing, research and development, engineering, production, etc. These core team members will have complete responsibility – and a good deal of resources – to design and develop the target product. They will be judged not by their immediate functional supervisors, but rather by a panel of senior managers well versed in the company's long-term strategy. Functional supervisors still have the job of making the divisions run smoothly. Such divided management can be chaotic and inefficient. As Figure 1.3–1 suggests, it is like parallel chemical reactions, with a good chance of synergy between functions. Above all, this form of product development is fast, and fast product development is believed to maximize profits, so that project management is currently the organization urged by most business consultants.

1.3–2 CORPORATE STRATEGY

Superimposed on its organization, a corporation will have strategic forces driving product development. Again, the driving forces are most easily described in terms of two limiting cases. First, corporations which look towards their markets for inspiration are said to be "market-pull." Corporations which emphasize extending their technology are said to be "technology-push." Examples of market-pull companies are common. W K Kellogg, the makers of breakfast cereal, are interested in new products from grain. They constantly assess the market for consumer wishes for new cereals or new grain-based snack foods. Honeywell make home thermostats, a major product because a significant fraction of the world's energy consumption goes for domestic heating. Honeywell are interested in any new products for home comfort that can complement their thermostats. Patagonia, makers of technical mountain-climbing equipment, now also make raincoats. They are pushing to expand their market: many more people need raincoats than ice axes.

Examples of technology-push companies are less common but can nonetheless be found among everyday names. W L Gore makes Goretex, that breathable film basic to high-quality raincoats. But Gore do not make raincoats: instead, they have used their basic material to make medical products, including arterial transplants. Exxon-Mobil has used its knowledge of petrochemical reactions to develop a series of new metallocene catalysts for polyolefins. Astra-Zeneca has used its experience with injectable therapeutics to develop delivery systems for different drugs. Interestingly, both "market-pull" and "technology-push" companies can use the same product design procedure. This procedure is described next.

1.4 The Product Design Procedure

Product design is a major topic both in disciplines like sales and marketing, and in technical professions such as mechanical engineering. Not surprisingly, schemes for this design procedure vary widely. Many are complex, especially with respect to the role of management. Many have features that seem specific to the particular subdiscipline that they represent.

The product design procedure used in this book is a simplification and generalization of those used in these other areas. It depends on four steps:

(1) *Needs*. What needs should the product fulfill?
(2) *Ideas*. What different products could satisfy these needs?
(3) *Selection*. Which ideas are the most promising?
(4) *Manufacture*. How can we make the product in commercial quantities?

The characteristics of this approach are discussed in the rest of this section. We shall see as we go along that the application of this template to the case of chemical products leads to several new and characteristic features of the design process.

1.4–1 HOW THE PROCEDURE ORGANIZES THIS BOOK

These four steps are the key to the organization of the first half of this book. Assessment of needs, the subject of Chapter 2, includes deciding on a standard for comparison – a benchmark – and on converting the qualitative needs to quantitative specifications. The benchmark chosen may be an existing product or an ideal. Needs must be as well defined as possible, and framed in technical terms, so that any specifications are definitive.

Finding ideas that might meet these needs is the next step in product design. Normally, we will wish to search for a large number of these ideas by all reasonable means. This search, the subject of Chapter 3, may include brainstorming by individuals and teams and synthesizing tangent compounds by combinatorial chemistry. Once these numerous ideas are identified, they must be screened using objective and subjective judgments, also described in Chapter 3.

At this point, we should have reduced the large number of fragmentary ideas for products to a short list of promising candidates. Typically, this reduction will be about a factor of twenty: if we start with a hundred ideas, we should have about five survivors. We must now select the best one or two for further design and development. If the characteristics of each of the surviving ideas were directly comparable, this would be easy. They normally are not. For example, we might be sure that one idea will work well but be expensive; a competing idea might be cheap but we may be unsure if it will work. Deciding between these ideas includes risk management, as described in Chapter 4.

Finally, we must manufacture the product, and estimate the costs involved. Parenthetically, this fourth step could correctly be called "process design." We have not done so because we found our descriptions then seemed more complicated than necessary. These manufacturing efforts, described in Chapter 5, are different from those expected for commodity chemicals, where we expect to use dedicated, optimized equipment which operates continuously. Here, we will normally use generic equipment, run in batch for a variety of specialty products.

This is different to traditional chemical engineering – and is exciting.

1.4–2 LIMITATIONS OF THE PROCEDURE

The four-step procedure outlined above is controversial. We should review the controversies now, so that we are prepared for exceptions and diversions in the practice of product design. The controversies cluster around three criticisms: that the procedure is not general, that management and not technology is the key, and that chemical product design is part of process design. Each controversy merits discussion; they are tackled below.

First, is the four-step procedure as outlined general? It is clearly a major simplification. Many business texts argue that such a procedure is universally applicable for any product in any industry. These texts are frequently written by business consultants eager to make money by applying their own standard template to specific problems. At the same time, many professional product developers argue that this or any procedure cannot represent the peculiarities of their own industry; that only those with particular interests can hope to be effective. While there is clearly some truth in this argument, these product developers may be like those who have denied that correlations of heat transfer could be used for food products because they were based on measurements for petrochemicals.

We believe that both sides of the debate have their merits. The four-step procedure used here is unquestionably an approximation. Certain techniques introduced in particular steps of the procedure can have value at other stages. For example, risk management, introduced in the selection step in Chapter 4, may have value in screening product ideas, explored in Chapter 3. It is unlikely that real product design will always be a simple sequential procedure as we suggest; iteration between stages is almost certain to be necessary. Still, we must begin somewhere, and the current procedure has been for us a sound and creative start. We suggest trying it; any necessary modifications quickly become obvious in specific cases. A framework in which the subject may be understood is an aid to learning in chemical product design just as an analogous template has been successfully applied for years in process design.

The second controversy is the claim that management, not technology, is key to product design. An irritating feature of most business books on product design is the extreme emphasis on the central role of management. The implication is that technology is always available if only the managers do their job properly

Table 1.4–1 *Process design vs. product design* All four steps of process design are contained in step four of product design.

Process design	Product design
1. Batch vs. continuous process	1. Identify customer needs
2. Inputs and outputs	2. Generate ideas to meet needs
3. Reactors and recycles	3. Select among ideas
4. Separations and heat integration	4. Manufacture product

(or at least do what the consultants say). These books on product design know no inconvenient constraints like the second law of thermodynamics or the meaning of Avagadro's number.

We believe that the application of technology is central to chemical product design. Product design governed only by management reminds us of a Sidney Harris cartoon showing a few managers and an engineer standing in front of a flip chart. Though the flip chart is covered with equations, pie charts, and organization charts, the engineer is pointing to one small box, which says:

"Then a miracle occurs."

The engineer remarks:

"I'm having trouble with this part."

On reading books on the management of product design, we can feel all too much like that engineer in the cartoon. In this book, we want to make sure that technology is central to chemical product design.

The third controversy about product design is the assertion that the subject is already covered as the process design part of the existing canon. This serious assertion is most easily tested by comparing our template for product design with an example of the type of intellectual hierarchy suggested for process design. One successful and powerful hierarchy, suggested by Douglas (1988), is summarized on the left side of Table 1.4–1. After we decide whether a process is batch or continuous, we then move on to flow sheets, which are almost always continuous. The initial flow sheets center on the stoichiometry. Our next level of the hierarchy, which adds the recycles, involves a discussion of the chemical reactions. Once these are established, we move on to the separation trains and finally to the heat integration. All of this makes for a good course.

If we want to emphasize product design, we need to go beyond this hierarchy. We cannot simply substitute the search for a product for drug delivery into the process design hierarchy. Instead, the four-step hierarchy suggested earlier is shown on the right side of Table 1.4–1. After first identifying a corporate need, we generate ideas to fill this need. We then compare these alternatives and finally decide on a prototype and its manufacture. The manufacturing includes all of the process design hierarchy.

Thus, the important steps in product design anticipate those of process design. Product design implies a focus on the initial decisions around the form of the product and implicitly de-emphasizes its manufacture. This shifts our efforts away from the common engineering calculations that have been our bread and butter for decades. The new emphasis includes subjects which have normally been left to those directly concerned with business. It is this combination of business and technology that is the subject of this book.

1.5   Categories of Chemical Products

The first half of this book, Chapters 2–5, is based on the assertion that the design of all types of chemical products can be usefully approached using a single template. This product development template works well, but it is applied differently to different types of chemical products. This is because these different products have different key variables and because they rest on different aspects of science and technology. In the second half of this book, we explore in detail how the different types of products are best developed. We emphasize which steps are the most difficult.

We have already identified four categories of chemical product: commodities, molecular products, microstructured products, and devices. The subject of Chapters 6–9 is how these different categories of chemical product differ and how this affects the way in which we approach their design, and the technological toolboxes which are most valuable.

Thus, the first half of the book approximates all chemical products as the same, while the second half examines the differences between categories of chemical products. This apparent contradiction can be resolved by remembering that the four-step design template we propose can never be more than a useful approximation. Different stages in the design tend to be emphasized for different types of product and we require different scientific and engineering tools to analyze them. We benefit by using the four-step procedure in approaching a design. At the same time, we consider the peculiarities of a particular product in utilizing this one-size-fits-all template, specifically in relation to the four categories of chemical product we identify. These two complimentary approaches to a design problem can be seen as being like the weft and the warp of a woven fabric – the combination of both being essential to give form, strength, and pattern to the material.

To give the second half of the book structure, we idealize chemical products as four types, based on the characteristic size scale which is critical to their performance, as shown in Table 1.5–1. The first, familiar type is chemical commodities like ethylene and ammonia, made in large quantities and sold at the lowest possible price: 131 million tons of ammonia were made in 2007, selling for around $300 per ton. Commodities are treated as continua: they are described without reference to any size scale. These commodities, which are the traditional focus of chemical engineering, are made using the tools of reaction engineering and unit operations. The chief uncertainty in their manufacture is their feedstock, which is most often a hydrocarbon obtained from politically unstable parts of the world.

Table 1.5–1 *Four types of chemical products* In the emerging chemical enterprise, the traditional commodities are supplemented by chemical devices, by molecular products, and by microstructures.

	Commodities	**Devices**	**Molecules**	**Microstructures**
Examples	Ethylene; ammonia	Artificial kidneys; home oxygen	Penicillin; Prozac	Sunscreen; surimi
Scale	Continuum	Meters	Nanometers	Micrometers
Key	Cost	Convenience	Discovery	Function
Basis	Reaction engineering; unit operations	Reaction engineering; unit operations	Chemistry	Recipe
Risk	Feedstock	Intellectual property	Discovery	Science

The three other types of chemical products are chemical devices, molecules, and microstructures. Chemical devices are basically miniature processes which accomplish a particular chemical transformation but at a small scale, for example for one person. An artificial kidney and an ultrafilter for drinking water are good examples. Another more complex example is a reactor for destroying nerve gas: here, the objective is less low cost than safety and convenience, making sure in a moderately sized batch reactor that the nerve gas is transformed into benign chemicals which are safe to discharge. The scale of these products is macroscopic, i.e. it is the device itself.

The third type of chemical product is molecules, most often exemplified by pharmaceuticals. The important scale of these products is nanometers, summarized as the specific chemical structure, normally with a molecular weight between 300 and 800 daltons, often with several chiral centers. Here, the key is not process cost or convenience but discovery, i.e. identification of the active compound in the first place. By comparison with commodity products, the amounts involved are miniscule. For example, the drug Zolodex, which in 2007 had sales of $800 million, is made at a rate of 44 kg per year. Processes for making these molecules are batch, in generic equipment; the processes are more like gourmet cooking than petroleum refining.

The final product type, microstructures, is also very different to commodity chemicals. These products include sunscreen and non-woven fabrics; they include surimi, the fish paste used to make artificial crab sticks. The important scale of these products is micrometers. This size is much larger than molecular, even larger than the largest drug molecules. The size is smaller than we are able to perceive directly, much smaller than our tongues or fingers. The key to this type of product is not its cost or discovery but its function. For example the success of a sunscreen depends on whether it makes our skin feel smooth and lets us tan easily and evenly; the success of a non-woven fabric depends on whether it feels soft and warm; surimi succeeds if it "tastes just like lobster." In each case, we have no interest in exactly what the chemical components are; we just care that the product fulfills its function. The key is tailoring the structure at the micrometer

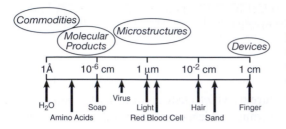

Figure 1.5–1 *Different chemical products have different critical sizes* These range from commodities, treated as continua, to chemical devices, with dimensions of centimeters.

level to give us the product properties we desire; we need to understand and control the relationship between microstructure and function. Because the state of knowledge of these structure–function relationships is often not well advanced, designing microstructured products can be challenging.

The categorization of product type by characteristic size scale is brought into focus in Figure 1.5–1, which shows the important length scales for the four types of chemical product and also of various well-known structures. Obviously, molecular products are described by molecular structures, described on the scale of Ångstroms. These structures have important complexities. For example, the drug thalidomide has two isomers. One of these has considerable therapeutic benefits, but the other was responsible for thousands of birth defects when it was given to pregnant women to alleviate morning sickness.

Microstructured products gain value from their micrometer-sized details, which are often not at thermodynamic equilibrium. One common example is latex paint. Under normal conditions, it is easily applied to make an effective barrier coating. However, if the paint is frozen, its microstructure is destroyed, and it is no longer useable. Chemical devices are much larger, with a size scale from millimetres to meters. They can be mini-chemical plants, made by scaling down larger facilities. Such scale-down, the converse of the usual chemical practice of scale-up, has its own problems. For example, flow tends to change from turbulent to laminar, giving lower pressure drops, which is good, but poorer mixing, which is bad.

Chemical commodities do appear in Figure 1.5–1, but are treated as continua. To be sure, they have a molecular structure; but this simple structure is usually ignored in design, replaced by other physical properties like viscosity or vapor pressure. This is why the periodic table rarely hangs on the wall of engineering classrooms, while it is ubiquitous in chemistry lecture halls. This concern with scale for all types of chemical products is basic to the discussions later in this book.

In Chapters 6–9, we discuss the characteristics of these four types of products. The summary for commodity chemicals given in Chapter 6 reviews the chemical process design well covered in other books as part of the discipline of chemical engineering. We give this review so all readers will have a common base, but we do not discuss this base in detail because it is easily accessible elsewhere. We detail the design of devices, molecules, and microstructures in Chapters 7, 8, and 9, respectively. Thus, in the second half of the book, we describe how the basic template of needs, ideas, selection, and manufacture can be applied to each of the four different chemical products.

1.6   Conclusions

Product design is the procedure by which customer needs are translated into commercial products. This procedure, which precedes process design, is especially valuable for molecular products, microstructured products, and chemical devices. Such products are an important focus of the present-day chemical industry, which is evolving beyond the commodity products that have been the emphasis in past decades.

In this book, the product design procedure is organized as four sequential steps. The first step, described in Chapter 2, is the identification of customer needs and the translation of the needs into product specifications. The second step, in Chapter 3, involves generating and winnowing ideas to fill these needs. In the third step, in Chapter 4, the best ideas are selected for commercial development. The last step, explained in Chapter 5, includes manufacture, regulatory requirements, and economics. The result is a template for chemical product design.

We must stress that management, especially senior management, is much more likely to be involved in product design than in process design. As a member of a product team, each engineer or chemist will be involved in a management review at each stage of the design process. This review will be critical; that is, the review will decide on whether the project should continue. To reflect this, in the "conclusions" section of each of Chapters 2 to 5, we will mention this review, and discuss what human interactions are likely. These human interactions are as important as the technology.

The second half of the book, Chapters 6–9, focusses on differentiating the four different types of chemical product – commodities, molecules, microstructures, and devices – on the basis of the size scale characteristic of each. How the design template described in Chapters 2–5 can be applied, and what science and engineering is most useful, is discussed for each type of product. Thus the two halves of the book give us synergistic ways of approaching the design of a chemical product, which are most effectively used simultaneously.

REFERENCES AND FURTHER READING

Biegler, **L. T.**, **Grossmann**, **I. E.**, and **Westerberg**, **A. W.** (2010) Issues and trends in the teaching of process and product design, *AIChE Journal*, **56**, 1120–1125.
Blessing, **L. T. M.** (2003) Future Issues in Design Research, in *Human Behaviour in Design*, **Udo Lindemann** (ed.). Springer Verlag, Berlin, pp. 298–303.
Blessing, **L. T. M.** and **Chakrabarti**, **A.** (2009) *DRM, a Design Research Methodology*, Springer Verlag, Berlin.
Cooper, **R. G.** (2001) *Winning at New Products, Accelerating the Process from Idea to Launch*, 3rd Edition. Addison-Wesley, Reading, MA.
Douglas, **J. M.** (1988) *Conceptual Design of Chemical Processes*. McGraw-Hill, New York, NY.

Graedel, **T. E.** and **Allenby**, **B. R.** (1996) *Design for Environment.* Prentice Hall, Upper Saddle River, NJ.

Kanter, **R. M.**, **Kao**, **J.**, and **Wiersema**, **F.** (eds.) (1997) *Innovation: Breakthrough Thinking at 3M, DuPont, GE, Pfizer, and Rubbermaid.* Harper Collins, New York, NY.

National Science Board (2003) *The Science and Engineering Workforce: Realizing America's Potential.* National Science Foundation, Arlington, VA.

Pahl, **G.** and **Beitz**, **W.** (2007) *Engineering Design, A Systematic Approach,* 3rd Edition. Springer Verlag, London.

Spitz, **P.** (1988) *Petrochemicals: the Rise of an Industry.* Wiley, New York, NY.

Ulrich, **K. T.** and **Eppinger**, **S. D.** (2007) *Product Design and Development*, 4th Edition. Irwin McGraw-Hill, New York, NY.

Wesselingh, **J. A.**, **Kiil**, **S.**, and **Vigild**, **M. E.** (2007) *Design and Development of Biological, Chemical, Food and Pharmaceutical Products.* Wiley, New York, NY.

2

Needs

Chemical product design begins by identifying customer needs, those unfilled wants which are the original spark for product development. The customers include both those who will buy the product and those who will use the product. Who these customers are and how their needs can be identified is the subject of Section 2.1. These needs are the starting point for our design. Sometimes, our product will be used primarily by consumers; in this case, the needs will often be described in non-scientific terms which are hard to quantify. In Section 2.2, we explore special problems associated with these consumer products, where we may wish to use subjectively assessed product attributes.

Needs are often vague, qualitative desires for solutions to ill-defined problems. To make these needs more definite, we seek particular specifications that our product must meet. Setting these specifications is explored in Section 2.3. Usually, the specifications will require continuing revision and re-evaluation. This revision can be greatly facilitated by using "benchmarks," which are often competing products that we hope to replace. The revision of specifications and the use of benchmarks are the topics covered in Section 2.4. By the end of the chapter, we will have a basis from which to begin designing a successful product.

2.1   Customer Needs

Elucidating customer needs involves three sequential steps: interviewing customers, interpreting their expressed needs, and translating these needs into product specifications. In each of these steps, we must be careful not to narrow the product definition prematurely. We will have to resist ideas from both ourselves and our colleagues, who will immediately identify potential new products and see some good ways to improve existing products. However, at this stage, we want to focus on stating a specific need, not on meeting that need.

2.1–1 IDENTIFYING CUSTOMERS

Our primary source in identifying customer needs should be the final users of the product. These users may not be those who buy the product from us; rather, it is those who will actually benefit from its chemistry. These users may not be individuals; they will often be organizations, including government agencies. Identifying these users, contacting them, and arranging times to discuss their needs is normally straightforward.

One group of customers are special and merit extra attention. These are the so-called "lead users," who depend very much on existing and competing products. These lead users will have needs which are well in advance of the marketplace. Accordingly, they will benefit most from any product improvements. Lead users are especially important in identifying needs for two reasons, both of which reflect their deeper experience. First, they often invent minor product improvements on their own. Second, they can usually clearly express what is wrong with existing products. When we can identify such lead users, we will find them an invaluable source of information.

EXAMPLE 2.1–1   CUSTOMERS FOR BARBIE DOLLS

One revolutionary toy, introduced in 1959, was the Barbie doll, marketed as a "teenage fashion doll." Based on a German sex toy, the doll is also controversial because it suggests an anorexic body type. Barbie originated in the inventor Ruth Handler's observations of her daughter:

> [My] daughter Barbara . . . and her friends always insisted on buying only adult female paper dolls. They simply were not interested in baby paper dolls. [The girls] were using these dolls to project their dreams of their own futures as adult women . . . This was a basic, much needed, play pattern that had never before been offered by the doll industry to little girls. Oh, sure, there were so-called fashion dolls, those who came with more than one outfit. But these dolls had flat chests, big bellies, and squatty legs – they were built like overweight six- or eight-year-olds. The idea of putting a prom dress on such a doll was ludicrous (Handler, 1994).

What was the inventor's insight?

SOLUTION

Ms. Handler saw that girls wanted dolls for at least two reasons. First, they wanted to pretend that they were mothers. Second, they wanted to pretend they were adults. This second insight made Ms. Handler wealthy and the doll's maker, Mattel, a major toy manufacturer. We note that in her most recent incarnation, Barbie is an engineer.

EXAMPLE 2.1–2 CUSTOMERS FOR CATARACT SURGERY

A cataract is a clouding of the crystalline lens of the eye that often develops with age. Remarkably, the lens can be easily replaced by surgery. To do so, the surgeon makes a small incision in the eye and removes the lens. A new plastic lens, which is rolled up to fit in a syringe needle, is then placed in the eye, where it stays permanently. The patient normally goes home on the same day.

If our company makes the plastic lens, is our customer the patient, the surgeon, or the health insurance company?

SOLUTION

On the most general level, it is all three. However, the key customer for our product is probably the surgeon. Of course, the lens must correct vision at a reasonable price. However, several companies make similar lenses which are going to cost similar amounts. The most successful lens will be the one that is easiest to implant. Once we have made an inexpensive, effective lens, we will find the surgeon will be the most important customer.

2.1–2 INTERVIEWING CUSTOMERS

The consensus in the business literature is that face-to-face interviews are the best way to discuss needs. Studies suggest that fewer than ten interviews may miss important information. More than fifty interviews results in little new information. We suggest the normal target should be about fifteen. In some cases, when you expect only one or two corporate customers, getting this many interviews may seem silly. We urge talking separately to many persons within a single corporate customer, for this often exposes different opinions which shape our product design.

Two alternatives to individual customer interviews are focus groups and trained test panels. Focus groups normally have a leader and perhaps eight panel members. Their discussions, which are sometimes videotaped, can supply a synergism leading to suggested innovations. Focus groups frequently show a smaller variance of opinion than a set of interviews of individuals. However, many in market research seem unconvinced that they are superior to individual interviews. Politicians in democratic countries often have a fondness for focus groups, with the aim of devising electorally safe policies. During Tony Blair's tenure as Prime Minister, government sometimes seemed to be run more by a focus group reporting to 10 Downing Street than by the Cabinet.

Trained test panels are most common in evaluating small differences in consumer goods. In these cases, the panels may be encouraged to use words with very specific definitions which can differ from those in popular use. For example, the words "papery," "rubbery," and "green" are used to describe flavor notes in coffee, notes which can be correlated with particular aroma chemicals. Test panels allow the reliable evaluation and categorization of smaller differences than groups of untrained consumers. Such evaluation often seeks to define "just distinguishable differences." These panels can guide consumer product improvements, but they are sometimes less useful for chemical products, where the primary customer is less often the consumer.

Before beginning the interviews, we need to decide what we want to obtain from them, both in terms of the product and for the benefit of our organization. This is best accomplished by having each member of the core team write out the project scope, the product's target market, and the key business goals. The project scope should be one simple sentence. The target market should include both the primary and secondary customers, and estimates of the sales that can be expected from each. The business goals should include both the timing of the new product and the organization's advantage in such an effort. After each member of the core team has completed this assignment, the entire team should reach a consensus for each item. This may seem silly; after all, each team member should know this information. In our experience, they will not. We are routinely astonished at the ignorance of engineers and scientists about business and the parallel ignorance of marketers about technology. For example, in the drug industry, development chemists may not know the expected sales of a new antibiotic; and those in marketing may not understand the cost savings possible with a whole broth extraction.

Once this consensus is reached, the core team should decide on a standard interview format to ensure a common starting point. Often effective questioning can start with a list:

> What do you do now?
> How do you use the existing product?
> What features work?
> What does not work?
> How do you buy the product?

We recommend that this list be simple and generic, without references to specific product ideas.

The interview's value depends on the interviewer's skill in eliciting useful responses. In many cases, the core team will want to seek extra help, most commonly from marketing. If possible, however, all core team members should participate in at least one interview. Members who do not do so tend to become passive, leaping into life only when they feel their own expertise is critical. In every interview, we will want to observe the following guidelines:

(1) *Encourage tangents.* These will force your thinking beyond the confines of your preconceptions.
(2) *Stimulate with alternatives.* Interviews often start fast and then stall. Have some alternatives, illustrated if possible, ready to re-start discussion.

(3) *Remove assumptions*. Often existing product use will be constrained, for example by cost or by a particular temperature range. Ask for responses where these constraints are removed.

(4) *Be alert for surprises*. Customers will often describe their dreams only after they urge extending existing products. These dreams offer chances of real innovations.

The result of the interviews will be a potpourri of responses, most of which will be incomplete. We now need to organize these.

2.1–3 INTERPRETING CUSTOMER NEEDS

The customers' needs, recorded in the interviews, will normally be a random collection, filled with redundancy and irrelevancy. Our challenge is to organize these needs as groups, and to edit them into a cogent list. In this effort, we may decide to drop stated needs, even when these are suggested by many separate interviews. For example, those interviewed may ask for what we believe is a perpetual motion machine or for a product well beyond our company's anticipated expertise. Dropping these needs is fine. However, we want to do so consciously, carefully remembering our assumptions to ensure that these omissions are appropriate.

We next want to rank our list of needs as "essential," "desirable," and "useful." We should believe that the new product must meet all the essential needs to be successful. We would like it to meet many of the desirable needs, especially if existing competitive products do not meet these needs. We acknowledge the existence of the useful needs, though we are not planning to design products explicitly to meet them. In many ways, the system is like that pioneered by the Michelin Travel Guides for ranking tourist sites. Sites with three stars are "worth a journey," sites with two stars are "worth a detour," and sites with one star are "interesting." We want to rank our needs to make our design journey profitable.

The ways in which we organize and rank needs differs depending on whether we seek to improve an existing product or to invent a new product. If we seek an improved product, we will know how the existing product is used. We will be able to define the essential, desirable, and useful product attributes without much effort. In doing this, we want to begin with the core team, discussing the needs until the team reaches consensus. Often, the team will decide that their ranking of needs requires an additional review with customers, especially with lead users. The degree to which this additional review is important will depend on how major the improvements are compared to the existing product. If we are seeking a new product, we may simply be forced to group the ideas by target market or by common function. For a new product, we almost certainly must return to the customers, perhaps a somewhat different group of customers, seeking to specify needs more tightly.

We illustrate these general ideas with examples.

Figure 2.1–1 *De-icing aircraft* Planes are currently sprayed with aqueous glycol to melt accumulated ice and snow. We seek a new fluid which can be more efficiently recycled than the glycol.

EXAMPLE 2.1–3   MEASURING BONE DENSITY

Our employer is interested in a device which could be used in the home to measure changes in bone density. Bone density is key to the assessment of osteoporosis. Such a device could provide a measure of the impact of the company's dietary supplements. At present, measurements are normally made by X-rays or ultrasound and require expensive equipment. We seek a simpler measurement which does not give an absolute reading of density, but only a relative indication.

What needs must our device meet?

SOLUTION

Our core team interviewed patients and physicians to come up with the following:
Essential. The device must measure weekly changes in some property proportional to bone density.
Desirable. The device must be small, the size of a laptop. It must be easy to use.
Useful. The device should cost less than $100 and should be likely to easily receive Food and Drug Administration (FDA) approval.

EXAMPLE 2.1–4   ALTERNATIVE FLUIDS FOR DE-ICING AIRPLANES

Minneapolis–St. Paul is a major airport with over 400 flights daily. This high traffic reflects the use of the airport as a hub by a major airline. The city is infamous for its cold winters. In the winter, snow can collect on planes as they wait at the gate for take-off. As shown in Figure 2.1–1, the snow is removed by spraying the planes with de-icing fluids like propylene glycol, which are discarded after use. These fluids are often discharged directly into groundwater, even though they can be toxic to humans and wildlife. In major airports like Minneapolis–St. Paul, the discharged de-icing fluid is sewered, causing a major burden on the local sewage treatment plant. These de-icing fluids cause major pollution.

Table 2.1–1 *Interview results for de-icing fluids* This is a synopsis of conversations with engineers responsible for de-icing aircraft.

What do you do now? "We no longer de-ice at the gate, because this was too difficult to control. Instead, when they are ready for take-off, aircraft are moved to a central location for de-icing. We collect the run-off from the de-icing in underground tanks and then slowly bleed it off to the sewage treatment plant."

How do you use the product? "We spray with a 70 °C solution of 50% water and 50% ethylene glycol ($HOCH_2CH_2OH$). We spray for ten minutes or till there is no snow visible, whichever is longer. We then spray with anti-icing fluids." (These are snow-melting hydrogels which adhere to the aircraft while it is waiting to take off, but are removed by shear during take-off. In this example, we ignore anti-icing fluids.)

What features of the product work? "It's a good product. It works even at –30 °C. It isn't volatile. It doesn't cause corrosion, like salt. It's hard to burn." (Once burning, it has a heat of combustion per kilogram around one third that of ethane.)

What features of the product don't work? "None. It has some odor, and some passengers get sick, but not many. The effluent contains about 2% ethylene glycol, which is toxic to fish. The environmental agencies say it is probably toxic to humans, but I'm not convinced it is. Still, they won't let us discharge it, and they're always threatening to shut us down."

How do you buy the product? "We get it through the State of Minnesota, who require bids. One company has the largest share but we always have at least two suppliers to ensure some competition."

Define specifications for alternative de-icing fluids which are environmentally less abusive because they are easily recycled.

SOLUTION

Because we already have a product which works well, we seek alternatives which not only work well but cause less pollution. Interestingly, our customers are not the airlines but the airports. In practice, the airports contract de-icing to local engineering firms. We should interview the engineering firms' employees who are directly responsible for the de-icing.

A synopsis of these interviews is given in Table 2.1–1, following the format suggested above. Additional information is available from other sources. First, all airports (in non-tropical regions) and airlines are worried about this pollution problem. Some airlines are considering a special hanger with infra-red heaters, an interesting idea for airplanes fully loaded with fuel. While North American airlines insist on virgin ethylene glycol, some European airports distil the effluent and recycle the glycol.

We next want to extract from the interviews the key characteristics which a perfect product would have. In this case, we conclude that the perfect product

(1)   is sprayable;
(2)   has a low volatility;
(3)   does not smell;
(4)   is not toxic to fish;

Table 2.1–2 *Characteristics of de-icing fluids*
This table organizes the more scattered topics in Table 2.1–1. A successful product will probably meet all essential needs.

Essential	The product must melt snow. It must be easily applied. It must be non-corrosive.
Desirable	The product is non-carcinogenic. It is non-flammable. It is easily recycled.
Useful	The product is inexpensive. It is water miscible. It is available from multiple suppliers.

---

(5) is not toxic to humans; and
(6) is easily recycled.

These characteristics imply no surprises.

However, Table 2.1–1 also shows that the interviews contain contradictions and redundancies. We want to remove these in our list of characteristics of a perfect product. For example, the product cannot be non-volatile and still smell. The smell makes some passengers sick, so the product is volatile. As a second example, the product should be non-toxic to fish, humans, and everything else. We do not need to list these characteristics separately.

The key product attributes chosen in this case, given in Table 2.1–2, merit a brief discussion. That the product should melt snow is obvious, but a good reminder that we must soon choose a useful temperature range ($-30\,^{\circ}$C to duplicate the glycol). Applying the product must be easy; spraying is one obvious alternative. Clearly, the product should not easily burn, but we have not made this essential because we feel that, with good practice, we can mitigate this risk. Similarly, it is obviously necessary to avoid poisoning passengers or operators, but this can be achieved both by the product itself having low toxicity and by its being used in a way to avoid contact with people – this will be facilitated if it is non-volatile. With this table, we are poised to seek ideas for this new, recyclable de-icer.

We can also benefit from considering the wide variety of de-icers already available, which are summarized in Table 2.1–3. In this consideration, we remember that each of these works by depressing the freezing point of water. The amount of freezing point depression ΔT is given by

$$\Delta T = \left(\frac{RT^2}{\Delta \tilde{H}_{\text{fus}}}\right) x_{\text{de-icer}}$$

where R is the gas constant, T is the temperature, $\Delta \tilde{H}_{\text{fus}}$ is the enthalpy of fusion of water, and $x_{\text{de-icer}}$ is the mole fraction of de-icer. Because all the quantities in

Table 2.1–3 Current airplane de-icers Some of these are used at the airports shown.

Chemical compound	US $/ton	MW	Moles of solute/moles compound	Moles solute/ US $	Min. temperature (°F)	Used	Corrosion	Environment
Sodium chloride	25	58.4	2	1370	−8	–	Very corrosive to metals	Environmentally benign
Ethylene glycol	1225	62.1	1	13	9	–	Mildly corrosive to some steel	Toxic to mammals
Propylene glycol	1400	76.1	1	10	−60	Minneapolis	Mildly corrosive to some steel	Small impact on aquatic O_2 demand
CMA (Calcium magnesium acetate)	850	449.0	9	24	8	–	Mildly corrosive to some steel	Environmentally benign
Urea	125	60.1	1	133	15	Chicago	Not corrosive	Toxic to aquatic life. High impact on aquatic O_2 demand
Aviform L50 (potassium formate)	1025	84.1	2	23	−50	Copenhagen	Mildly corrosive to some steel	Environmentally benign
Safeway KA (potassium acetate)	645	98.1	2	33	−24	General Motors	Mildly corrosive to some steel	Environmentally benign

the parenthesis are fixed, the key is $x_{\text{de-icer}}$, which is highest for highly soluble compounds of low molecular weight. Ionization is useful if it does not cause excessive corrosion. These factors result in the complex comparison shown in the table.

EXAMPLE 2.1–5 "SMART" LABELS

Your company currently manufactures the labels that are printed to attach to food in supermarkets. For example, the labels attached to packaged chicken give the weight, the price per kilogram, the price of the package, and the date by which the chicken should be eaten. While your labels are a successful product, you want to improve them so they tell the consumer more about the chicken. For example, the date on such a "smart" label might change if the chicken were stored frozen.

Your core team has decided that your goal is to make labels that let consumers judge product quality. The team has also decided to focus on the food and pharmaceutical areas, building on company experience with its current labels.

Identify and organize customer needs for these new smart labels.

SOLUTION

To investigate this new product, we cannot easily use our standard questions, suggested above, because we are not improving an existing product, but developing a new one. We do not have good general guidelines. Instead, we begin by assembling several peers and collecting their thoughts. Then we organize these ideas, at least in a preliminary fashion.

Our efforts to identify needs are partially summarized in Table 2.1–4. While this list is abridged, it is not edited. In this abridgement, we did not exclude obvious tangents or irrelevancies. For example, idea #13 implies that the chemistry used for smart labels could be used for better condoms. Now it is true that there could be a major market for condoms that gave their users greater pleasure by, for example, releasing sexual stimulants. Such condoms might be more widely used, and hence reduce the spread of the HIV virus. However, your company is unlikely to jump from food labelling into condom manufacture. This irrelevant idea has value only in waking up any who are dozing.

Once we have a collection of needs like those in Table 2.1–4, we must organize them. In cases as unstructured as this one, we are probably best to organize them not around business topics but around intellectual topics, as shown in Table 2.1–5. From this outline, we can target product needs consistent with our company's business interests. We could choose to develop labels which use thermally triggered, irreversible reactions as our initial focus. We could choose as secondary needs labels which detect spoilage or are consumer activated. However, choosing product needs for final development will clearly require taking these organized needs back to our customers for further definition.

Table 2.1–4   *Customer needs for "smart" labels* These unedited needs are collected from customer interviews.

1.   We need labels to tell if the chicken has spoiled.
2.   Similar labels would be useful for ground beef.
3.   Smart labels should say if ice cream has ever melted.
4.   They should say if the chicken ever got warm.
5.   The best labels would be stamped onto meat, like the current USDA inspection stamps.
6.   Labels on canned goods would be good, too.
7.   Canned-goods labels should detect botulism.
8.   Stick-on labels are better.
9.   Labels should remind you when drugs go bad.
10.   Some labels could release drugs slowly.
11.   Nicotine patches are one good smart label.
12.   Could labels also release good smells?
13.   This technology could be used to make better condoms.
14.   Dairy goods should have some way to show their temperature history.
15.   These labels could respond to lactic acid, so you could tell if the milk had spoiled.
16.   You tell if things are spoiled by smelling them.
17.   Labels for cream should tell you just before the cream turns sour.
18.   Labels for milk should too.
19.   Eggs do not spoil much.
20.   Labels for frozen foods could say if they have ever been thawed.
21.   Beer labels should say if it is at the right drinking temperature.
22.   Wine labels could, too.
23.   Labels for fish should indicate that it has no smell.
24.   We need labels for mussels.
25.   How can color-blind people use labels that change color?
26.   Eggs should be thrown away when their stated shelf life expires.
27.   Pharmaceuticals stay useful long past their stated dates.
28.   Prescription drugs are usually finished, so there is no need for smart labels.
29.   Could the labels be activated by the customer?
30.   These would be like "Post It" notes which tell you when you must eat specific foods.
31.   These must be activated by the consumer.
32.   They could be part of tamper-proof packaging, activated automatically by opening the package.
33.   Some labels could be irreversible: once they change, they should not change back.
34.   The most important question is, "Has this food ever been warm?"
35.   These labels are especially important for caterers.
36.   Labels could respond to pH changes as well as temperature changes.

This is a good example. As a reader, you can teach yourself much more about evaluating customer needs by repeating this example than by reading the text. At this stage of product design, you will almost certainly get results different from those given above. Indeed, when we have repeated our interviews, we also got somewhat different results. These differences will disappear as we continue with the product design.

**Table 2.1–5 *One possible organization of "smart" label
needs*** This table organizes the needs listed in Table 2.1–4. The
numbers in parentheses refer to that table; needs #13, 19, 25,
26, and 35 are omitted.

I   *Temperature*
    A. Actual temperature assessed
        Beer and other foods (#21, 22)
    B. Temperature history assessed (#14, 34)
        Implies irreversible reactions (#33)
    C. Target: foods (#3, 4, 20)

II   *Spoilage*
    A. Anticipate spoilage (#17, 18, 30)
    B. Detect spoilage (#15, 16)
    C. Targets: foods (#1, 2, 23, 24) and drugs (#9, 27, 28)

III *Customer-activated labels (#29, 31, 32)*

IV   *Tangents*
    A. Label materials (#5–8)
    B. Controlled release (#10–12)

2.2   Consumer Products

Often the needs of chemical products are easily evaluated with conventional scientific instruments. We can use an ammeter to measure the current in a battery
which has been shorted out to determine how long a battery separator takes to
shut down. We can use a balance to measure the ice melted by a particular mass
of chemical. While these measurements may require judgment, they are familiar.

In contrast, many consumer products have important characteristics which
may be more difficult to measure using conventional instruments. We may want
to measure the "smoothness" of skin so that we can develop superior cosmetics.
We may want to determine the "softness" of fleece during development of new
winter parkas. We will normally be less confident about these attributes, and so
may have trouble developing appropriate specifications.

2.2–1 EVALUATING CONSUMER ASSESSMENTS

We want to discuss the special problems for these consumer products. To understand what consumers want, we must assess their needs. Doing so requires a scale
of measurement. There are two such scales: "hedonic," which is what customers
like; and "intensity," which is what they feel. We can understand the difference
from some examples. The first is from the fairy tale, "Goldilocks and the Three
Bears." Remember that Goldilocks judges Papa Bear's bed as "too hard," Mama
Bear's bed as "too soft," and Baby Bear's bed as "just right." Goldilocks is using
a hedonic scale.

More complex examples are shown in Figure 2.2–1. The top part of this figure
gives customer assessments vs. sugar concentration. For the hedonic scale shown

(a) Sweetness vs. Sugar Concentration

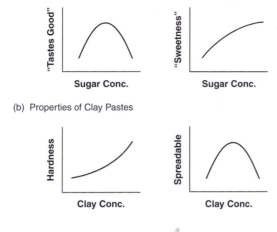

(b) Properties of Clay Pastes

Figure 2.2–1 *Customer assessments of (a) "sweetness" of sugar solutions and (b) "hardness" of clay pastes* The assessments can reflect preferences ("tastes good") or intensity ("sweetness"). (Clay data in (b) from Bourne, 2002.)

at the upper left, low sugar concentrations are judged not sweet enough, and very high sugar concentrations are judged too sweet. The maximum at intermediate concentrations is sometimes called the "bliss point." Not surprisingly, this point is culturally dependent: Europeans like food sweeter than Chinese, and the obese prefer more sugar than the slender.

The intensity data in Figure 2.2–1(a) show that higher sugar concentrations are perceived as sweeter. In contrast with the hedonic scale, these results do seem to be largely independent of culture or of personal condition. However, they are not necessarily linear, as shown both by the sweetness data and the clay paste data at the bottom of the figure.

The data shown in Figure 2.2–1(b) are for clay pastes: customers were asked to assess hardness (left) and spreadability (right). Both of these are intensity scales, and they are linked – if we get the correct hardness we can obtain optimum spreadibility. Once the relationship is established, we should need only assess hardness, which is easier for customers to do. If we had asked customers to assess "feels good on the skin," we would have obtained a hedonic scale for the clay pastes, which we speculate would look like the spreadibility curve.

These complexities may partly be because of the scales chosen to measure customer responses. Three different kinds of scales are used to assess consumer needs. The first is a simple comparison test. For example, we could ask customers for a skin cream, "Is sample A creamier than sample B? Is sample B smoother than sample C?" This type of test is fast and simple, suitable for simple product improvements, but it does not supply a model for further developments. It can be checked for consistency. In other words, if customers judge A creamier than B, and B creamier than C, then they should judge A creamier than C. We should check this by asking.

The second kind of scale is an ordinal or category scale. Professors' grading of their students is one good example: each professor decides which students deserve A grades, which B grades, and so on. Another common ordinal scale, shown in Figure 2.2–2, is used by physicians to judge pain. We might present

Table 2.2–1 *Assessments of softness for non-woven fabrics* These data show advantages and demerits of an ordinal scale from "1" to "10."

Person	Fabric 1	Fabric 2	Fabric 3
1	8	7	6
2	9	5	2
3	8	8	2
Average	8.3	6.7	3.3

Figure 2.2–2 *Assessments of pain* This chart, a fixture of doctors' offices, is an intensity scale.

consumers with a selection of non-woven fabric samples and ask them to judge "softness" of each sample from one to ten. "One" would mean a very rough fabric; "ten" would be as soft a sample as the consumer could imagine.

Ordinal scales are somewhat more difficult to implement than comparison scales, but they provide more information. They are probably the most common type. They give data that require analysis. To illustrate some of the problems in this analysis, consider the results for three customers and three non-woven samples shown in Table 2.2–1. Two points stand out. First, no samples are graded "1" or "10." This is the common result: everyone being tested assumes rougher or smoother samples could possibly exist, even if they were not presented. Thus, a scale from "1" to "10" is really a scale from "2" to "9." Second, person #1 judged all three samples to be almost the same, so the number of persons really making assessments is two, not three. The number making useful assessments is smaller than expected. We will discuss other details of the analysis below.

The most difficult scales to use, and the ones capable of producing the most information, are ratio scales. As an example, imagine we are evaluating six different cookies. We ask each customer to choose one cookie as a standard and

Table 2.2–2 *Assessments of "floral" notes in two perfumes* These data illustrate averaging of assessments with ratio scales.

Person	Scent 1	Scent 2	Average	\<Assess 1\>	\<Assess 2\>
1	10	20	$\sqrt{200}$	0.71	1.41
2	50	100	$\sqrt{5000}$	0.71	1.41

to evaluate the others relative to this standard. This evaluation could include answers like;

- Cookie #2 is three times sweeter than the standard.
- Cookie #3 is half as sweet.

Thus, this scale makes the ranking of an ordinal scale more specific.

The data for ratio scales are evaluated as follows. First, we define n_{ij}^0 as the evaluation of product "i" by consumer "j". Then we calculate the assessment n_{ij} of that consumer relative to his average score for all "l" samples

$$n_{ij} = \frac{n_{ij}^0}{\left(\prod_{i=1}^{l} n_{ij}^0\right)^{1/l}} \tag{2.2-1}$$

This removes the effect of choosing different standards. Next, we find the average score for each cookie by averaging over all "k" consumers:

$$\bar{n}_i = \left(\prod_{j=1}^{k} n_{ij}\right)^{1/k} \tag{2.2-2}$$

These are the results we seek. The use of geometric averages in this analysis is a consequence of ratio scaling.

A simple example of these ideas is shown in Table 2.2–2 for a "floral" aroma in perfume. One perfume is judged half as floral as the second, but the way in which this can be represented is unclear. We can make this clear using the equations above. While in this simplest case the results are obvious, they are much less so for eight perfumes and twenty consumers, a more typical experiment.

2.2–2 CONSUMER VERSUS INSTRUMENTAL ASSESSMENTS

We next want to reformulate these consumer needs as more quantitative specifications. The qualitative needs may well be trite: we already know that consumers want crunchier cereal, softer woollens, and smoother skin. The question is how much improvement will be noticed and valued.

Making the jump from "better" to "how much better" is normally easier if we can relate specific attributes to particular scientific parameters. For sure, we can continue to guess new product formulations, and then run back to our consumers to ask which formulation they like best. Such repeated consumer evaluation can be effective, but it is slow and expensive. In product design, where reducing development time is a constant goal, such a strategy should be avoided. Instruments are better.

The connection between consumer attributes and instrumental measurements is normally sought empirically. Often, this connection is exactly what we would expect. For example, "thick" soups have high viscosity and "thin" soups have low viscosity. This implies that assessments of "thick" should correlate inversely with assessments of "thin" but directly with viscosity measurements. This is supported by experiment. In the same way, assessments of "sweetness" correlate with measured sugar concentrations and judgments of "sourness" are proportional to citric acid concentration.

Sometimes, the connection between consumer attributes and instrumental measurements is unexpected. For example, we might expect that a breakfast cereal's "crunchiness" would be fairly directly related to the cereal flakes' fracture mechanics. It is not. Instead, it correlates best with the sound released during chewing. In the same way, we might expect that a beer's "smoothness" would be related to its viscosity. It is not. It is associated with the force of contact lubrication. This is why beers sold as "smooth" often have many very small bubbles. These small bubbles apparently reduce the coefficient of friction on the tongue and hence the related contact lubrication force.

The uneven correlation between consumer assessments and instrumental measurements has spawned an enormous number of proprietary instruments which try to imitate consumer assessments more exactly. Some of these are bizarre. One famous example attached a pair of dentures to an Instron machine. To use such an instrument, one would place a sample of, for example, beefsteak between the dentures and measure the force vs. distance that the beef is squeezed. The result would be a measure of the beef's rheology and hopefully of attributes like its tenderness.

Occasionally, we can see how consumer assessments and instrumental measurements are quantitatively related. For example, imagine that you pick up a metal spoon and a wooden spoon from a kitchen drawer. The metal spoon feels colder than the wooden one. In fact, both spoons are at the same temperature – the kitchen's temperature. Thus your perception is not of the spoons' temperature, but of your skin's temperature. If you estimate how your skin's temperature is influenced by the thermal diffusivity of the spoon, you can predict how much colder the metal spoon feels compared with the wooden one.

In the case of many important consumer products, purely empirical relationships between consumer and instrumental assessments have been painstakingly established by companies interested in improving product quality. Take the case of coffee. Over two hundred aroma molecules, important in coffee flavor and smell, have been identified. Each has been related to one or more words used by expert panels to describe coffee. Word like "citrus," "nutty," "papery," "green" are correlated to specific molecules by placing an expert's nose at a sniffing port on the end of a gas–liquid chromatography column. The molecule is identified by its residence time in the column and simultaneously described by the expert. To get meaningful results, this must be repeated many times by many different experts. After years of effort, major coffee companies have built multidimensional maps associating molecules with flavor and aroma qualities. These companies are able to go a long way in characterizing their products by the purely scientific methods of chromatography and mass spectrometry, without the expensive, time consuming, and inconsistent intervention of human experiences. This major advantage will be discussed in detail in Section 9.2.

Two further examples of relationships between consumer attributes and instrumental measurements conclude this section.

EXAMPLE 2.2–1 TASTY CHOCOLATE

One of the attractive characteristics of good chocolate is its smooth "melt-in-the-mouth" quality. Some chocaholics even claim a pleasant cooling of the mouth as the fat crystals in the chocolate absorb latent heat and melt. Sometimes chocolate exhibits a powdery tan-colored layer called "blooming," especially when it is old and has been stored at fluctuating temperature. While such chocolate has identical ingredients and chemical composition to good chocolate, it has an unpleasant powdery texture in the mouth.

Chocolate manufacturers employ expert panels to assess the quality of their chocolate. These testers are extremely competent – some can even identify the country of origin of the cocoa beans, in much the same way an expert oenologist can tell the terroir of a fine wine. The testers do a good job. However, it would be quicker, cheaper, and perhaps more reliable if we could augment their assessments using scientific instruments. How can we do so?

SOLUTION

Chocolate gets its "melt-in-the-mouth" sensation from the melting of cocoa butter crystals. Cocoa butter is a triglyceride, similar to olive oil, margarine, and animal fat. While triglycerides are the commonest form of naturally occurring fat, most have a poorly defined melting point because they contain a range of different fatty acids. Cocoa butter is unusual in having a narrow range of composition; it mainly consists of symmetrical triglycerides with oleic acid (unsaturated) in the 2-position and palmitic or stearic acid (saturated) in the 1- and 3-positions. This results in a well-defined crystal structure and a sharp melting point. However, cocoa butter can crystallize into six different crystal forms, each with its own melting point. The structures of forms IV, V, and VI are shown schematically in Figure 2.2–3. Which one is formed depends on the exact preparation. This is why confectionery chefs worry so much about the cooling rate and stirring speed when preparing chocolate dishes from the melt. Crystal form V is the desirable one, with a melting point of about $34\,^{\circ}\text{C}$, just below mouth temperature. It is easy to produce form IV by cooling molten cocoa butter without stirring; however, form IV has a melting point of about $28\,^{\circ}\text{C}$ and does not produce the desirable

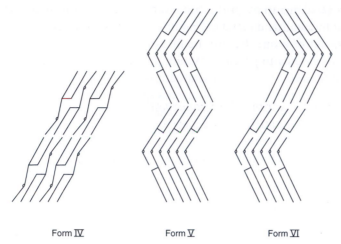

Form IV   Form V   Form VI

Figure 2.2–3 *Schematic structures of cocoa butter* In chocolate, form V tastes better than form IV. Form VI is "bloom."

mouth feel of form V. Form V is more stable than form IV, but is much harder to produce, requiring a prolonged period of gradual cooling while shearing, a process known as "tempering." Form VI, the most stable form with a melting point a little over 36 °C, is the cause of "blooming."

Thus, chocolate is an example of kinetics competing with thermodynamics: the thermodynamically more stable structures are kinetically harder to reach, presumably because the molecules are more tightly packed in their crystal structures. Cooling and stirring – "tempering" – is a delicate procedure because we are attempting to produce neither the kinetically favored form, nor the thermodynamically most stable one, but an intermediate between the two.

Chocolate testers are spending a lot of effort in identifying crystal forms of cocoa butter. We can do this more reliably and more quickly by using differential scanning calorimetry (DSC). In this method, we measure the rate of heat flow as the temperature is raised: as the crystals melt, energy is absorbed in the form of latent heat, without a concomitant temperature increase. We identify the presence of different crystal forms by the presence of different melting points in DSC. If we wish to be more quantitative, we can use the more sophisticated and expensive technique of powder X-ray diffraction, which shows how much of each crystal form is present. We can then assess product quality and so optimize manufacturing and storage procedures.

EXAMPLE 2.2–2   TEXTILE COATINGS WHICH FEEL SILKY

Our company makes a variety of textiles whose sales are steady but flat. As a result, we have formed a team searching for coatings or additives – "finishes" – which will provide additional value to our products. We are especially interested in making fabrics which feel "silky," "soft," and "supple," because these are attributes of considerable commercial value. These attributes are currently achieved by coatings of silicone elastomers.

Our company is willing to support a two-year project to understand attributes like "silky," and to use this understanding to develop intellectual property suitable for the entire textile line. Before the first meeting of our core team, we are asked to recommend how to proceed. We should discuss four specific aspects of the project:

(1)   What consumer attributes are important?
(2)   How should these attributes be measured?
(3)   What physical properties of the finishes are important?
(4)   How will consumer attributes and physical properties be related?

Our recommendations will be used as a starting point for the core team's discussions. What should we recommend?

SOLUTION

We want to outline a strategy for a project on textile finishes which make a textile feel "silky." These finishes will be applied as solutions or sprays, much in the way silicone elastomers are currently applied. These silicone elastomers will be our benchmark, the standard to which our new formulations must be compared.

Our strategy involves three complementary steps which should be pursued simultaneously. These are the choice and measurement of consumer attributes, a model of consumer perception, and identification of relevant physical properties. These are discussed below.

Consumer attributes. We will initially focus on four consumer attributes: "soft," "silky," "supple," and "smooth." The first three are mentioned in the project statement; the last, "smooth," is often mentioned in the textile literature. All these are part of what tailors call "hand," and are used to describe fabric texture.

We will measure consumer reactions by giving small groups of consumers samples of treated and untreated fabric. These samples should be chosen to have radically different physical properties: different weights, flexibilities, sheen, fiber sizes, and so forth. At present, we will only ask the consumers for a ranking: "Sample A is silkier than sample B"; "sample C is more supple than sample D;" etc.

While we understand that other measurements of consumer reactions – like ratio scaling – will provide more information, we are interested only in the simplest analysis for now. In this analysis, we are especially interested in how attributes depend on each other. For example, if the ranking for "silky" is the same as that for "smooth," then we may be able to treat these as synonyms.

A model of perception. To organize our thoughts, we suggest assuming that the attributes are associated with different physical effects. To start, we assume that "smooth" is associated with friction forces caused by rubbing the fingers across the fabric. We assume "supple" is associated with forces from bending larger amounts of fabric between the fingers. We suspect that "soft" is similar, but with smaller fabric samples squeezed normally. We guess that "silky" is a combination of "smooth" and reflected light, i.e. of tactile and visual perception.

These assumptions suggest that we need a model describing how the forces on the fingers are affected by the geometry and the physical chemistry of the textile. The geometry will include the size, spacing, and weave (if any) of the yarn in the textile. The model should be as simple as possible, but emphasize the impact of surface treatment rather than the ultimate properties of the fabric array. For example, it should emphasize changes in friction forces caused by a silicone spray rather than the force required to tear the fabric apart. At this stage, we may only want the model to identify the relevant physical properties, rather than to predict quantitative variations.

Relevant physical properties. We feel that the most important physical properties are the coefficients of friction for fiber–skin and fiber–fiber contact. We also want to check the amount of reflected white light. For both sets of properties, we are most interested in the values before and after chemical treatment.

We expect other physical properties will be less important. These include viscosity of the treatment chemicals, Young's modulus of the fibers, and the yield strength of the yarns. While these physical properties may be important in the overall characteristics of the textiles, they seem to us less relevant to the changes caused by surface treatments. However, we are uncertain about a third group of properties, including how hygroscopic and elastic the yarn is. We will need to continue to check these properties as our model for perception evolves.

2.3   Converting Needs to Specifications

The customer needs that we discover during interviews are often qualitative. These needs may include trivial product improvements and unrealistic product dreams. Their assembly and editing will be dominated by those whose strengths lie in marketing. Those on the core team whose strengths are in chemistry and engineering can be good critics or cheerleaders, but they will depend on others for detailed expertise.

Our next step is to convert these qualitative needs into particular product specifications. For example, if we want to make the airplane de-icers, we must decide on the degree of freezing-point depression, the speed of melting, and the efficiency of any recycle. Marketing is now less important and chemistry and engineering are paramount.

We can benefit from a cogent strategy for setting specifications. The experienced designer may use this strategy merely as a checklist for his own creativity. The novice may use it as a convenient way just to get started. The strategy we suggest has three steps:

(1)   Write complete chemical reactions for any chemical steps involved.
(2)   Make mass and energy balances important in product use.
(3)   Estimate any important rates which occur during product use.

Many of you with technical training will recognize that this strategy is a précis of the undergraduate curriculum.

Our rationale for these three steps merits discussion. Carefully writing out the equations of any chemical reactions may seem silly, especially because the reactions in some chemical mixtures may be incompletely defined. Many chemists will find the exercise trivial. Nonetheless, in our experience, this step is important to force consideration of mass versus moles. We have found that those without chemical training often write specifications in terms of mass, though the actual chemical phenomena imply moles. These confusions are not always limited to non-chemists. We vividly remember one petrochemical research group carefully developing a partial oxidation catalyst based on a hemoglobin analogue, only to realise that the analogue's large molar volume meant that it would have 1000 times too few active sites per volume to be practical. The first strategic step should always be checking the stoichiometry.

The second step is identifying the important mass and energy balances in the system. Such a thermodynamically grounded step establishes what can happen. It will give the maximum volume change, or the final temperature after an adiabatic reaction, or the size of any process recycles. In making these balances, we seek the simplest non-trivial solution. All physical properties, like densities and heat capacities, are taken as constant. All solutions should be assumed to be ideal. Like the first step, this will not take long.

The third step, the rate processes, should again make as many simplifying assumptions as possible. All relevant transport coefficients, like the thermal conductivity and the diffusion coefficient, should be taken as constants. The viscosity should be assumed Newtonian unless non-Newtonian behavior is essential to the specifications. Chemical reactions should be taken as first order or zero order. Reaction rates should be capped: they cannot go faster than diffusion control, even at high temperatures. At this stage, we want a simple answer.

Once these steps are complete, we will often want to choose a "benchmark," the standard against which our new products will be judged. Often, this benchmark will be an existing product, either the market leader or the product which we judge has the greatest intrinsic merit. Sometimes, the benchmark will be a hypothetical future product, perhaps one which we think our competitors could develop. In any case, we now set a standard: will our new product be better than our benchmark?

We should recognize that in choosing benchmarks, we are taking a serious risk. The risk is that we will skip the product design process and jump straight to our best guess of a solution. We will not finish carefully setting specifications, searching for new ideas, selecting rationally among these ideas, and developing our chosen product. This risk is real. To minimize it, we urge the product development teams to downplay their favorite guesses when they seek product ideas in the next stage. At the same time, we urge revising specifications with any extra information which seems important to the team.

EXAMPLE 2.3–I MUFFLER DESIGN

Automobile mufflers (in Britain, "silencers") rust from the inside out. They do this because after the car is driven, the muffler contains exhaust gases, including

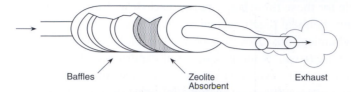

Baffles

Zeolite
Absorbent

Exhaust

Figure 2.3–1 *Car mufflers with reduced corrosion* Putting a zeolite
adsorbent in the muffler increases its life.

water vapor. When the muffler cools, the water condenses and corrodes the inside
of the muffler.

One clever route to avoiding this problem is to put a small bag containing
hydrophilic zeolite in the muffler, as shown in Figure 2.3–1. This adsorbs the
water vapor, preventing liquid condensation and so dramatically reducing corrosion. When the car is restarted, the hot exhaust gases heat the zeolite and drive
off the adsorbed water. The zeolite is then ready to adsorb more water when the
engine is stopped.

We are considering making a muffler which has this feature. How much water
will we need to adsorb? How fast should the adsorption be?

SOLUTION

This problem is simply an exercise in stoichiometry. Assume the muffler's volume
is about 5 litres, and that it is about 70% voids. The basic reaction in the engine
is:

$$C_8H_{18} + \frac{25}{2}O_2 \rightarrow 8CO_2 + 9H_2O$$

If the engine is run with 10% excess air, then the exhaust concentration at complete combustion is easily shown to be 0% C_8H_{18}, 2% O_2, 75% N_2, 11% CO_2,
and 12% H_2O. Thus:

$$\left(\frac{5\,L}{22.4\,L/mol}\right)0.7\left(\frac{0.12\,mol\,H_2O}{mol\,total}\right)\frac{18\,g}{mol} = 0.34\,g\,H_2O$$

Adsorbing this small amount can prolong the life of the muffler. We want the
adsorption to be prompt, but it need not be faster than the time that the muffler
takes to cool. Thirty seconds is probably a reasonable starting point.

EXAMPLE 2.3–2   BETTER THERMOPANE WINDOWS

In hot or cold climates, windows must not only allow light entry but must also
reduce heat transport. Such windows, shown schematically in Figure 2.3–2, are
typically made of two 0.3 cm panes of glass separated by a 1.6 cm gas gap. The
gap is filled with a gas and glued with a sealant. Because the gas should have a low
thermal conductivity to be an effective insulator, freons like CCl_2F_2 were used in
the past. Since freons have been phased out, argon is the industry's choice.

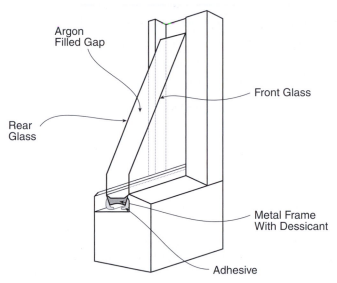

Argon Filled Gap

Front Glass

Rear Glass

Metal Frame With Dessicant

Adhesive

Figure 2.3–2 *Thermopane windows* We seek a better adhesive with lower water permeability.

However, the problem with the windows is also their appearance. This appearance relies on the sealant used to glue the glass panes together. Water vapor diffuses through this sealant into the gas gap between the window's panes. This water vapor can then condense when the weather cools, fogging the window. To avoid this, manufacturers put a dessicant between the window's panes, but this becomes saturated with time. In addition, the argon can leak out faster than air leaks in, causing low pressure so the panes bend together. This can distort the view seen through the window.

Suggest specifications for a better window sealant.

SOLUTION

Before setting specifications, we define the needs more completely. The most important customers for this sealant are the manufacturers of thermopane windows, whom we will want to interview. We may also want to talk to some building contractors to discover if window fogging depends on how the window is installed or where in the house it is located. We probably will gain less from talking with homeowners, even though they are the eventual buyers of the windows.

The questions we want to ask are straightforward:

(1)   What are the requirements of the sealant?
(2)   What sealants have already been tried?
(3)   What is the temperature change that the window encounters?
(4)   What is the window lifetime?

While other questions are helpful, they are not as central.

Table 2.3–1 *Possible gases for thermopane windows* Finding a replacement for freon has proved difficult.

Gas	Molecular weight	Thermal conductivity W/m °K
Air	29	0.024
Carbon dioxide	44	0.015
Argon	40	0.016
Krypton	84	0.003
Xenon	131	0.002
Freon	121	0.004

The specification for the sealant depends on the answers to the questions given above. Our interviews report that of the hundred sealants tried to date, silicone rubber gives the best seal. However, this sealant has a water permeability of 550×10^{-6} cm^2/sec, six hundred times greater than butyl rubber, which is 0.9×10^{-6} cm^2/sec. Unfortunately, butyl rubber doesn't stick to glass. Thus we suggest seeking a sealant with a ten times lower water permeability than silicone rubber but with adhesion not less than half that of silicone rubber.

In passing, we note that we want to minimize the heat flux q across the window, subject to a given temperature difference. This is

$$q = \left[\frac{1}{\dfrac{l}{k_{glass}} + \dfrac{l'}{k_{gas}} + \dfrac{l}{k_{glass}}} \right] \Delta T$$

where l and l' are the thicknesses of the glass and the gas layer, respectively, and k_{glass} and k_{gas} are the relevant thermal conductivities. The parameter k_{glass} is about 0.50 W/mK, and k_{gas} is about 0.02 W/mK. Thus, the thermal resistance of the glass doesn't matter. The thermal conductivities of possible gases are shown in Table 2.3–1. Because the thermal conductivity is proportional to the inverse square root of the gas's molecular weight, freons were the best choice. While noble gases are a possible alternative, radon is toxic and krypton is expensive. Argon is the industry's current choice, though its thermal conductivity is only 20% less than that of air. Given the problems with argon's high permeability through sealants, we should consider urging the abandonment of the industry's specification of argon and using dry air or carbon dioxide.

EXAMPLE 2.3–3 WATER PURIFICATION FOR THE TRAVELLER

People travelling into wilderness areas require drinking water. Often, water sources like streams and ponds are contaminated by viruses and bacteria. A particular problem in North America is giardia, a protozoan potentially present in 90% of water sources. (It is carried by animals as well as humans, which explains its wide spread.) Giardia results in severe intestinal problems, with very unpleasant symptoms including noxious farts, which are particularly unfortunate when sharing a tent.

Interviews with potential users (hikers, mountaineers, soldiers, equipment suppliers) might reveal the following list of needs in a water purification device:

> Produces safe water
> Is light/small
> Is fast acting
> Has a long lifetime
> Requires no power source
> Is cheap and re-usable
> Improves odor/flavor

Write specifications for a water-purification product.

SOLUTION

This example is different to the two just discussed because we are now considering a consumer product. For this product, the two steps of chemical reactions and thermodynamics suggested above are not helpful. There is a rate process at work: how fast do people lose water? The answer is about five litres or one gallon per day. Indeed, the original definition of a gallon derived from the daily allowance of water a farmer had to provide to each laborer.

We might aim to design for groups of two to four, for trips of up to two months. We need to be able to purify 2000 L of water before our product fails. We need around 20 L per day, but probably we would like to produce it all quickly, for example for cooking, so we want a flow rate of say 1 L/min. Because the product must be carried up mountains it must weigh less than 1 kg and occupy less than 1 L volume.

For safety, we can look to legislation: US health regulations require 99.9% removal of bacteria and protozoa in surface water treatment. Note we have focused here on bacteria and protozoa. While this is appropriate in the North American wilderness, travellers in third world countries are also threatened by waterborne viruses, passed on through human waste. Here we have an example of different specifications for different markets. Price? Climbers will pay in excess of $100 for a good ice axe, so one might expect they would be willing to fork out this much for clean water. The product must be effective over a full range of temperatures at which water is found and at altitudes up to 7000 m: 0–40 °C and one third to one atmosphere pressure.

Our final specifications might take the following form:

> Has a capacity of 2000 L
> Has a production rate of 1 L min^{-1}
> Removes 99.9% of bacteria and protoza
> Costs less than $100
> Has an operating range of 0–40 °C, and 0.3 to 1 atm
> Improves odor/flavor

(a)

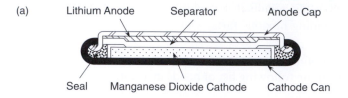

Lithium Anode   Separator   Anode Cap

Seal   Manganese Dioxide Cathode   Cathode Can

(b)

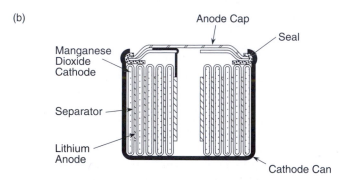

Anode Cap

Seal

Manganese Dioxide Cathode

Separator

Lithium Anode

Cathode Can

Figure 2.3–3 *High-performance batteries* The coin cell (a) is drawn at larger scale than the cylindrical cell (b).

We are now in a good position to start thinking of ways to achieve these specifications.

EXAMPLE 2.3–4   PREVENTING EXPLOSIONS IN HIGH-PERFORMANCE BATTERIES

Modern electronic devices like cellular phones and laptop computers require powerful batteries in order to function without frequent recharging. The result, shown schematically in Figure 2.3–3, is batteries with ever increasing power per volume and power per mass. Putting more and more energy into smaller and smaller packages means that the batteries are becoming bombs. Indeed, the president of one battery manufacturer had his mobile phone blow up while he was talking to his wife.

Obviously, there is a need for safe batteries which shut down before they blow up. How fast does the battery need to shut down?

SOLUTION

In answering this question we consider the most powerful batteries available, which are based on lithium. We choose a common type of lithium ion battery as our standard, using lithium cobalt oxide as the cathode and graphite as the anode. Our results will be applicable to other types of lithium ion batteries, since they all have similar energy densities and principles of operation. We also choose a typical laptop computer as a standard application. Such computers typically use a battery with about 100 W hr energy.

We begin by writing out the basic reactions during discharge. At the cathode:

$$\mathrm{Li}_{1-x}\mathrm{CoO}_2 + x\mathrm{Li}^+ + x\mathrm{e}^- \rightarrow \mathrm{LiCoO}_2$$

At the anode:

$$\mathrm{Li}_x\mathrm{C}_6 \rightarrow 6\mathrm{C} + x\mathrm{Li}^+ + x\mathrm{e}^-$$

The overall reaction is thus:

$$\mathrm{Li}_{1-x}\mathrm{CoO}_2 + \mathrm{Li}_x\mathrm{C}_6 \rightarrow \mathrm{LiCoO}_2 + 6\mathrm{C}$$

This reaction is reversed during charging. The electrodes are separated by an electrolyte consisting of a lithium salt and an organic solvent, which allows the transport of lithium ions. The concentration of lithium ions in the electrolyte might typically be 2 M. Under normal operation, the lithium ion concentration is nearly constant between the electrodes.

We now turn to the mass and energy balances. A battery in a laptop typically contains two electrodes around 3 mm thick, separated by a membrane separator. The membrane separator, which is around 20 µm thick, contains about 30% pores filled with 2 M Li^+ ions in the electrolyte. The battery is usually about 20 cm square, so its total volume is around 240 cm^3.

The adiabatic temperature rise if the battery is shorted out is found from the first law of thermodynamics:

$$\Delta U = Q + W$$

where ΔU is the internal energy change, Q is the heat, and W is the work. Because the battery is adiabatic, Q is zero. Because the battery has a constant volume, W is zero. Because the battery contains no gas, ΔU is about equal to the enthalpy change ΔH. Thus:

$$\Delta U = 0 \approx \Delta H_{\mathrm{rxn}} + V\rho \hat{C}_V \Delta T$$

where ΔH_{rxn} is the heat of reaction under the initial conditions, V is the battery volume, ρ is its average density, $\hat{C}_V$ is its average specific heat capacity (note that for solids and liquids $\hat{C}_V \approx \hat{C}_p$), and ΔT is the temperature rise caused by the short. Consistent with our goal of keeping our analysis simple, we assume that the battery is initially at 25 °C, the normal reference temperature for chemical reactions.

The heat of reaction is the energy in the battery or (–100 W hr). An average value of $\rho \hat{C}_p$ is about 3 J/cm^3 K. Thus,

$$0 = -100 \frac{\mathrm{J\ hr}}{\mathrm{sec}} \left(\frac{3600\,\mathrm{sec}}{\mathrm{hr}} \right) + 240\,\mathrm{cm}^3 \left(\frac{3\mathrm{J}}{\mathrm{cm}^3\,\mathrm{K}} \right) \Delta T$$

and

$$\Delta T = 500\,°\mathrm{C}$$

Before this temperature rise is reached, the solvent will have vaporized and the battery will have exploded. We may be in trouble. This completes our summary of the mass and energy balances.

Next, we turn to calculating the rate at which the battery heats up when it is shorted out. When the battery is shorted, the electrochemical reactions shown above are controlled by the rate of lithium diffusion from one electrode to the other. Thus, the time taken to discharge a shorted battery is roughly given by:

$$\begin{pmatrix} \text{amount Li} \\ \text{in battery} \end{pmatrix} = \begin{pmatrix} \dfrac{\text{Li}^+ \text{flux}}{\text{area time}} \end{pmatrix} \text{area (time)} = j_1 A t = \left(\dfrac{D \Delta c_1}{\delta} \right) A t$$

where j_1 is the lithium ion flux, D is its diffusion coefficient, δ is the thickness between the electrodes, Δc_1 is the lithium ion concentration difference, and A the battery area.

We can calculate the amount of lithium in the battery from the energy stored and the voltage at which lithium ion batteries operate, about 3 V:

$$\text{Energy} = \text{Voltage} \times \text{Charge}$$

$$\text{Charge} = 100 \frac{\text{J hr}}{\text{sec}} \left(\frac{3600 \, \text{sec}}{\text{hr}} \right) \Big/ 3\text{V} = 120\,000 \, \text{C}$$

Consistent with our approximate approach, we use a typical diffusion coefficient, $10^{-5} \, \text{cm}^2/\text{sec}$. Thus:

$$\left(\frac{120,000 \, \text{C Li}^+}{96,500 \, \text{C/mole}} \right) = \left(\frac{10^{-5} \, \text{cm}^2/\text{sec}}{20 \times 10^{-4} \, \text{cm}} \frac{2 \, \text{mole Li}^+}{10^3 \, \text{cm}^3} \right) \times 0.3 \times 400 \, \text{cm}^2 \times t$$

$$t \approx 1000 \, \text{sec}$$

However, we must shut down the battery before any vaporization begins. Because we start at 25 °C, we can only stand a temperature change of perhaps 50 °C. Thus:

$$t = \frac{50}{500} \times 1000 = 100 \, \text{sec}$$

The customer's need is to not be blown up. The resulting product specification is to shut down the battery within about a minute.

EXAMPLE 2.3–5 SPECIFICATIONS FOR COMFORTABLE ATHLETIC SHOES

Men's shoes are usually uncomfortable. Those which are open, like sandals, do not provide good support. Those like running shoes, which do provide good support, are usually laced tightly so that sweat cannot escape. Thus good support means wet feet, as suggested by the interview in Table 2.3–2.

Our company is interested in developing new shoe designs which both provide support and keep feet dry. Our core team expects that the new designs will center around three areas. First, we are interested in more breathable materials for the shoe uppers, i.e. a better Gortex. Second, we are interested in desiccants

---

Table 2.3–2 *An interview on how can running shoes be improved.*

What do you do now?
I wear either running shoes or leather dress shoes. In the summer, I will wear sandals for a
day or so, but longer periods make my feet hurt.

How do you wear the shoes?
In the normal way. I wear socks, and I keep the shoes laced. I do wear orthotics in the dress
shoes, but not in the running shoes.

What's wrong with your current shoes?
My feet sweat a lot so they get wet. Summer or winter, rain or shine, my socks are damp and
my feet are wet. I get athlete's foot a lot.

Do any other products help?
No. Foot desiccants work for less than half an hour. Heavyweight socks stay drier longer than
lightweight socks but not enough to make much difference. Short socks, coming just to the
edge of the shoe, do help some, but then my feet get really cold in winter.

Do you have any suggestions?
I've wondered if shoes with a small electric fan would work. The battery for the fan could fit
in the shoe's heel.

---

which adsorb water. Third, we wonder about a fan installed in the shoe's heel and
powered by the wearer's movement.

The team wants to write specifications and establish a benchmark for development of this product. The specifications and the benchmark depend on experiments which show that one foot sweats about 5 g of water per hour. Our goal is
to keep the relative humidity in the shoe at 60%, expecting that the humidity in
the surrounding air is 40%. The typical shoe area for mass transfer, which should
not include the shoe's sole, is around 120 cm^2.

We need general criteria which must be met by any design, and other criteria
specific to the three areas. Set the specifications which the product must exceed
to be successful.

SOLUTION

In discussing the specifications needed, we first describe general requirements
which any design must meet. We also suggest a physical model which facilitates
comparisons between different designs. We then describe specifications for the
three particular design solutions, depending on a vapour-permeable membrane,
a desiccant, and a small fan.

General specifications. Any improved shoe design must have three key characteristics. First, it must keep the air in the shoe at about 60% relative humidity for
eight hours. This humidity is the maximum felt to be "comfortable." Second, the
new design must remove the 5 g H_2O produced per hour by a typical foot. Third,
the new shoe should not weigh more than 10% of the weight of an existing shoe.
If that shoe weighs 500 g, then the weight increase must be less than 50 g.

Benchmark. The best commercially available shoes are based on the microporous perfluoronated polymer film Gortex. Because of its perfluorination, this

material is hydrophobic, impermeable to liquid water because of surface tension effects in its 0.2 μm pores. The water vapor transmission rate of Gortex, given as 28,000 g/m^2 day, allows us to calculate its permeance $D\varepsilon/l\tau$:

$$j_1 = \frac{D\varepsilon}{l\tau}\Delta c_1$$

where j_1 is the water vapor flux across the film and Δc_1 is the water concentration difference, which is the value at 100% relative humidity (RH) minus zero. Note that the permeance $D\varepsilon/l\tau$ is a combination of the diffusion coefficient D, the film thickness l, the void fraction ε and the pore tortuosity τ. If we assume the feet are at 30 °C, then the concentration of water vapor at saturation is about 3×10^{-5} g/cm^3. The permeance of Gortex is given by

$$\frac{28,000 \text{ g/m}^2\text{day}^{-1}}{10^4 \text{ cm}^2/\text{m}^2(24 \times 3600 \text{ sec/day})} = \frac{D\varepsilon}{l\tau}(3 \times 10^{-5} \text{ g/cm}^3 - 0)$$

$$\frac{D\varepsilon}{l\tau} = 1.08 \text{ cm/sec}$$

This value is the standard of the current technology, the one which we must out-perform.

Thin-film designs. We now turn to the additional specifications for thin films which could outperform our Gortex standard. We seek a thin non-porous poly-mer film which will not be penetrated by water liquid and which has a higher permeance than our benchmark. To see what we need, we parallel the calcula-tion for Gortex using values in the problem statement.

$$j_1 = \frac{DH}{l}\Delta c_1$$

$$\frac{5 \text{ g}}{120 \text{ cm}^2 (3600 \text{ sec})} = \frac{DH}{l}(0.6 - 0.4) \times 3 \times 10^{-5} \text{ g/cm}^3$$

$$\frac{DH}{l} = 1.9 \text{ cm/sec}$$

In these equations, DH is the permeability, the product of the diffusion coef-ficient D of water in the polymer, and the solubility H of water in the poly-mer. This calculated value, which is what we need, assumes 60% relative humidity for air in the shoe and 40% relative humidity for air outside the shoe.

The permeance is only twice that available from Gortex. It may be possible to design a better shoe based on diffusion through a thin film, but it will be difficult.

Desiccant designs. A second group of designs uses desiccants. These absorb water to keep humidity in the shoe at some specific limit, for example, at 40% rel-ative humidity. The desiccant could be contained within the shoe's insole, much like the activated carbon inserts to reduce foot odor. The desiccant would need to be regenerated, perhaps in a microwave oven.

The challenge with such designs will clearly be the desiccant mass. The general specifications given above show that, over eight hours, we need to absorb 40 g of water. This is close to the 50 g maximum weight gain allowed by our specifications. If we are serious about these designs, we will need to make sure that the limits placed by these changes in weight are feasible.

Pumps. The third general area of design involves supplementing water loss by diffusion with water removal by convection. In its simplest form, we imagine a pump, which blows fresh, drier air through the shoe. The pump could be powered either by a battery or by walking. We must specify the required air flow in and out of the shoe, Q. The air will enter at the ambient relative humidity of 40%, and leave with the maximum humidity for comfort of 60%. Again, assuming that the saturated water concentration is 3×10^{-5} g/cm^3 at the foot temperature of 30 °C, we have

$$\frac{5\,\mathrm{g}}{3600\,\mathrm{sec}} = Q \times (0.6 - 0.4) \times 3 \times 10^{-5}\,\mathrm{g/cm^3}$$
$$Q = 230\,\mathrm{cm^3/sec}$$

This large flow will be difficult to achieve. We are in trouble with all three options. To increase our chances of success, we must seek a mechanism beyond diffusion, chemical adsorption, or forced convection, which are basic to the detailed specifications developed here.

2.4 Revising Product Specifications

The strategy above – stoichiometry, overall balances, and rates – leads to preliminary product specifications. We will normally be able to obtain good estimates of these. However, such estimates often have three serious shortcomings. The first shortcoming is that our initial specifications may be blatantly unrealistic. They may suggest that we need materials which are excessively expensive. They may require huge flows or huge concentrations. They may imply elements which are not in the periodic table. Such shortcomings mean that we must revise the product specifications or abandon the product design.

Parenthetically, this type of revision is described idiomatically with different images in different companies. Sometimes, it is called a "sanity check" or a "gut check." Often it is described as a "back of the envelope calculation." One company calls this revision a "chicken test," an apparent reference to a fabled test of aircraft engines made by the Canadian government. The test involved tossing frozen chickens into the running jet turbines. The British navy have a similar test to check out the viability of new marine impeller designs; they throw a railway tie into the propeller and reject those designs which fail to survive intact.

The second shortcoming of the strategy in the previous section is that it does not make a careful comparison with existing products, and especially with our benchmark. In many cases, we will have a good idea about what particular improvements we want, and we will have several specific ways in which we want to make these improvements. We can gauge the significance of these improvements by comparing our specifications with the benchmark. The comparison may show that our new specifications are inadequate or overly ambitious. If so, we should revise them.

The third shortcoming results from presuming a completely unregulated society, where we can make any desired product without interference. In our current industrial society, this is not true. We will want to respect broader social needs. We will not want to kill our employees or poison our customers. We won't want to pollute our communities or produce unnecessary waste. We do not want our products to violate laws imposed by government to safeguard its citizens.

These societal constraints will produce a second set of needs often phrased as governmental regulations. Most often, such regulations are not capriciously invented but are developed with governmental oversight by groups of stakeholders, including customers, manufacturers, and environmental groups. Though the results are widely disparaged, they are hard to develop and merit sympathy for those who have struggled to establish them.

We illustrate the revision of specifications to account for these shortcomings in the examples which follow. The use of a benchmark is vividly illustrated in the first example given below, which seeks a de-icer for winter roads which causes less corrosion. (We know, we had another de-icing problem in an earlier section; but this book was partially written in Minnesota.) Roads are currently de-iced with rock salt. Thus, rock salt is an obvious benchmark, and one which is an enormous help in revising our specifications.

In other cases, we may be developing a new product for which no direct competition exists. We can sometimes benefit by looking for products which perform similar but chemically different functions. The second example in this section shows how this "similar function" strategy works. We seek a liquid which selectively absorbs nitrogen from natural gas. We currently have no such liquid. However, other liquids are available which do absorb other chemical species from other gas mixtures. Thus, as a benchmark, we choose an aqueous solution of monoethanol amine, which is used to absorb carbon dioxide from gas mixtures found in petrochemical processing. Comparing our hypothetical nitrogen absorber with monoethanol amine helps us to revise specifications.

The third and fourth examples describe the challenges of governmental regulations. In the third, we discuss regulation of persistent organic pollutants, released into the environment and then accumulated in the human body in potentially dangerous concentrations. The fourth example discusses the labelling of sunscreens and what the labelling implies. All four examples illustrate the difficulties of revising product specifications.

EXAMPLE 2.4–1   DE-ICING WINTER ROADS

In winter, roads are spread with a mixture of sand and salt to improve traction and melt ice and snow. While this treatment works well above –20 °C, the salt causes significant environmental damage. It corrodes cars about four times faster than water alone. It weakens bridge decks and parking ramps, sometimes causing their collapse. It can pollute local water wells.

Not surprisingly, government agencies frequently look for alternatives to salt which are less environmentally abusive. Using salt as a benchmark, they suggest the following specifications for the alternative chemical:

(1)   It should melt ice over a similar temperature range.
(2)   It should melt a comparable amount of ice per kilogram as salt does.
(3)   It should cause less corrosion per kilogram than salt.

Two alternatives to salt which are frequently suggested are urea and calcium magnesium acetate (CMA).

Compare the performance of salt with these alternatives as a means of revising product specifications.

SOLUTION

To make this comparison, we must first consider why chemicals cause ice to melt. The basic phenomenon is freezing point depression. When any solute is dissolved in water, the chemical potential of the water molecules is lowered. As the temperature is lowered, this chemical potential rises faster than that of ice does, because the entropy of the solid is lower than that of the liquid: remember that $(\partial G/\partial T)_p = -S$. When the chemical potential of the water in solution becomes higher than that of ice, the water freezes. To put these ideas on a more quantitative basis, we remember that for a freezing point depression of ΔT,

$$x_2 \approx \left(\frac{\Delta \tilde{H}_{\text{fus}}}{RT^2} \right) \Delta T$$

where R is the gas constant, T is the freezing temperature, $\Delta \tilde{H}_{\text{fus}}$ is the enthalpy of fusion of water, and x_2 is the mole fraction of additive. For example, for a depression of freezing temperature of 10 K,

$$x_2 = \left(\frac{6 \times 10^3 \, \text{J/mol}}{8.31 \dfrac{\text{J}}{\text{mol K}} (273 \, \text{K})^2} \right) (10 \, \text{K})$$
$$= 0.097$$

For salt, this is about equal to 18 g NaCl per 100 g H_2O. This result is useful for revising the first two product specifications given above.

Developing a basis which is similarly useful for the third, corrosion-based specification is more difficult. The corrosion rate will normally involve two or more sequential chemical steps. In the limit where surface chemical reactions are controlling, the rate of corrosion will be given by the Butler–Volmer equation, which for small over-potential η is:

$$j_1 = \frac{j_0 \eta F}{RT}$$

where j_0 is the "exchange current density," i.e. the reaction rate at equilibrium, and F is Faraday's constant. Unfortunately, we have no easy ways of guessing

---

Table 2.4–1 *Relevant properties of three possible de-icers* Sodium chloride (NaCl) is the benchmark. Urea is sometimes used on airport runways because of reduced corrosion. Calcium magnesium acetate (CMA; $(CH_3COO)_4CaMg$) is one alternative favored by some government agencies.

Basic compound	NaCl	Urea	CMA
Solubility g/100g H_2O	36	100	40
Species in solution	Na^+, Cl^-		$Ca^{++}, Mg^{++}, 4CH_3COO^-$
Molecular weight	58.5	60	300
Moles particles/mole compound	2	1	6
Mole fraction at saturation	0.18	0.23	0.13
Moles particles/mass compound	34	17	20
Cost per metric ton, US$	20	130	800
Moles particles per dollar	1700	130	25

---

either j_0 or η in this situation, so this result does not help us. However, we can estimate the fastest possible corrosion rate, which will occur when paint or rust limits how fast the de-icing chemical can reach the surface. This rate is the diffusion-controlled limit:

$$j_1 = \left(\frac{D}{\delta}\right) c_1$$

where D is the diffusion coefficient of the chemical across a rust layer of thickness δ. Note that c_1 is the molar concentration, not the mass concentration. Similarly, in the freezing point depression, the important variable was the molar concentration.

We can now compare the expected performance of the three possible chemicals. While salt and urea are straightforward, CMA requires some explanation. This material, which is not readily available commercially, is made by reacting dolomite limestone and acetic acid

$$CaMg(CO_3)_2 + 4CH_3COOH \rightarrow (CH_3COO)_4 CaMg + 2CO_2 + 2H_2O$$

The limestone sells for around $20/ton; the acetic acid is about $0.37/lb; so the material cost is almost $800 per ton. Clearly, this cost is dominated by the acetic acid, and can be reduced if waste acetic acid is available.

The results of the comparison of salt, urea, and CMA are given in Table 2.4–1. Right away, we see that salt will be difficult to beat if the initial specifications are used. In particular, the first specification says that an alternative chemical should melt ice over a similar temperature range to salt. From the equations above, we see that the maximum freezing point depression is proportional to the maximum mole fraction, i.e. the mole fraction of saturation solubility. From Table 2.4–1, we see that these mole fractions are roughly comparable, so we seem to be on the right track.

However, the second specification is that a fixed mass of chemical should melt the same amount to salt. Here, the alternatives stumble. While urea has a similar molecular weight as salt, it does not ionize, so a mass of urea gives a solution whose mole fraction is about half that of salt. In the same way, CMA produces only half the mole fraction as salt, or in the terms used in the table, 20 vs. 34 particles per mass. Salt is better.

The third specification, regarding corrosion, is the hardest to evaluate. The corrosion caused by urea will be modest, because urea does not produce ions. Salt and CMA do produce ions. When corrosion is diffusion controlled, the corrosion rate is proportional to the ionic concentration. But ice melting is also proportional to ionic concentration. Thus any ionic chemical will accelerate corrosion and melt ice in roughly proportional amounts.

Finally, we turn to chemical cost. While cost was not included in our original specifications, it will certainly have a role in our final decision. What will matter is the ice melted per dollar spent, or in chemical terms, the particles delivered per dollar spent. As Table 2.4–1 shows, salt is way out in front.

Thus, if we still seek an alternative chemical, our specification requires significant revision. The exact nature of this revision depends on additional information. Even without this, we can see alternatives requiring creative thinking. For example, we are immediately struck by the high cost of the alternative chemicals. Urea is already a commodity, so we are unlikely to be able to cut its cost much. On the other hand, CMA's price is dominated by the acetic acid cost. To cut this cost, one imaginative group creatively fermented a mixture of lime and garbage in cement mixers which normally sat unused in the winter. They then used the fermented garbage as a de-icer, a solution better suited to rural areas.

As a second example, we can remember that our real objective is to clear the road, not necessarily by melting the ice. If we can just get the ice to debond from the road surface, then we can remove it with a snowplow. This suggests replacing the specification of ice melted per mass of chemical with ice removed per mass in some standard test.

EXAMPLE 2.4–2 SCRUBBING NITROGEN FROM NATURAL GAS

In the next few decades, a major fraction of the world's energy is expected to come from natural gas. The best natural gas is largely methane, with small amounts of other hydrocarbons. Future natural gas may be much less pure, containing large amounts of carbon dioxide and nitrogen.

These impurities need to be removed. One easy way is cryogenic distillation, but this is prohibitively expensive. Carbon dioxide can be removed by absorption into aqueous solutions of amines, using packed towers like that in Figure 2.4–1. There is no similar method of absorption for nitrogen.

Thus, the need is for a liquid or liquid solution which absorbs nitrogen but not methane. There is currently no liquid with this property, so there is no obvious benchmark. Instead, we choose monoethanol amine, the standard absorbent for carbon dioxide. Our first attempt at product specification seeks a new liquid solution for nitrogen which would be used as the amine solutions are used for carbon

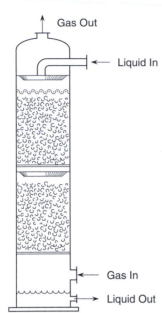

Gas Out

Liquid In

Gas In

Liquid Out

Figure 2.4–1 *A Packed tower for capturing carbon dioxide* We want to use similar equipment to remove nitrogen from low-quality natural gas.

dioxide. This new solution would have similar loading, react with similar kinetics, and use similar equipment to that used for CO_2. Such similarity is sometimes called a "transparent technology" because the operator running the process does not need to know if she is removing CO_2 or N_2.

See if these specifications are reasonable, and revise them if necessary.

SOLUTION

We can begin by looking at the stoichiometry and the mass balances for carbon dioxide. The key reaction is

$$CO_2 + H_2O + 2RNH_2 \rightleftharpoons (RNH_3)_2 CO_3$$

where R is, for example, $HOCH_2CH_2$. If the aqueous solution contains 10 wt% amine and reacts completely, then the amount of CO_2 absorbed will be

$$10\% \left(\frac{1 \, mol \, CO_2}{2 \, mol \, amine} \right) \left[\frac{44 \, g \, CO_2/mol}{61 \, g \, amine/mol} \right] = 3.6 \, \% \, CO_2$$

To meet our specifications, we would like our new liquid to absorb a similar amount of nitrogen.

To see if this specification is reasonable, we need to anticipate the chemistry of nitrogen complex formation. The complexes are most likely going to be organometallic, possibly with a porphyrin structure aping that in hemoglobin. These chemical compounds are likely to be large, with a molecular weight around 500 g/mol or more; hemoglobin itself consists of four sub-units each with a molecular weight of around 17,000 g/mol, although the active heme group is much smaller than this. They are unlikely to be soluble in water to greater than 10 wt%. Animals' oxygen-absorbing systems achieve pigment concentrations in

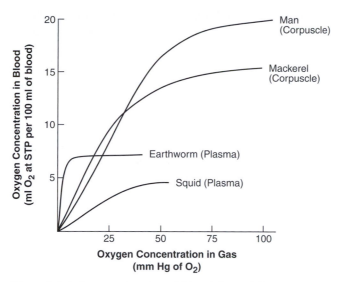

Figure 2.4–2 *Oxygen capacity of blood from different animals* These data let us set specifications for similar compounds that would absorb nitrogen.

their blood of up to about 15 wt% as shown in Figure 2.4–2. Because the evolutionary pressure to maximize the oxygen-carrying capacity of blood has been intense, we are most unlikely to be able to develop a heme analogue with a higher solubilty; if anything, 10 wt% represents an optimistic value. Thus, if the organometallic compound forms a 1:1 complex,

$$10\% \left(\frac{1 \, mol \, N_2}{1 \, mol \, complex} \right) \left[\frac{28 \, g \, N_2/mol}{500 \, g \, complex/mol} \right] = 0.6 \, \% \, N_2$$

Like it or not, the nitrogen scrubbing solution is going to be more dilute. For comparison, the oxygen-carrying capacity of human blood is around 0.026%, so achieving even 0.6% nitrogen will be challenging.

The relative dilution in our absorber, a consequence of reaction stoichiometry, has a significant effect on the rates of absorption as well. There are two types of rate processes which are important. First, the new complex must react with the nitrogen quickly and reversibly. Ideally, we would like complex formation to be as fast or faster than the amine–CO_2 reaction. Equally, we would like to have the complex quickly decompose when we increase the liquid temperature slightly. To discover if this can happen, we must synthesize our complexing agent and make absorption and desorption experiments. The results of these experiments will determine the height of the packed towers which we will require for absorption and stripping.

The second type of rate process concerns not the height but the diameter of the packed tower. To explore this, we return to our carbon dioxide benchmark. Remember we have already calculated that if the amine solution is completely saturated with CO_2 then we have a liquid which is 3.6 wt% CO_2. Imagine that we have a gas flux G of density ρ_G and a liquid flux L of density ρ_L flowing countercurrently in a packed tower of cross section A. As a basis, we assume that we have a total gas flow equal to 100 mol/min. This feed stream contains 80 vol% CH_4 and 20 vol% CO_2. Such a feed will have an average molecular weight of 21.6 g/mol. If 90% of the CO_2 is removed to produce an amine solution which is 50% saturated, then the mass flow of amine solution LA may be found from

$$(\text{mass } CO_2 \text{ leaving in liquid}) = (\text{mass } CO_2 \text{ absorbed from gas})$$

$$0.50\,(0.036)\,LA = 0.90\,(0.20)\,\frac{100\,\text{mol}}{\text{min}}\left(\frac{44 \times 10^{-3}\,\text{kg}}{\text{mol}}\right)$$

$$LA = 44\,\text{kg/min}$$

The operation of the packed tower depends critically on the "flow parameter," which for this feed is

$$\frac{LA}{GA}\sqrt{\frac{\rho_G}{\rho_L}} = \frac{44\,\text{kg/min}}{\dfrac{100\,\text{mol}}{\text{min}}\left(\dfrac{0.0216\,\text{kg}}{\text{mol}}\right)}\sqrt{\frac{21.6\,\text{g}/22.4 \times 10^3\,\text{cm}^3}{1\,\text{g/cm}^3}}$$

$$= 0.63$$

This value, in the middle of the normal design range, can be used to determine the tower's cross-sectional area.

We now turn from the benchmark, carbon dioxide, to our new target, nitrogen. As before, we imagine that we have the 100 mol/min feed which is now 80% CH_4 and 20% N_2, and 90% of the N_2 is removed to produce a liquid which is again 50% saturated. The mass flow of this new product solution is

$$0.50\,(0.006)\,LA = 0.90\,(0.20)\,\frac{100\,\text{mol}}{\text{min}}\left(\frac{28 \times 10^{-3}\,\text{kg}}{\text{mol}}\right)$$

$$LA = 170\,\text{kg/min}$$

Now the flow parameter is quite different

$$\frac{LA}{GA}\sqrt{\frac{\rho_G}{\rho_L}} = \frac{170\,\text{kg/min}}{\dfrac{100\,\text{mol}}{\text{min}}\left(\dfrac{0.0184\,\text{kg}}{\text{mol}}\right)}\sqrt{\frac{18.4\,\text{g}/22.4 \times 10^3\,\text{cm}^3}{1\,\text{g/cm}^3}}$$

$$= 2.6$$

This value is higher than the normal design range, and may require packed towers of larger cross sections than used for the amines.

Thus, the original specification, that the absorption of nitrogen closely imitates that of carbon dioxide, will be difficult to achieve. Because complexing the nitrogen will require an absorbent of high molecular weight, the nitrogen concentration in the absorbing liquid will be less than that for the carbon dioxide. This means that relatively more liquid will be needed to remove the nitrogen, which in turn suggests that the packed-tower geometry will change. This is by no means a fatal flaw for nitrogen absorption. It does mean that the specifications for the two absorptions will be different. Amine scrubbing of CO_2 may not be a good benchmark for potential N_2 absorption products.

EXAMPLE 2.4–3   IDENTIFYING PERSISTENT ORGANIC POLLUTANTS

Persistent organic pollutants (POPs) are synthetic organic chemicals which are released into the environment as combustion products, insecticides, and waste streams. Examples are dioxins, DDT, and PCBs (polychlorinated biphenyls, formerly used as non-conductors in electrical transformers). These toxic species can accumulate in the lipids (i.e. the fats) in our body, potentially to toxic levels. They accumulate from the water we drink, the air we breathe, and the food we eat. This "biomagnification" is especially important for animals at the top of the food chain, like us humans.

Commonly, the risk posed by each POP is judged by approximating all the body's lipids as octanol and all the blood and tissue as water. The key parameter for the danger of a toxin is then taken as its octanol–water partition coefficient K_{OW}, defined as

$$K_{OW} = \frac{\text{concentration in octanol}}{\text{concentration in water}}$$

Small values of K_{OW} are felt to be good, while large values are bad. This has been conventional wisdom, especially for fish. However, some studies (*Science* **317** (2007)182–183, 236–239) suggest that, for land animals, K_{OW} is less significant than the corresponding partition coefficient between octanol and air K_{OA}

$$K_{OA} = \frac{\text{concentration in octanol}}{\text{concentration in air}}$$

Their arguments are impressive but incomplete.

Which of these parameters provides a better criterion for environmental risk?

SOLUTION

Governmental regulations identify a toxin which is most likely to biomagnify by means of its octanol–water partition coefficient, or K_{OW}. Values of K_{OW} between 10^5 and 10^8 are believed to be especially dangerous. Smaller values of K_{OW} mean that the compound is effectively excreted; larger values suggest that the compound is so insoluble in water that there isn't enough there to cause problems in the first place.

The use of K_{OW} does predict biomagification in fish. However, it works much less well for mammals, including humans. An alternative is the octanol–air partition coefficient K_{OA}, closely related to the product of K_{OW} and a Henry's law constant. K_{OA} is useful for animals which feed on land rather than in water. For example, lichens accumulate persistent organic pollutants from air; caribou eat lichens and so biomagnify the pollutants' concentrations; and artic wolves eat caribou and so further magnify these concentrations.

The values of K_{OA} which are ecologically important are different to the values of K_{OW}. For humans, values of K_{OW} between 10^3 and 10^8 are said to be potentially dangerous if K_{OA} is also greater than 10^7. Presumably, smaller values of K_{OA} mean that the pollutants will not be persistent because they will be eluted by respiration. Still, the range suggested as dangerous by these values of K_{OW} and K_{OA} is much larger than that previously accepted. The range probably includes at least one third of the organic compounds made commercially, that is, over 4000 different chemical species.

These results imply that trace chemicals in the environment may be considerably more dangerous than has been believed in the past. If this is true, it is an implicit threat to the chemical industry – and to the chemical profession. However, we note that the current parameters are equilibrium concepts, that both K_{OW} and K_{OA} are based only on thermodynamics. Thermodynamics cannot be the only important parameter, and environmental engineers recognize that rate processes must also be important. To make this more specific, they note that the uptake rates in either the lungs or the intestine will probably not vary much with K_{OW} or K_{OA}. This uptake rate will be given by

$$\begin{pmatrix} \text{uptake} \\ \text{rate} \end{pmatrix} = k \left[\begin{pmatrix} \text{concentration} \\ \text{in environment} \end{pmatrix} - \begin{pmatrix} \dfrac{\text{concentration in body}}{K_{OW} \text{ or } K_{OA}} \end{pmatrix} \right]$$

where k is an overall mass transfer coefficient based on a driving force in the environment. From other studies of mass transfer, we expect k will be about the same for all pollutants. Because the values of K_{OW} and K_{OA} are large for the compounds of interest, we expect that the term containing these values will be small. Thus we expect that uptake rates will largely be independent of the values of K_{OW} and K_{OA}. Still, the suggestion that high values of K_{OA} are an important indication of toxicity should be taken seriously in future environmental regulations, at least until more data are available.

EXAMPLE 2.4–4   REVISED SUNSCREEN SPECIFICATIONS

The Food and Drug Administration (FDA) has suggested a new, revised sunscreen-rating system, originally proposed in a different form on May 12, 1992. This system would replace the SPF ("sun protection factor") ratings used for current products with a new system based on four stars (one is lowest and four is highest). The new regulations are, in part, a response to a consumer fraud lawsuit brought in California against sunscreen manufacturers.

Humans tan because UV light stimulates the production of the pigment melanin; they burn because UV light destroys cells in the outer layer of skin, called the epidermis. The UV light can also cause premature skin aging and melanoma, one of the most common skin cancers. Somewhat arbitrarily, UV light is divided into three categories: UVA, from 315–400 nm, has the lowest

Table 2.4–2 *Common chemicals used in sunscreens.*

Generic name	Structure	Comments
p-Aminobenzoic acid (PABA)		• For UVB • Often esterified
Cinnamic acid		• For UVB • Usually esterified
Benzophenone		• For UVA
Salicylic acid		• Largely replaced by others • As acetate, aspirin
Avobenzone		• Proposed for UVA • Not currently allowed
Dihydroxyacetone (DHA)		• Not a sunscreen • Used for "tan-in-a-bottle"
Ensulizole		• For UVB

energy and is assumed to produce melanin and, hence, be responsible for tanning; UVB, from 280–315 nm, is blamed for burning; and UVC, at shorter wavelengths and so still higher energies, is largely screened by the atmosphere and so is ignored.

Sunscreen products most commonly try to block UVB and allow permeation of UVA. They have a different goal than sun blocks which try to prevent penetration of all sunlight. Sun blocks, which are the familiar white coating on the noses of pool lifeguards and cricket players, are usually suspensions of particles of ZnO or TiO_2. Sunscreens are also different to "tan-in-a-bottle" products, usually based on dihydroxyacetone. These effect a chemical reaction with the skin, producing a tan color.

In contrast, sunscreen products use compounds like those shown in Table 2.4–2. These are commonly based on aromatic species; similar structures are often carcinogens themselves. Note that the actual molecules shown in the table are usually modified, most often by esterification. They are combined with detergents and oils which may keep the skin hydrated and flexible. Still, compounds like those in the table are the key to a sunscreen's effectiveness.

Until recently, sunscreen products were described by a single number called an SPF ("sun protection factor"). In popular terms, this number is believed to be the increase in the time necessary to cause sunburn. For example, if you normally burn in ten minutes and you apply a sunscreen with an SPF rating of ten, then you should not burn before 100 minutes. At present, the FDA allows products sold without prescription ("OTC" products) to carry SPF ratings from 2 to 30. Products judged still more effective can be labelled 30+. Note that this type of labelling implies *in vivo* testing, that is, testing on actual humans under standardized conditions. The FDA also requires supplemental *in vitro* testing. The regulations are now judged misleading.

What should the regulations be?

SOLUTION

There are three major changes in the new proposed regulations. First, future sunscreen products will carry a warning: "UV exposure from the sun increases the risk of skin cancer, premature skin aging, and other skin damage. It is important to decrease UV exposure by limiting time in the sun . . . " The comparison with the warning on cigarettes is striking. For cigarettes, the product itself is dangerous; for sunscreens, the product allows the consumer to reduce risk.

The second major change complicates the labelling by giving separate ratings for UVB and UVA. For UVB, the high-energy rays which cause burning, the same rating "SPF" system used previously is extended to products labelled as high as 50+. The UVB rating is supplemented by a second rating for UVA protection. This new UVA rating, which would be displayed with the same prominence as the SPF rating, would be from one to four stars. These stars would represent UVA protection, indicating "low," "medium," "high," and "highest." Keep in mind that UVA is responsible for tanning so that a product with a high SPF rating and one star would imply no burning and fast tanning. A high SPF rating and four stars would be a sun block.

The third major change in the FDA regulations is the rejection of a single *in vitro* test and a commitment to continuing a combination of *in vivo* and *in vitro* tests. This is in spite of the FDA's recognition of the wide variation in human skin, and their careful, but apparently only partly successful, efforts to describe this variation. It is in spite of the fact that though consumers are encouraged to apply products "liberally," "generously," and "evenly," they often use considerably less than the 2 mg/cm^2 recommended. Consumers also apply the products less frequently than is required for good protection.

Thus current *in vivo* testing, with its standard generous and frequent applications, would seem of limited value in describing actual product safety. A reliable *in vitro* test has been developed (*Federal Register* **72**(165) (August 27, 2007) L19095). This test measures the absorption of 2 mg/cm^2 product applied to a collagen membrane and placed between a xenon lamp and a spectrophotometer. By measuring absorption vs. wavelength and time, the test measures not only the product's SPF but also its effective stability. Such a test provides a scientific, objective criterion for judging a new sunscreen's efficacy.

The FDA should require warnings on sunscreen products. They should retain the SPF ratings in the short term but move to minimum standards for UVB and a four-star rating for UVA in the longer term. They should reconsider the decision not to rate products solely on the basis of *in vitro* tests.

2.5 Conclusions and the First Gate

Product development begins with the identification of customer needs. While the customers may be individual consumers or large corporations, their needs are effectively explored through interviews with fifteen to twenty individuals. Once these needs have been collected and rank-ordered, they are best evaluated by the product development team, which should include representatives of marketing, research, engineering, and manufacturing. This team must reach a consensus on how the needs are translated into preliminary product specifications. Developing these specifications can be guided by first writing out all chemical reactions, by estimating mass and energy balances, and by assessing the rates of key processes. This is the point where the scientists and engineers on the development team first have a major input.

The preliminary specifications will need revision. This revision is facilitated by using a standard "benchmark," which is frequently a competing product. The revised specifications must get a critical analysis to see whether they make sense. The team must reflect on whether the specifications excessively restrict the possible solutions, as they did in the road de-icing example. In addition, the specifications should be consistent with corporate strategy, if possible building on existing corporate strength. Once the specifications are complete, we are ready to start seeking ideas for their achievement.

The first management review of the product's development takes place at this point. Such a review is sometimes called a "gate," to suggest that the review will decide whether or not to continue the project. The core team will prepare both a written report and an oral presentation. Normally, the audience will be senior-level managers, often well above the immediate supervisors of the core team.

The decision on whether to continue development will, at this early stage, usually be positive. After all, the core team will probably have been together for only a few weeks. Moreover, the impetus for the project may have come from the same senior managers making the decision, who may be unwilling to see their ideas abandoned without more scrutiny.

In fact, management studies suggest that this review – this "first gate" – tends to be too casual, and that its conclusion should more frequently be negative. These management studies point to the high cost of cancelling projects later in the product design procedure, and to the importance of quick checks on a spectrum of alternatives. The studies urge being carefully critical at this early stage.

We believe that there are two ways in which chemists and engineers can aid this early decision. The first way is to be ruthlessly objective but without being destructive. This is a skill which comes naturally to some; others become easily involved in championing a particular perspective.

The second way in which chemists and engineers can be effective is in teaching management about the science behind the specifications. For example, we must explain that freezing point depression depends on the mole fraction of ions, and not on the mass of chemical. That is why we have so stressed specifications in this chapter. It is through specifications that those with scientific training can most help to decide whether product development should be continued or stopped. Product specifications are the key to this first management gate.

Problems

1. *New product ideas*. Go into a food store, a drugstore, or a hardware store to identify three new potential products. Describe in a sentence or two what need each new product would fill.

2. *Cappuccino powder*. Your company is a major manufacturer of instant coffee powder. You wish to extend your product portfolio to include instant cappuccino. Define needs and set specifications for this new product.

3. *A varnish for country cabins*. Develop a list of needs for a clear, long-lasting varnish to be used on exposed wood in weekend cabins in remote areas of Minnesota. The cabins are rustic, exposed to extreme weather, and surrounded by lakes, which breed the Minnesota state bird, the mosquito. Classify these needs as essential, desirable, and useful. Decide how much improvement the new varnish should have compared with existing products.

4. *Collecting rainwater*. We wish to ensure a pure, inexpensive rainwater supply for a single-family home of 200 m^2. Identify and rank the needs of such a system.

5. *A new ice cream*. A small ice-cream company makes fresh ice cream for a local market and wishes to increase its sales. Interviews with customers result in the responses shown in Table 2.5A. Add your own opinions and those of friends. Then rank these responses to determine the best route to proceed.

6. *Water in space*. The United States proposes sending astronauts to Mars. One major problem on this mission is the quantity of water required. Because this is too great to carry, a highly efficient recycling system will be essential. Determine the needs for the water recycling system if such a Mars mission is to be successful. What should be used as a benchmark? Does the Mars mission seem viable?

7. *A heat-pack trigger*. Heat packs are used for relief of minor aches and strains or for the provision of warmth on cold days. They are made from sodium

---

Table 2.5A *Suggestions for a new ice cream.*

What is your favorite flavor?
"Mint chocolate chip in a cone, or chocolate or vanilla with chocolate sauce in a bowl."
  Rebecca, aged 22
"Italian limone." Jo, aged 12
"Pistachio and almond. Hard ice cream in a cone." Maria, aged 56
"Chocolate with chocolate chips." Monica, aged 22
"Green." Henry, aged 10
"I like two types together, like chocolate and vanilla. Always in a bowl. Sorbet if the weather
  is hot." Bronach, aged 37
"Soft vanilla ice cream in a cone." Harriet, aged 40

What is good about ice cream?
"It's sweet but doesn't fill you up too much – good in a restaurant." Rebecca, aged 22
"It's cold and it tastes good." Jo, aged 12
"It cools you down. It's not too sweet. Texture in creamy matrix." Maria, aged 56
"It's comfort food. It makes you feel good." Monica, aged 22
"It's sweet and cold." Henry, aged 10
"It's all about fulfillment of need. I only eat it at home. It's sweet and tasty." Bronach,
  aged 37
"Texture and flavor." Harriet, aged 40

What is bad about ice cream?
"It drops down your clothes. At the theatre, you can't get it out of the corners – the spoons
  are rubbish. I don't like it with bits in." Rebecca, aged 22
"Nothing." Jo, aged 12
"It's fattening. Look at poor Elvis." Maria, aged 56
"You always eat too much. It's expensive. It doesn't keep properly in the freezer."
Monica, aged 22
"It melts too fast." Henry, aged 10
"Sometimes, it's too sickly. It's too cold and makes your teeth feel funny." Bronach, aged 37
"It's either too hard or already melted" Harriet, aged 40

Would you like bacon-and-egg or mustard ice cream?
"Yuk. Marmite or saffron ice cream are great though – but they don't taste like marmite or
  saffron." Rebecca, aged 22
"No, but pumpkin or fig are delicious." Maria, aged 56
"I don't think so. Brown bread ice cream is the best." Monica, aged 22
"Disgusting ideas." Bronach, aged 37

---

acetate trihydrate. The mixture has a melting point around 60 °C. Below this temperature, the thermodynamically stable state is sodium acetate trihydrate crystals. The heat pack can be regenerated by heating it above 60 °C (for example, in a pan of boiling water) and then cooling without crystallization. In the absence of a nucleating agent, the sodium acetate trihydrate remains metastable in the liquid form well below its melting point. Thus, even at 0 °C, the contents of the heat pack remain liquid. When the heat pack is "triggered," crystallization begins, releasing latent heat. This causes the heat pack to warm to its saturation temperature, around 60 °C, at which crystallization continues at the rate required to provide the heat necessary to maintain that temperature. Thus, a constant, moderate temperature is achieved.

The device works well. The problem is the trigger. In current technology, this is a flexible metal disc containing microscopic cracks on its surface. To initiate nucleation, the disc is flexed. This usually works, but not always. More seriously, the disc sometimes triggers nucleation without being flexed, meaning that such heat packs are unsuitable for military applications where reliable performance is critical. Finally, because the trigger is metal, the heat packs cannot be regenerated in a microwave. Discuss how the current trigger works and write specifications for an improved trigger.

8. *Better dental fillings.* Many adults in developed countries have "fillings" in their teeth. These gray-colored amalgam insertions repair regions where the original hydroxyapatite, a calcium phosphate, has decayed. The fillings, often required during adolescence, last decades. Recently, customers have chosen tooth-colored fillings, or as dentists call them, restorations. These newer restorations are often ceramic, which can be colored and wear well. However, the ceramics' hardness means that they can fracture easily: they don't absorb shock well. In addition, they are so hard that they may cause the bone to which they are attached to demineralize.

    As a result, manufacturers have developed systems based on polymeric resins, especially acrylics. Originally, these resins behaved very differently than the amalgams they replaced, so dentists had trouble using them. Now, resins are made to handle like amalgams. Also, the original resin systems wore too rapidly, even when filled with 50 μm silica. Now, resin systems use 0.06 μm silica for wear and 0.8 μm silica to adjust color.

    Your company manufactures a variety of inorganic colloids of controlled size and shape. You are interested in exploring the use of your products in dental materials. How should you begin to explore this possible use?

9. *Ecologically benign detergents.* A recent newspaper article discussed a variety of detergents described as "ecologically friendly." In many cases, this seems to be because they are based on vegetable oils rather than petroleum. In the past, "ecologically friendly" detergents frequently had very little detergent *per se*, but extremely high concentrations of sodium hydroxide. These largely seem to have been withdrawn.

    However, the remarkable part of the article is its almost complete absence of substance. This is not the fault of either the author or the paper; there is no consensus about what an "ecologically friendly" detergent is. Develop criteria for this case. Suggest what information is missing that will allow these criteria to be applied. Then write a short, sound article on this subject.

REFERENCES AND FURTHER READING

Astarita, **G.**, **Savage**, **D. W.**, and **Bisio**, **A.** (1983) *Gas Treating with Chemical Solvents.* Wiley, New York, NY.

Atkins, **P. W.** and **Julio de Paula** (2010) *Physical Chemistry*, 9th Edition. W H Freeman, New York, NY.

Bard, **A. J.** and **Faulkner**, **L. R.** (2000) *Electrochemical Methods: Fundementals and Applications*, 2nd Edition. Wiley, New York, NY.

Beckett, **S. T.** (2008) *The Science of Chocolate*, 2nd Edition. Royal Society of Chemistry, London.

Bourne, **M.** (2002) *Food Texture and Viscosity*, 2nd Edition. Academic Press, Englewood Cliffs, NJ.

Fogler, **H. S.** and **LeBlanc**, **S. E.** (2007) *Strategies for Creative Problem Solving*, 2nd Edition. Prentice Hall, Upper Saddle River, NJ.

Handler, **R. M.** (1994) *Dream Doll*. Longmeadow Press, Stamford, CT.

McGrath, **M. E.** (1996) *Setting the Pace in Product Development, A Guide to Product and Cycle-Time Excellence*, Revised Edition. Butterworth-Heinemann, Boston, MA.

Rosenau, **M. D.** Jr., **Griffin**, **A.**, **Catellion**, **G. A.**, and **Anschuetz**, **N. F.** (1996) *The PDMA Handbook of New Product Development*. Wiley, New York, NY.

Sato, **K.**, **Arishima**, **T.**, **Wang**, **Z. H.**, **Ojima**, **K.**, **Sagi**, **N.**, and **Mori**, **H.** (1989) Polymorphism of POP and POS. *Journal of the American Oil Chemists' Society* **66**, 664–674.

3

Ideas

Once we know the specifications for our target product, we need some ideas to meet these specifications. In fact, all we really need is one good one. Finding this idea is sometimes discussed by reference to the children's fairy tale, "The Frog Prince." You will remember the story: a somewhat vain princess who is walking in the woods promises her hand in marriage to a frog in return for some simple service. The frog performs the service and then shows up to claim his bride. The distraught princess submits ungraciously and then is astonished to discover that, after she kisses him, the frog turns into the prince of her dreams. Freudian psychoanalysts have enjoyed interpreting this story.

For us, the story is the antithesis of product design. The chance of our beginning with one frog of an idea and making it into a prince is remote. Our chance of success is much greater if we can screen a large number of ideas. In terms of the fairy tale, we need to behave like a more modern princess, and kiss a lot of frogs. Exactly how many if we want to find a prince? While estimates in different businesses vary, most experienced product developers suggest we need around one hundred ideas to find one winner. We need to kiss one hundred frogs to find one prince.

This large number of ideas can come from a variety of sources. Many ideas come from individuals. These include customers, competitors, consultants, and members of the product development team. As explained in Section 3.1, generating these ideas should be as free and unrestricted as possible. This is the time for craziness, for off-the-wall notions, and for asking, "What if?"

In some cases, the ideas generated by all these groups will be insufficient. In these cases, we will often seek additional chemical ideas sparked by the methods described in Section 3.2. We can look at natural products, checking if folk medicines are really effective. We can use the automated synthesis and analysis of combinatorial chemistry to explore many more compounds than has been possible with traditional synthetic techniques. These methods will increase the chemical possibilities.

Once we have generated a large number of ideas, we will want to winnow the best ideas from the large number assembled. Typically, we want to start with perhaps a hundred ideas and wind up with five. We suggest doing this in two stages. As described in Section 3.3, we will first sort the ideas, removing redundancy. We will drop those groups of ideas which are inconsistent with our corporate strategy, or which do not build on our corporate strengths. We will nervously drop ideas which seem pure folly, knowing that some of these may contain seeds of innovation. This sorting of ideas will normally leave perhaps twenty survivors.

The second method of pruning our ideas, described in Section 3.4, is a more aggressive screening. This screening tries to cut the ideas from twenty to around five. We feel that one good route for this step is to use a concept screening matrix. In this procedure, we evaluate the general characteristics of each of our ideas – chemical understanding, engineering requirements, etc. – and compare weighted averages of these evaluations. The result of these screenings will be strong, practical ideas, any one of which should work. With these strong ideas, we will be ready for selecting the best idea, as described in Chapter 4. For now, however, we focus on getting good product ideas.

3.1 Human Ideas

The consensus in product design is that we will normally need between twenty and several hundred ideas to get one winning product. These estimates vary with the particular industry. Du Pont suggests that they need around three hundred initial concepts to get one commercially viable product. 3M feels they can develop a winning product from only ten ideas. Zeneca and Pfizer suggest that they need around one hundred ideas per success. While these different estimates certainly reflect differences in what is meant by a new idea, the conclusion is clear: we will need a lot of ideas.

Generating many ideas requires answering two questions. First, who are the sources of our ideas? Second, how do we get these sources to give us ideas? Exploring these questions is the subject of this section.

3.1–1 SOURCES OF IDEAS

Because we need so many ideas, we will accept them from any sources we can find. However, because some groups will be willing to spend much more time trying to help us, they will usually be our main sources.

One major source is the product development team itself. This team will normally have representatives who have made, used, and been frustrated by existing products which we are trying to replace. They will be quick to see advantages and demerits of any new concepts. Their professional careers depend on the team's success, so they have a large stake. They will be an excellent source.

The second group are the product's potential customers. These customers are those who will directly benefit from the new product's characteristics. The most important customers are the so-called "lead users." These customers will frequently already have tried to modify the product, improving its utility for their particular goals. While such improvements may be of narrow scope, they will often identify new ways to meet specific needs. A related group to the customers are competitors because both are interested in products which meet the same need. The competitors' marketing efforts may supply clues to their own plans.

The third area to look for ideas is the literature. The trade literature, and indeed trade shows, are the best source of information about current products beyond the product team and current customers. The archival literature, i.e. the peer-reviewed publications of scientists and engineers, can provide the secrets of new products, but only if the product team has experts who critically read this literature. The patent literature is also valuable. While many in management argue that patents are paramount, we feel that chemical patents are of limited value because they are often incomplete. We remember the anguished reports of chemists who, during the First World War, were trying to manufacture ammonia using the Haber process. They concluded that however Haber made ammonia, it was not with the catalyst described in the patent.

Other persons who can be good sources of ideas are product experts, private inventors, and consultants. Experts are those with particular knowledge of the products which we want to make. Those retired from our organization or from a competitor may be especially helpful. Private inventors can be an important resource, especially for innovations which go beyond the boundaries of our current thinking. Dealing with private inventors is tricky: in many cases, their pet ideas may be impractical, but may spark different and important product ideas from the product development team. As a result, many companies keep careful written records of all dealings with private inventors to ensure all are clear on who invented what.

Consultants are the most difficult group to characterize because they are so diverse. Those who supply special services, like schemes for product development or innovation encouragement, seek to catalyze ideas from the organization's employees rather than produce ideas of their own. They can be valuable in limited doses. University professors can be frustrating consultants. The product development teams usually forget that professors are scholars, interested in truth and in education. They are the intellectual descendents of the monks who guarded society's knowledge. As such, they are often excellent critics but poor innovators.

3.1–2 COLLECTING THE IDEAS

The most direct and effective method of collecting ideas is to ask the various groups listed above to write ideas down and send them in. Writing them down is important, for writing forces an objectivity which can spark improvements. If the ideas that we seek depend on chemical processes, a flow sheet can help. If the ideas include chemical synthesis, then guesses about synthetic routes and mechanisms are helpful. The ideas collected in this way form the core of our product ideas.

Asking the team, the customers, and a few experts for written submissions will rarely generate the hundred or so ideas which we normally seek. To get more, we will commonly assemble three or more groups of five to eight persons, and ask them to suggest still more ideas. Each group should have a formal leader who runs the session. Such a "brainstorming" session will work best under a few rules:

(1) *Use a common format.* Have all groups cover the same topics in more or less the same way.
(2) *Generate ideas freely.* Do not be worried that some ideas have problems.
(3) *Eschew ownership.* Do not worry about which idea is whose, or whether the suggestion is competent.
(4) *Encourage eccentricity.* Do not squash weirdness, even if it suggests the impossible.

While it may take thirty minutes to get started, most groups will be able to generate ideas under these rules for an hour or two.

During idea generation, the group should keep a written record of its progress. While the leader can sometimes do this, a separate scribe is usually better. The scribe often keeps his record on large sheets of paper which are then posted around the room. These posters are an enormous stimulant, because the group can refer back to old ideas whenever anyone thinks of improvements.

However, all brainstorming sessions will tend to stall, usually after about an hour. While the group's productivity will drop, its creativity may actually rise, because the obvious avenues have been exhausted. Thus, most sessions probably should be kept going somewhat past the point where the group's members want to stop. To keep them going, we find four stimuli are especially useful:

(1) *Invite criticism of ideas generated by other routes.* This produces comments like, "That's stupid, but if we went to an aliphatic side chain, then…"
(2) *List all assumptions made in the specifications; then dismiss each in turn.* This sparks statements like, "Well, I guess we don't really need a solvent if we use an aerosol…"
(3) *Use analogies.* Everyone remembers that heat conduction and mass diffusion are described by the same mathematics, but fewer remember that free energies can sometimes be changed by pH instead of temperature.
(4) *Probe opposites.* This leads to comments like, "Reverse osmosis uses a water-selective membrane to separate a lot of water from a little salt in sea water. What if we use a salt-selective membrane?"

The purpose of these stimuli is to restart the group, encouraging their creativity in less-explored directions.

3.1–3 PROBLEM-SOLVING STYLES

In this brainstorming, we will probably see the emergence of different problem-solving styles. Extensive psychological research, especially that by Kirton (2003), suggests that it is useful to divide intellectual development into two styles: adaption and innovation. Adaption is problem solving using existing or closely related

technology. Innovation is problem solving using apparently unrelated information. The choice in the psychological literature of the word "innovation" is unfortunate, for this word in the vernacular is often taken as a synonym for "creativity." Innovation may be creative, but adaption can be equally creative.

Problem-solving styles are often evaluated by psychological testing. More simply, you can guess your own problem-solving style by considering how your contribution to a project might be seen. If you are called a "nerd," a "drone," or a
"bean counter," then you are probably an adaptor. If you are called a "flake,"
a "weirdo," or a "loose cannon," you are probably an innovator. We need both
adaption and innovation. How much of each depends on our objectives, e.g. on
the business we are in.

Not surprisingly, successful professionals within a single discipline tend to have
similar styles, but different professions encourage different styles. For example,
at one extreme, accountants are adaptors: the last thing we would want is an
innovative tax accountant. At the other extreme, successful entrepreneurs are
often innovative. Other professions fall in between these limits, in more or less
the order expected: chemists are more innovative than engineers, who in turn are
more innovative than elementary school teachers. Please remember than innovative is *not* a synonym for creative.

We also believe that there is an additional style beyond those of adaptors and
innovators – that of a product "champion." A champion wants the product to
work, whether it seems to make sense or not. He is often less deeply trained
and less objective than either adaptors or innovators. He just wants the product to be successful. While he knows that a successful product will make money
and advance his career, he also believes in the product, with an almost religious
fervor.

We can make these three problem-solving styles more understandable by giving three examples. As an adaptor, we choose Pierre de Montreuil, the thirteenth-
century architect of Sainte Chapelle and of the famous rose window in the south
transept of the Cathedral of Notre Dame, both in Paris. Pierre worked for over
30 years, carefully extending the empirical knowledge of how to use stone to
build higher and higher, "to the glory of God." While the builders of this period
knew no civil engineering, they came closer and closer to the limits of the stone
they were using. Occasionally, these adaptors went too far: in AD 1248, trying
at Beauvais to increase the height of the nave 50% beyond Notre Dame, they
built a cathedral which collapsed. Most of the time, however, they adapted and
extended earlier successes to complete the magnificent structures which remain
today.

Our example of an innovator is Michael Servitus, a fifteenth-century physician
and monk with strong but unusual religious beliefs. Servitus became convinced
that Christianity was a wonderful religion, except that there was no Trinity and
Christ wasn't divine; otherwise the religion as practised in his day was fine. Like
most innovators, he was sure he was right. He went to Geneva to convince John
Calvin, who had him burned at the stake, a not unusual fate for innovators. They
can have great ideas, but they can be a pain.

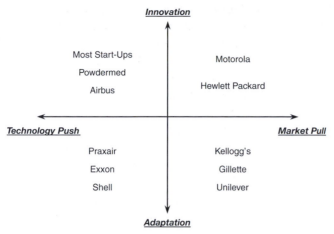

Figure 3.1–1 *Corporate problem-solving styles* Different companies tend to develop different problem solving styles.

A good example of a champion is Lorenzo de Medici, "Lorenzo the Magnificent." Lorenzo died in 1492, making his death easy to remember. More than any other single person, he was responsible for making Florence a powerful city-state, and promoting a cultural flowering. His motives and methods were certainly objectionable: he lied, cheated, and murdered, probably at least partly for personal gain. But he was important as a champion of Florence, and was certainly partly responsible for her ascension in politics and the cultural renaissance she harbored.

Just as individuals and professions tend to have a problem-solving style, so do corporations. While this may be just a consequence of the most common profession employed, we suspect that it is often a deeper characteristic of the particular corporation. When we combine this style with the forces responsible for product development, we get company profiles like those suggested in Figure 3.1–1. Try plotting corporations of your own experience on these coordinates. We feel that most universities fall in the lower left-hand corner of this figure: they are pushed by technology but feature largely adaptive scholarship. We find a company like 3M especially illuminating. Their high development costs force them to be innovative, but they respond both to market pull and technology push. As a result, they would appear as a line across the top of the figure.

We understand that many find the idea of inventing new products intimidating, and that they expect to have difficulties. That has not been our experience. We are often startled by the number and quality of new ideas generated by all types of trained and untrained groups. Students, who are our most frequent subjects, are especially good, possibly because of their tolerance for looking foolish. We are also startled that generating good new ideas does not seem to correlate with age, training, intelligence, or economic achievement. Apparently, idea generation is a form of creativity which everyone can do.

3.1–4 EXAMPLES OF UNSORTED IDEAS

The results at this point will be long lists of raw ideas. Five examples, given in Tables 3.1–1 to 3.1–5, illustrate the scope involved. The first four of these examples will be discussed in more detail in the later sections of the book. Table 3.1–1 suggests ways to do laundry more efficiently, either by redesigning current washing machines or by developing new types of detergents. While some of these ideas seem bizarre, we were amused to learn that one of our graduate students saves time and money by running his laundry through the clothes dryer three times more often than through the washing machine. Table 3.1–2 is a broad search of new methods of drug delivery. Since this list was made, at least ten of these ideas have been produced commercially. These speculations will be used in Section 3.3 to illustrate how ideas can be efficiently sorted. Such sorting can reduce the number of ideas from perhaps a total of one hundred down to twenty.

The next two lists of ideas will be used in Sections 3.3 and 4.1 to illustrate how ideas can be more critically screened. Table 3.1–3 lists ideas for a new lithographic ink to replace an ink which contains a carcinogenic solvent and which requires large quantities of the same solvent to clean the presses. The new ink must offer an opportunity for reduced pollution. Table 3.1–4 is a partial list of ideas for treating high-level radioactive waste containing ^{137}Cs. This isotope, produced in the manufacture of atomic weapons, is especially dangerous because it is water-soluble. Finally, Table 3.1–5 lists targets for the application of dipstick technology. One commercial example is the pregnancy test, in which a dipstick develops a blue line when wet by a pregnant woman's urine. A second example, developed but not yet commercially available, is a small card which, when held against a piece of fish, changes color in response to the freshness of the fish.

Later, we will give details of the processes for dealing with the information in these five tables; for now, we want only to illustrate typical results of the idea-generation process.

3.2   Chemical Ideas

In many cases, the ideas resulting from the strategies given above for generating product ideas will show real promise in meeting the customer needs. The ideas may be devices for chemical change, like better catalytic converters or cheaper kidney dialysis machines. They may involve particular chemical compounds, like the crown ethers for radioactive cesium extraction. In some cases, however, the general strategies will not help much, because we do not know which chemical compounds we want to make. This may be especially true for pharmaceuticals.

When we do not know our target compounds, we can use common chemical methods as a stimulus towards generating ideas. Two such methods are discussed here: natural-product screening, and combinatorial chemistry. Natural-product screening takes advantage of the variety of active chemical species present in nature. Plants, animals, fungi, lichens, marine organisms, and microbes are all rich sources of new chemical species with specific functions which could fulfill our

Table 3.1–1   *Ideas for new laundry detergents causing less pollution* These speculations will be used in Section 3.3 to illustrate the sorting of ideas.

1. Wash without soap.
2. Throw the clothes away.
3. Use less soap and less water.
4. Use a more effective soap.
5. Add enzymes to detergents.
6. Add dead cells to detergents.
7. Add live cells to detergents.
8. Mop up dirt with particles.
9. Use specific chemical interactions.
10. Improve the washing machine.
11. Recycle the soap.
12. Filter bigger detergent particles.
13. Make larger micelles.
14. Make emulsions out of the soap.
15. Grow microbes on dirty clothes.
16. Attach soap to particles, facilitating recycle.
17. Imitate dry cleaner agents.
18. Use a fine adsorbent.
19. Cook clothes under N_2.
20. Air out clothes as washing substitute.
21. Prevent soiling with antistatic coatings.
22. Wash until semi-clean.
23. Remove odor without removing dirt.
24. Wash with base, converting sweat compounds into soap.
25. Split objectives of clean, color-fast, and odor.
26. Ultrafilter dirty water.
27. Imitate dry cleaning.
28. Get a new dry-cleaning solvent.
29. Make a home dry cleaner which is sealed.
30. Use supercritical CO_2 for dry cleaning.
31. Use another supercritical solvent.
32. Wash with Fuller's Earth.
33. Dry clean with chlorine-free solvents.
34. Grind the clothes up and remake them.
35. Recycle the surfactant using a pH change.
36. Recycle the surfactant exploiting its cloud point.
37. Make a detergent which precipitates on command.
38. A detergent which forms many phases.
39. Wash clothes with dry shampoo.
40. Clean ultrasonically.
41. Shine with a UV light (to sterilize?).
42. Use pressure waves.
43. Cook clothes in high-pressure water.
44. Freeze clothes; shake off dirt.
45. Calcine dry shampoo to make it pure.
46. Adsorb detergent in clay.
47. Use ultrafiltration.
48. Dry cleaning recycle is distillation.
49. Flocculant aid for detergent.

Table 3.1–2 Ideas for new drug delivery system Note the table has been slightly abridged, so not all numbers appear here.

	Ideas	Device	Specific ailment	Drug	Comments
1.	Antibiotic-coated implants	Implantable	Infection, inflammation		Coating dissolves into tissue, artificial joints, etc.
2.	Antibiotic delivery through skin	Topical–transdermal			Small device attached to skin to discharge at precise intervals of time
3.	Asthma patch, nano controller to measure heart beats	Topical–transdermal	Asthma attack	Corticosteroids	
5.	Bathing system	Topical–transdermal			
7.	Biosensor to detect alien chemicals	Implantable			
8.	Birth-control pill	Oral–tablet	Pregnancy		Taken monthly instead of daily, stored in fatty cells; slowly released
10.	Botox mask	Topical–transdermal	Wrinkles (skin care)		Diffusion through pores in mask
12.	Caffeine patch	Topical–transdermal	Drowsiness, caffeine withdrawal		Military market
13.	Cancer recognition injections	Intravenous	Cancer		Injected at young age, polymeric system, senses cancers
15.	Carbon spheres injection	Intravenous	Birth control, hyper-activity	Ritalin, birth control	Drug elutes slowly into bloodstream.
17.	Eye drops	Topical–eye	Cornea shape		
18.	Skin lotion	Topical–transdermal	Hair removal		
19.	Chemotherapy patch	Topical–transdermal	Cancer-tumor		Used for chemotherapy applications
20.	Cholesterol absorber	Gut	High cholesterol		Cholesterol absorbed in gut
22.	Coated cancer pill	Oral–tablet	Cancer		Only tumor cells can break down the coating of drug
23.	Coated scissors	Topical–hair	Split-ends		Product seals hair ends to prevent damage
25.	Concentration gradient patch	Topical–transdermal			Facilitate constant drug delivery

No.	Name	Route	Application		Description
26.	Controlled release wafer	Implantable	Cancer-tumor	Chemotherapy drug	Site specific, allows for longer treatment intervals, shorter rest intervals
28.	Di-block copolymer micelle	Implantable	Tumor		Functionalized on hydrophilic end with a protein; reaction with specific cells
31.	Drug-coated artificial joints	Implantable	Inflammation		Reduces inflammation and accelerates healing
32.	Drug delivery via food	Oral		Antibiotics	Food produces antibiotics, instead of pill
35.	External blood filter	External			Small volume of blood bypasses the heart through external filter where drug acts
36.	Eye contact	Topical–eye	Vision, protection		Covers entire eye, protects against scratching, filters UV light
37.	Flu inhaler	Inhalation	Flu		Inhaled version of flu shot
38.	Fluoride releasing gum	Oral	Dental		Activated by saliva, strengthens teeth, treats area weakened by sugar
39.	Flux-controlled release capsule	Oral–microcapsule			Porous, non-dissolvable micro-structure; drug release via diffusion
41.	Drug-coated condom	Topical–transdermal	Sexually transmitted diseases (STDs)		Treats STDs
42.	Drug-coated sock	Topical–transdermal	Athlete's foot		Insert inside sock
44.	Glue with antibiotic	Topical–transdermal	Inflammation	Antibiotics	Seal cuts, drug released over few days, keeps down stink
45.	Hair gel	Topical–hair	Hair care		Release hold using a special solvent or light
48.	Hormone suppressant pill	Oral–tablet	Overeating		Triggered by hormones in the body, prevention instead of treatment
51.	Implantable headache device	Implantable	Chronic headaches		
52.	Implantable microprocessor	Implantable			Contains multiple microcapsules; electric pulse stimulates breakdown of one

(cont.)

Table 3.1–2 (cont.)

	Ideas	Device	Specific ailment	Drug	Comments
53.	Implantable osmotic pump	Implantable			Deliver two drugs at one time
54.	Implantable spinal device	Implantable	Arthritis		
55.	Implantable vapor pressure piston pump with heat patch	Implantable			Temperature makes piston release drug
56.	Insulin microcapsule	Oral–microcapsule	Diabetes	Insulin	Permeability of outer polymer dependent on concentration of sugar in bloodstream
57.	Insulin patch	Topical–transdermal	Diabetes	Insulin	Continuous level of insulin
59.	Insulin pump with sugar sensor	Implantable	Diabetes	Insulin	Sensor provides feedback to pump to determine rate
60.	Invisible needle	Intravenous	Prevent patient fainting at sight of needle		Looks like a large pen, not visible to patient
61.	Knee brace drug delivery	Topical–transdermal	Swelling		Delivers drug from brace through the skin
62.	Laser activated pain relief	Implantable	Pain		External laser activates chemical reaction
65.	Long-lasting patch	Topical–transdermal	Asthma		Replaces daily inhalers
66.	Longer-lasting asthma inhaled medicine	Inhalation	Asthma		Minimize number of times inhaler must be used
68.	Magnetic property devices	Implantable			Magnet can attract drug to certain locations in body
70.	Metal nanoparticles	Intravenous	Cancer		Functional groups attach to cancer cells; drug released when a certain wavelength used
71.	Microcapsule	Oral–microcapsule			Drug released as layers of biodegradable polymer degraded; multilayered; drug in between layers

No.	Name	Type	Condition	Drug	Description
72.	Microcapsule	Oral–microcapsule			Big enough to be trapped in a tumor but small enough to travel in body
74.	Microscission drug delivery	Topical–transdermal		Aluminum oxide	Gas stream removes top layer of skin and creates holes
75.	Microscopic mechanical device	External			Release drug payload through external mechanism
77.	Migraine patch	Topical–transdermal	Headache		
79.	Multi-layered time release pill	Oral–microcapsule			Layers degrade as pill travels through body
82.	Needleless pain relief device	Intravenous	Pain		
83.	Numbing patch	Topical–transdermal	Dental	Novacain	Numb gum area for dental work
84.	Oral decongestant with delivery device	Inhalation–nasal	Nasal congestion		Device delivers drug to nasal area
85.	Orthodontic device	Oral device	Osteoporosis	Calcium	Device contains saliva-soluble calcium supplies for constant delivery, slow release
86.	Over-the-counter painkiller inhaler	Inhalation	Pain		Vaporized dose, fast-acting headache medicine
87.	Over-the-counter painkiller patch	Topical–transdermal	Pain, arthritis	Aspirin	Localized dose
89.	Pain-relief patch	Topical–transdermal			
91.	Pancreatic enzyme device	Implantable	Cystic fibrosis	Pancreatic enzymes	Device enlarges to release enzymes, triggered by environment of the intestine
93.	Patch with microneedle	Topical/injection			Delivers drug through direct injection
94.	Peristaltic pump	Implantable	Shin splints		Force absorbing material excreted by system
95.	Permanent gastrostomy tube	Implantable	Obesity		Surgically place in skin, leads to stomach

(cont.)

Table 3.1–2 (cont.)

	Ideas	Device	Specific ailment	Drug	Comments
96.	pH-activated pill	Oral–tablet	Acid reflux–indigestion		Pill coated with substance that will be activated when stomach pH changes
97.	Pituitary gland device	Implantable	Thyroid disease		Electrical stimulation of pituitary, brain, and thyroid
98.	Portable insulin device (patch)	Topical–transdermal	Adolescent diabetes	Insulin	Constant delivery of insulin
100.	Pump and catheter drug delivery system	Implantable	Chronic pain		Surgically implanted; externally programme dose
102.	Slow-release holding vessel	Implantable		Depression drugs	Gradual continuous drug release
103.	Slow-release polymer-coated implant	Implantable	Cancer		Drug transported through polymer film, refilled with injections into polymer
104.	Streamlined inhalers	Inhalation	Anything		For all injected drugs
105.	Nitroglycerine patch	Topical–transdermal	Chronic angina		Pressure activated
107.	Time-release eye contact	Topical–eye	Vision, protection		Replace eye drops, drug release
108.	Time-release pill	Oral–microcapsule			Larger shell contains small capsules, release at various rates
109.	Time-released capsule	Oral–microcapsule	Deep-vein thrombosis	Coumadin	Release into the venous side
110.	Time-released insulin pill	Oral–microcapsule	Diabetes	Insulin	Digestible polymer loaded with drugs
111.	Watch for drug delivery	Topical–transdermal			Releases drug through skin based on the time

Table 3.1–3 *Ideas for a new lithographic ink with reduced solvent emissions* More chemical details of the current ink are given in Example 3.3–2.

1. Don't use a solvent.
2. Switch solvents.
3. Clean the press with robots.
4. Change the press.
5. Use an electrostatic ink.
6. Use a laser printer instead of the current design.
7. Change ink chemistry.
8. Recycle all of the solvent.
9. Clean the press with a high-pressure spray.
10. Extract the solvent from the rags used to clean the press.
11. Do the whole process in a clean room.
12. Isolate all equipment.
13. Clean the press less often.
14. Clean the press in a fume hood.
15. Print more checks at a time.
16. Mix the current solvent, methylene chloride, with other solvents.
17. Have workers wear a self-contained breathing apparatus.
18. Use a solvent mixture.
19. Use a solvent which dissolves the ink.
20. The solvent in the ink should differ from the cleaning solvent.
21. Use a non-volatile solvent.
22. Use partial cleaning of specific components of the press.
23. Steam clean the press.
24. Clean the press with air.
25. Put the press in a car wash.
26. Clean the press by brushing.
27. Clean the press by burning.
28. Make the lithography more like a jet printer.
29. Don't use checks.
30. Use a disposable press.
31. Use oil to trap the solvent.
32. Make checks by photocopying.

needs. Combinatorial chemistry uses robotics to provide a first pass at screening thousands, even millions, of compounds which may have the desired product properties. The chemical compounds discovered by these methods can provide ideas for those working on the project. A synopsis is given in the following paragraphs.

3.2–1 NATURAL-PRODUCT SCREENING

The first route to new chemical ideas is to look for possible sources in nature. During the past century, this has been a major source of complex chemical species with higher molecular weights. The pharmaceutical industry has benefited especially from mimicking nature: aspirin and opium, quinine and colchinine, caffeine

Table 3.1–4 *Ideas for treating high-level, water-soluble radioactive waste* This partial list of submitted ideas is used in Section 3.4 to demonstrate the concept screening matrix.

1. Store waste in bedrock.
2. Separate cesium with irreversible sodium titanate ion exchange.
3. Cesium extraction out of caustic solution.
4. Better process control for cesium tetraphenylborate precipitation.
5. Zeolite ion exchange.
6. Stabilize cesium tetraphenylborate precipitate with palladium catalyst poison.
7. Electrochemical separation.
8. Fractional crystallization to remove solvent.
9. Make a ceramic of waste.
10. Precipitate as cesium tetraphenylborate and vitrify quickly.
11. Electrochemical membrane support.
12. Potassium precipitation before cesium precipitation.
13. Cesium ion exchange with acid regeneration.
14. Build more storage tanks to hold waste.
15. Stabilize cesium tetraphenylborate precipitation.
16. Hollow-fiber extraction of cesium.
17. Remove water from salt tanks for more storage capacity.
18. Precipitation at reduced temperature and storage as cesium tetraphenylborate.
19. Ion exchange on glass; vitrification.
20. Concrete formation of total waste (grout).
21. Simulated moving bed adsorption to separate cesium.
22. Inject in ground; vitrify *in situ* with nuclear explosives.
23. Concentrate cesium in microorganisms.
24. Adsorption in sodium titanate.
25. Regenerable ion exchange.
26. Cesium extraction in acidic solution.
27. Magnetic particles which adsorb cesium.
28. Electrochemical nitrate destruction.
29. Fluidized-bed ion exchange.
30. Cesium separation and concrete formation (grout).
31. Cesium absorption in $MnFe(CN)_6$.
32. Flocculate tetraphenylborate precipitate.
33. Salt washing plus fractional crystallization.
34. Electrically regenerated ion exchange.
35. Reversible adsorption using crown ethers on inert substrate.
36. Reduce explosion risk for cesium tetraphenylborate precipitate.
37. Use smaller size of cesium ion for separation.
38. Total vitrification of waste.
39. Make storage tanks safer.
40. Separate interstitial liquid, evaporate to dryness, and vitrify.
41. Buy additional benzene-release permits.
42. Precipitate potassium before cesium.
43. Selective crystallization.
44. Properly designed ion-exchange processes.
45. Properly designed precipitation processes.
46. Alternative precipitation chemistry.
47. Countercurrent exchange with sodium titanate.
48. Electrodialysis.
49. Salt dehydration and vitrification.

> **Table 3.1–5** *Ideas for the application of dipstick technology* At this point we are not focused on inventions which could actually accomplish these tasks. We are just looking for areas where the technology would be beneficial, for further analysis and development. A pregnancy test is an example already available commercially.
>
> | | |
> |---|---|
> | Fish freshness | Allergens |
> | Chicken freshness | Hair cleanness |
> | Beef tenderness (from color) | Body odor |
> | Blood cholesterol (HDL, LDL) | Body temperature |
> | Temperature history of meat | Water quality for drinking |
> | Organics in water | HIV |
> | Blood sugar | Water hardness |
> | Lead in paint | Asbestos |
> | Air quality (especially for asthma) | Total aromatic exposure |
> | Radioactivity (film badges) | Sniffing explosives at airport |
> | Prostate function | Milk sourness |
> | Fridge contents spoilage | Caffeine levels in tea/coffee |
> | Skin dryness | Drug overdoses |
> | Sun intensity | Total smoke inhalation |
> | Bug-repellant concentration | Loss of lung capacity |
> | Pap smears | Fat metabolized during workout |
> | Total exposure to sun | Pesticide exposure |
> | Total exposure to noise | Total exposure to insecticides |
> | Relative humidity | Skin oiliness |
> | Nerve agents (on battlefield) | Scalp dryness |
> | Corn-on-the-cob sweetness (freshness) | Nappy wetness |
> | Diaper odor | Alcohol content in drinks |
> | Alcohol in breath (is it safe to drive?) | Third-hand smoke (for hotel room) |
> | Breath odor | Chlorine in pools |
> | Water for houseplants | Sterility (is the toilet clean?) |
> | Carbon monoxide | Fertilizer for houseplants |
> | Formaldehyde (from soft furnishings) | Pollen count |
> | Malaria detection | Cancer detection |
> | Ice-cream smoothness (ice-crystal size) | Second-hand smoke |

and codeine all originate from natural sources. Other industries which rely on specific chemical activity have also benefited from natural products or their analogues. For example, stevioside is a low-calorie sweetener derived from *Stevia rebandiana*, commonly called cua-hê-hê (sweet herb). Sunillin is a plant-based anti-fungal pesticide. Bulletproof vests are based on a synthetic polyamide. The hydrogen of each amide group in the polyamide hydrogen bonds to the oxygen of an amide group in a neighboring chain: the result is interlocking helices, much like a miniature rope, very strongly bound because of the large number of hydrogen bonds. Strength comes from this strong bonding, and flexibility from the ability of the helices to stretch. Such vests imitate spider silk, a natural polyamide with large numbers of regions consisting of repeating glycines. Glycine is the amino acid with hydrogen as a side group (i.e. the smallest of the amino acids). The polyglycine regions pack and hydrogen bond in exactly the same way as the synthetic polyamides of body armour. While the man-made imitation is good, it is not as good as spider's silk, so research continues to more accurately imitate the spider.

There are four ways in which natural products may be used to produce active chemical species:

(1) If the active ingredient is expensive or impossible to synthesize, it may be isolated directly from an organism. Vinchristine, a highly effective treatment for childhood leukaemia which is isolated from the Madagascan periwinkle (*Catharanthus Roseus*), is one example. It requires 53 tons of Madagascan periwinkle leaves to isolate 100 g of vinchristine, which is worth $22,000.

(2) A precursor may be isolated from a natural product and then used as a building block for a more complex molecule. Diosgenin, the active molecule in the first oral contraceptive, was originally synthesized entirely in the laboratory, but its economic production depends on a suitable precursor extracted from the Mexican yam (*Dioscorea floribunda*). A yam seems an unlikely candidate to have been responsible for a social revolution on the scale of birth control!

(3) The natural product can be slightly modified to produce a different material. For example, the cholesterol-lowering drug lovastatin, isolated from a fungus, can be made longer acting if one particular hydrogen is replaced by a methyl group, yielding simvastatin. Infamously, morphine, derived from the opium poppy, is diesterified to make heroin.

(4) The active ingredient may be identified in a natural product, but then used as a model for a chemical synthesis of an identical or similar molecule. Reserpine, a drug used to treat hypertension, was first identified in the Indian snakeroot (*Ravolfia serpentina*), used in traditional Ayuvedir medicine in India as a tranquilizer, but is now produced entirely synthetically.

Although the first group is historically the most significant, the fourth category is the target of many modern investigations, particularly using microorganisms.

The exploitation of microorganisms merits further discussion. We start by seeking samples of new microorganisms. Fewer than 0.1% of bacteria and fungi present in the soil have been tested; yet penicillin, cyclosporin (the immunosuppressant which makes transplant surgery possible), and lovestatin are just a few of the products which were derived from fungi. One way to find these is to send a graduate student armed with a trowel off to a tropical country for a few weeks. The student will return with bags of soil samples which can be cultured to seek new, unusual microbial species. Alternatively, we can look closer to home: the mold from which current penicillin cultures are descended was found growing on a cantaloupe in the garbage of a supermarket in Peoria, Illinois.

As another approach, we can begin with a culture which we already know produces active chemical species. We then stress the microbe either by chemical treatments or, more classically, by high doses of radiation. These doses should be so high that they will kill most of our culture. We will then grow the few survivors to see if the stress of radiation has rearranged their DNA beneficially.

We will often test these new microorganisms by culturing them within an existing colony of the target pathogens. In a few cases, we will find that our new microbes will kill target pathogens growing close by, presumably because the new strain is producing a toxic chemical. Seeing such behavior is said to have been the stimulus which led Alexander Fleming to the discovery of penicillin. Thus, we can discover which of our candidate microbes are producing toxins of value.

We next need to determine the chemical structures of these toxins. To do so, we will normally depend on a combination of chromatography, mass spectrometry, and nuclear magnetic resonance. Hopefully, we will discover the chemical structure in sufficient detail so that we will be able to compare synthesis of the target species by chemical methods or by fermentation with our new microorganism. If we decide on fermentation, we can try to dramatically increase the production (the "titre") by inducing mutation in our new microorganism.

In addition to microorganisms, we can seek active chemical species in plants. This rich source of ideas is hardly touched. While there are on the order of half a million species of flowering plants, only 5000 of these have been extensively investigated for active chemicals, and most of those originate from temperate regions. The potential of aquatic organisms has also recently started to be realized: discordomolide (a powerful immuno-suppressant) and marinovar (an antiherpes agent) were extracted from a Bahaman sponge and a Californian marine bacterium, respectively.

A final approach to natural-product screening is to investigate the traditional uses of natural products, primarily as medicines. This is not a new method. In 1785, William Wittering discovered that the ingestion of dried foxglove (a member of the genus *Digitalis*) eased dropsy, a condition caused by the heart's failure to pump properly. He reported, "I was told this had long been kept secret by an old woman in Shropshire, who had sometimes made cures after the more regular practitioners had failed." Today, two components of the foxglove – digitoxin and digoxin – are prescribed to cardiac patients to regulate and strengthen their heartbeats.

At present, this natural-product screening centers on conserving the knowledge of traditional healers, who may be remnants of dying cultures in tropical countries. For example, flavanone is a topical anti-inflammatory extracted from tree bark in traditional Samoan medicine. Another example, the adrenergic blocker yohimbine, is extracted from *Pavsinystalia johimbe*, the source of a traditional Indian aphrodisiac.

However, since 1990, interest in natural-product screening has waned. The reasons for this loss of interest are complex. Many in the larger drug companies felt that the number of successes had dropped, so that synthetic chemistry offered better prospects. Critics of these drug companies argue that interest shifted from curing disease to relieving symptoms: after all, the critics assert, the companies would make less money from curing malaria than from medication taken daily to lower blood cholesterol. Still others argue that because natural products are often highly dilute mixtures, many promising molecules are present in such small amounts that they are hard to identify and to test.

3.2–2 COMBINATORIAL CHEMISTRY

This is the chemical equivalent of using a sledgehammer to crack a nut. Though the sledgehammer may be inelegant it is unquestionably powerful! Very often we are faced with a chemical problem we know we want to solve: a drug to attack a specific protein, a catalyst to speed a known reaction, or a poison specific to a microbe. We might, however, have no idea how to achieve this result. There are thousands of potential solutions out there: we have no chance of testing them all other than via combinatorial chemistry.

The core idea behind combinatorial chemistry is to identify possible active ingredients or molecular fragments and to test all of them robotically, in all possible combinations. Clearly both the combining of ingredients and the testing of activity needs to be automated for this to be practical. The method has particular value for biochemical problems. Most biological molecules are by their nature sequences of a limited range of alternatives. DNA is a code written as a sequence of four bases, an alphabet with only four letters. Proteins are a linear sequence of twenty naturally occurring amino acids. Naturally occurring fats are almost all triglycerides: only the chain length and degree of saturation of the fatty acids vary. Clearly it is a conceptually easy matter to synthesize all possible DNAs, polypeptides, or fats and to test them on the biological systems of interest. Testing for biological activity is well established and lends itself well to automation.

As an example, imagine that we want to investigate the efficiency of hexapeptides for affinity for the μ-opioid receptor. Houghten and co-workers (1999) identified a potential library of 52,128,400 hexapeptides which might have the desired affinity. Even for a robot this represents a lot of work. They therefore decided to structure the problem. First, they tested 400 alternatives, in which only the first two amino acids were varied. The most efficient of these was then taken as fixed, and successive amino acids were used to optimize efficiency, i.e. a further four tests of 20 alternatives each. This strategy thus paralleled the method of steepest descent used in chemical-process optimization. Encouragingly, the molecule identified by this particular search has the sequence occurring naturally in proteins which stimulate this receptor. This demonstrates the possibilities of the combinatorial method.

However, at this point, the consensus is that unfocused combinatorial chemistry has failed. Large pharmaceutical companies do continue some studies of low-molecular-weight synthetic compounds which are analogs of known drugs. While the success rates are typically below 0.001%, any molecules which are found are usually easy to make. The best chance for future success may be a combination of combinatorial and natural-product screening techniques, possibly using metagenomics. These techniques are made much more feasible because of advances in high-performance liquid chromatography (HPLC) and mass spectroscopy. The resulting molecules are easier to evaluate because of recent advances in nuclear magnetic resonance (NMR).

We illustrate these ideas with an application to fuel-cell catalysis.

EXAMPLE 3.2–1   FUEL-CELL CATALYSTS

An elegant example of combinatorial methods used in catalytic chemistry is given by Reddington *et al.* (1998). The problem is to optimize the composition of a catalyst for methanol fuel cells. High-surface-area Pt–Ru catalysts are the current technology, but waste about 25% of the fuel's energy. By chemical analogy, we want to consider the other platinum group metals as additives. The problem is to test efficiently many catalysts of different composition and with different combinations of elements.

How can this be done?

SOLUTION

To make these tests, Reddington *et al.* built a 645-member electrode array, including the five pure elements, 80 combinations of two elements, 280 ternaries, and 280 quaternaries, by using a modified ink jet printer to spray dots of mixed metal salts. The dots were dried and reduced to the metals. To test each combination, the authors used a fluorescent molecule which luminesces in acid but not base. (H^+ is produced as part of the fuel cell's catalytic cycle.) On testing, the most effective catalyst simply lit up the brightest.

The results are fascinating. A quaternary alloy (Pt(44)/Rh(41)/Os(10)/Ir(5)) was found to be the most efficient catalyst. It was significantly more efficient than the commercial binary, although the surface area achieved was only about half. Other attractive catalysts were identified in different regions of both ternary and quaternary space. The most efficient ternary (Pt(62)/Rh(25)/Os(13)) lies in a ternary region bounded by inefficient binaries, Pt–Os and Pt–Rh. We would not intuitively expect high activity from this ternary composition. These results could never have been achieved by conventional catalytic testing: the amount of work required would be just too great and the results show that a rational or intuitive approach would have failed.

3.3   Sorting the Ideas

Generating possible solutions to a need is fun, but produces a hodgepodge of ideas. These ideas are incomplete. They are not so much like frogs which must be screened to find a prince; instead, they are like fragments of frogs. Somehow, we must compare these fragments and choose the most promising candidates. We will probably also want to combine some of our better ideas to produce something closer to a whole frog.

The situation seems to us much like the screening of candidates for an open professorship. If we advertise such a position, we will get at least one hundred applications. We want to narrow this number to about five whom we wish to interview. We hope that at least one of the five will be acceptable. However, we do not have the time to carefully evaluate all the applicants. How do we go from one hundred applications to five?

One possible strategy uses two separate stages. First, we could eliminate those who are unqualified. This could include candidates without a doctorate or without publications and research funding over a long career. We could eliminate recent Ph.D. graduates from universities without much research effort. We could eliminate non-citizens or older applicants, though such actions could be illegal. We could eliminate those in a particular sub-specialty, like polymeric materials, feeling that our department is already well represented in that area. We would use this strategy almost without thinking, just to cut the number of applicants down to perhaps twenty. These twenty we would carefully study to choose the five for interviews.

The situation here is similar: we will have perhaps one hundred ideas, and we want to choose the best five. Just as we might do for faculty candidates, we will make this choice in two stages. First, we will sort and prune these ideas, which will normally reduce the number to around twenty. Second, we will use matrix-screening methods to choose the best five. Unlike the case of faculty candidates, we also have the opportunity to combine the best points of more than one of our ideas – sorting ideas also involves combining them. The sorting is the subject of this section, and matrix screening is covered in the next section.

3.3–1 GETTING STARTED

The first step is simply to prepare a list of all the ideas. The list will almost certainly contain considerable redundancy, which is easily removed. The redundancy most often will occur because some ideas are more general and some are more specific. For example, in an effort to recover orange juice flavor during juice evaporation to make a concentrate, one idea could be:

"Remove the flavor from the vapor."

A second idea could be:

"Use a selective membrane to concentrate flavor from the vapor."

The second idea is a specific example of the first.

The overall list will also contain some ideas which seem folly and which should be pruned. This pruning is much trickier than removing redundancy. Some ideas will simply be irrelevant, perhaps recorded incorrectly or just not thought through in the turbulent brainstorming session. They are easy to drop. Some ideas may prove just plain wrong, perpetual motion machines in disguise. However, some ideas may be incorrect but still contain innovations which we do not want to lose. As a result, we suggest a bin of random thoughts which we keep aside from the rest of the sorting, and periodically recheck. Removing redundancy and folly will typically cut the number of ideas by about a third, from around one hundred to about seventy. We need to organize these ideas further in an effort to get down to the targeted twenty.

3.3–2 "THE MATERIAL WILL TELL YOU"

The next step is to organize the ideas into categories, but the hard part is to know what form the categories should take. In this effort, the best guide is to remember what good writing teachers say: "The material will tell you." By this, the teachers mean that the structure will never be the same for any two sets of ideas. The organization of ideas for concentrating orange juice may be as different from that for ideas for a reusable detergent as from the structure of an essay on "Macbeth." The ideas themselves must be the basis of any organization.

In many cases, we will find the organization of the ideas is obvious, and jumps out to all members of the core team. Be careful that the organization does not simply reflect the training of the core team. In one case, a team of mostly engineers insisted on organizing the ideas around unit operations. In so doing, they overrode team members trained in chemistry who urged organizations based on chemical mechanisms. The latter scheme would, in this case, have provided more insight.

The normal rules of outlining apply to this effort. First, you want to use around five main headings. These five heading should be roughly equal in importance. If you have many more headings, examine them carefully to see if some cannot be combined. While sometimes they cannot, make the effort.

Second, subheadings should be special cases of each main heading. There should rarely be more than four subheadings; if there are, consider combining these further. There should never be only a single subheading; if there is, then the subheading should probably replace the main heading.

In the outlining, we should remember a guideline sometimes taught in elementary school, called "the Rule of The Table." The guideline argues that an outline is like a group of dinner tables. Each heading is like the top of the table, covering everything under it. The subheadings are like the table's legs. Three or four legs are the best number for a table. A few good tables do have two legs ("harvest tables"), but very few good tables have six or eight legs. Thus, says this guideline, we should make the sections of our outline like tables. While this may sound simplistic, we have found this rule helpful.

Once the outline is made, it should be carefully edited. Most commonly, the editing will focus on three areas. First, the outline will often expose gaps. For example, in an evaluation of reducing pollution from airplane de-icing, we might have a heading "Extract the de-icer from water for recycle." This really should be a subheading under "Recycle the de-icer," which may not have been one of our original ideas. Such a heading could have subheadings like "Extract the de-icer", which we got, and like "Adsorb the de-icer," which we missed. We may decide to add this new "Adsorption" idea.

The second area for editing is to prune unpromising or politically impossible ideas. For example, one group of ideas may call for inventing a new type of chemistry, and we may have little in-house chemical skill. In this case, this group of ideas should probably be dropped. Similarly, our company may make centrifuges, and we may have ideas for membrane separations. Pursuing the membrane ideas would be a completely new direction for our company, and hence imply deciding to shift corporate strategy. Such a shift is probably beyond the mission of our product development team.

The third area for editing is to acknowledge different patterns of thinking in different individuals. Some persons cannot organize ideas in a linear structure like that used here, and may prefer circular outlines or pie charts or some other structure. Some may prefer decision trees. In particular cases, organizations may have adopted particular development schemes. Any of these can work. Which we use does not matter. Just use those which let us cut the number of ideas from our original hundred down to twenty.

EXAMPLE 3.3–1   REUSABLE LAUNDRY DETERGENTS

Our start-up company has raised considerable resources to seek pollution-preventing, environmentally benign chemical technologies. Without many specifics, the company's prospectus promises "reusable detergents." These should be more attractive ecologically than many existing "environmentally friendly" detergents, some of which have high sodium hydroxide concentrations. A few consumer surveys have generated product ideas like those shown in Table 3.1–1.

Sort these ideas so that we can begin to choose those which are most promising for development.

SOLUTION

The ideas in the earlier table break into the four groups shown in Table 3.3–1. The numbers shown in parentheses refer to the particular ideas in Table 3.1–1. These four groups represent different regions on the intellectual triangle in Figure 1.2–1. For example, those under heading "I. New soap" are close to the "chemistry" corner of the triangle. "IA. Chemistry" is nearest this corner. "IB. Easier to recycle" also depends to some extent on chemical engineering. The heading "II. New washer design" is going to be dominated by mechanics, and be relatively independent of chemistry.

This organization of ideas shows immediately that our new company needs to choose between several, very different strategies. The first heading implies making a new soap, and so makes sense for a company like Proctor & Gamble, who already make conventional detergents. It would be a major new initiative for a company like Whirlpool, whose current business centers on making washing machines. Whirlpool has already explored the topics under "II. New washer design." Interestingly, the washing machines available in Europe are dramatically more expensive and slower than those available in North America.

Table 3.3–1 *Sorted ideas for reuseable laundry detergents* The numbers in parentheses refer to the ideas given in Table 3.1–1.

I.   New soap (4, 9, 11, 37)
   A. Chemistry
      1. Base (24)
      2. Powders (8, 16, 18, 32, 39, 46)
      3. Biochemical (5, 6, 7, 15)
      4. Emulsions (14)
   B. Easier to recycle (48)
      1. Size (12, 13, 14, 16, 26, 47)
      2. Temperature (36, 38, 45)
      3. pH (35)
      4. Other chemistry (49)

II. New washer design (3, 10)
   A. More mechanical energy
      Ultrasonic and pressure waves (40, 42)
   B. Thermal energy
      1. Cooking (43)
      2. Freezing (44)
   C. Light (41)

III. Improved dry cleaning (1, 27, 28, 48)
   A. Altered equipment
      1. Sealed (29)
      2. Supercritical (30, 31)
   B. New solvents (33)
      1. Gases (19, 20)
      2. Liquids (30)

IV. New directions
   A. Disposable clothing (2, 34)
   B. Soil-resistant clothing (21)
   C. Altered mores (22, 23, 25)

However, European machines use both less water and less detergent, and hence cause less environmental intrusion.

EXAMPLE 3.3–2 NEW METHODS OF DRUG DELIVERY

The second example of idea generation was undertaken by a consumer products company curious about developing platforms for delivering a wide variety of drugs. The objective was not the drugs themselves, but to explore ways in which old drugs could be made more effective. The results of these efforts were given above, in Table 3.1–2.

Organize these results to help the company choose areas for development.

Table 3.3-2 *Organized ideas for drug delivery* The list sorts ideas by their method of administration. Ideas for different methods of drug injection (13, 15, 28, 70) were ignored as inappropriate for this company.

I.   Oral delivery (pills)
   A. Slow release – in fatty tissue (8),
      from gum (38), as nanocapsules (39, 56)
   B. Degradation (71, 79, 108)
   C. pH-activated (96)
   D. Absorbent in gut (20)

II. Topical delivery (on the body's interfaces)
   A. Transdermal patches (2, 10, 12, 19, 25, 41, 42, 61, 65, 77, 33, 87)
   B. Ocular (eye drops, contacts) (17, 36, 107)
   C. Inhalers
      1. Lung (37, 66, 86)
      2. Mucus membrane (84)

III. Implantable delivery (requiring surgery)
   A. Released by diffusion (1, 15, 31, 103, 26, 53, 85, 102)
   B. Released on demand (52, 55, 68, 94, 97, 100)
   C. Feedback control (3, 39, 51, 91)
   D. Diagnostic (13)

SOLUTION

Because these ideas scatter broadly, many organizations are possible. The organization given in Table 3.3–2 uses the method of delivery as a grid for the sorting. By far the largest number of ideas are based on diffusion of the drugs across the skin (topic IIA.1). Some of the most interesting are based on implantable devices, where the drug can be released passively, with feedback control, or by the patients themselves. In this case, the company involved flirted with designing improved inhalers, but eventually decided not to enter this market area.

EXAMPLE 3.3–3   A POLLUTION-PREVENTING INK

A printing company prints personal checks with a lithographic ink containing the carcinogenic solvent methylene chloride (CH_2Cl_2). Workers at this company also clean the presses by wetting a shop rag with the same solvent, and scrubbing down the press. This procedure works well. The trouble is that much of the methylene chloride evaporates and so risks workers' health and censure from the environmental authorities. Also, the soiled rags have been reclassified as a hazardous waste, so that the cost of their disposal almost equals the cost of buying the solvent in the first place.

   The company clearly needs to use a different ink, one which has less negative environmental impact. Some ideas for this ink were shown in Table 3.1–3. Sort these ideas to identify those most worth pursuing.

> **Table 3.3–3 *Sorted ideas for a pollution-preventing ink*** The numbers in parentheses refer to the ideas suggested in Table 3.1–3.
>
> I.   *Improve current printing*
>   A. Change press (4)
>   1. Isolate press (3, 11, 12, 14, 17)
>   2. Use laser printer (6)
>   3. Use photocopying (32)
>   B. Change cleaning
>   1. Less often (13, 15)
>   2. Other solvents (9, 23, 24, 25)
>
> II. *Use a new solvent*
>   A. Change CH_2Cl_2 operation
>   1. Recycle (8)
>   (i) Extract (10)
>   (ii) Spin dry (new)
>   2. Burn (27)
>   3. Freeze (new)
>   B. Replacement of CH_2Cl_2 (2, 20)
>   1. Non-volatile solvent (21)
>   2. Oil as solvent (31)
>   3. Solvent mixtures (16, 18)
>
> III. *Solvent-free ink chemistry (1, 7)*
>   A. Electrostatic ink (5)
>   B. "Solvent which dissolves ink" (19)
>
> IV. *Don't use checks (29)*

SOLUTION

The ideas easily break into four groups, as shown in Table 3.3–3. Again, the numbers in parentheses refer to the original sequence of ideas, which in this case are in Table 3.1–3. The first group in Table 3.3–3 involves changes in the printing presses. Because the company does not want to make the enormous capital investment involved in changing the presses, this group is deferred until other alternatives have been explored.

The second group of ideas involves either containing the solvent or using a different solvent. These ideas are the easiest to implement, and hence the most tempting for further development. The third group of ideas implies the invention of a new ink, a more major effort than the substitution of a new solvent. We will examine this option more carefully in Section 4.1.

The final idea, "Don't use checks," may initially seem foolish; but consider the explosion in electronic money transfers. The company may decide that electronic data processing, which replaces handwritten checks, is like the automobile, which replaced the horse-drawn buggy. If so, then printing checks may be like making buggy whips. Thus this fourth idea should be carefully considered in the idea screening process described in Section 3.4.

Table 3.3–4 *Sorted ideas for "Lotus Leaf" and "Gecko Foot" technology*
These ideas are broader and less developed than those in the other tables in this section.

I.   Self-cleaning surfaces (Lotus Effect)
    A. Glass
        Office buildings, homes, anti-fog for windows, anti-foam beer glasses
    B. Metal
        Cars, pipes
    C. Plastics
        Boats
        Building materials, tiles, kitchen worktops
        Packaging for food

II. Products which stay clean (Lotus Effect)
    A. Personal care
        Cosmetics, eye glasses, visors, bike goggles, etc
    B. Clothing
        Shoes, ties that stay clean, socks that stay dry, waterproofs
    C. Housewares
        Drinking glasses, ice-cream scoop, cooking dishes, sieves
    D. Electronics
        Electronic circuit boards, TV/computer screens

III. Products with high adhesion (Gecko Effect)
    A. Glues
        Strong but reversible
    B. Velcro analogues
        Shoes to climb walls, cross-country skis without wax
    C. Dust rag

EXAMPLE 3.3-4 THE LOTUS EFFECT

The self-cleaning properties of the lotus leaf and the adhesive properties of the gecko foot represent natural phenomena which depend on stuctures at the nanometer length scale. In both cases, they are a consequence of regular "pillars" protruding from a surface, combined with specific surface chemistry: hydrophobic in the case of the lotus leaf, hydrophilic for gecko feet. Recent advances in nanotechnology have allowed the replication of similar structures in man-made materials. Regular protrusions or indentation of a few tens of nanometers up to a few microns can be routinely produced on polymer or metal surfaces at perhaps a few hundred dollars per square meter. Given these advances, it may not be long before "lotus leaf-" and "gecko foot-" based products reach the consumer market. Consider where time and money for development are best spent.

SOLUTION

Table 3.3–4 gives a sorted list of ideas to exploit this new technology. The original list from brainstorming included around 80 ideas. Note that these ideas are much less well developed than those in some of the other examples. Their manufacturing details are unknown, and their eventual market value is unexplored. They will require more work to form a basis for effective products.

3.4   Screening the Ideas

Organizing the fragments of ideas which we have generated can greatly clarify our thinking, reducing the number of concepts for more quantitative consideration. For simple product designs, the idea sorting may suggest only one or two strong ideas for new products. If this is the case, we can skip directly to the "Selection" procedure described in Chapter 4.

In many cases, however, our search for new products will have generated a large number of promising fragments of ideas. For many industries, we have suggested that this will be around one hundred. Sorting through these ideas to remove redundancy and inconsistency with corporate goals, and combining some of the ideas will normally cut this number to twenty. We need to cut the number further, for we will not normally have the resources to make the quantitative calculations necessary for selection between all of these. We need a basis for qualitative judgments to further reduce the selection.

We explore one method for achieving this in this section. To compare different product ideas, we recognize that our comparison must include a wide variety of criteria. It will include purely objective questions like: "Which of these two absorbents has a greater capacity?" and "Which battery has a greater power per mass?" Some objective comparisons will also involve cost. We will normally prefer a catalyst which has only half the activity but one tenth of the cost of our original benchmark.

At the same time, our comparison of different product ideas should often include more subjective criteria. We will be less interested in a new non-woven fabric if it is less "wearable." We will normally prefer a marginally more expensive product which we judge is "safer." In more complex cases, we will be making compromises between two conflicting criteria. For example, we may be hard-pressed to decide between different home air purifiers whose performance and cost go up together.

3.4–1 STRATEGIES FOR IDEA SCREENING

There are clearly many possible strategies by which we can screen our ideas for product design. Our easiest approach to screening the ideas is to look at the headings in the outline, and choose the best candidate under each heading. We then use these candidates in the selection procedures in the next chapter. This simple strategy can work well if the product designs are simple extensions of existing technology. It does have the significant risk that the two best ideas will be under the same subheading. This strategy will have major risk if there are many, very different product designs.

A more effective strategy is to choose the most important factors by which we want to evaluate the product. These factors will often include at least some of the following:

(1) *Scientific maturity*. We will prefer designs based on scientific knowledge which we already have and understand.
(2) *Engineering ease*. We will prefer designs which imply straightforward engineering like that already used in established manufacturing.
(3) *Minimum risk*. We do not want to take unnecessary chances. At least, we want to know what our chances of success are.
(4) *Low cost*. We may want a rough estimate of the relative cost of our ideas.
(5) *Safety*. We want to identify which products are inherently safer or more dangerous than our benchmark.
(6) *Low environmental impact*. We will tend to choose products which cause less pollution.

Other important factors may be more subjective. Examples are "The product should be quiet," or "The product should be comfortable." The list of Needs established in Chapter 2 will be a good source of these factors.

We typically seek five or fewer of these attributes which we feel are most important. This choice is best made by consensus, with the entire core team working together. Asking the individual members of the core team for their choices and averaging these choices usually does not work well. In seeking this consensus, the members of the team need to be careful not to compete, not to feel that their chosen attributes will be "winners" or "losers." We are not sure how to ensure this. We do recognize that some individuals have personalities which catalyze rational judgment, just as other individuals can precipitate polarization and win–lose arguments. We also recognize that some companies prefer much longer lists of attributes. In our experience, these longer lists are less helpful.

Once the key factors are identified, we need to assign weighting factors to them. These weighting factors should be normalized:

$$\sum_{i=1}^{n} \omega_i = 1 \tag{3.4-1}$$

where ω_i is the weighting factor of factor i, which is one of a total of n attributes. In our experience, the core team will reach a consensus on these weighting factors more quickly than on the choice of the factors themselves. We suspect that this may be because the discussion about which factors are important implicitly includes a discussion of the relative weights.

With these weighting factors in hand, we now evaluate the key ideas in our outline on the basis of some scale. The easiest scale for most core-team members ranges from a low score of one to a high score of ten. First, we assign an average score of five to our benchmark product, and we grade each of our product ideas relative to this benchmark. When we are through grading, we have a group of scores s_{ij} for each attribute "i" and each idea "j". We then calculate the total score for each idea:

$$\text{Score}\,(j) = \sum_{i=1}^{n} \omega_i s_{ij} \tag{3.4–2}$$

The ideas with the highest scores are those we will use for the next stage of product design, selection.

A numerical scoring system such as that just described can be useful. Certainly, when it comes to the more detailed comparisons of the Selection stage we are likely to want this quantification. At this stage it is sometimes adequate to use a more qualitative scoring system, such as one or two plusses or minuses. We really just need an indication of which ideas are likely to perform well or badly with respect to each of our criteria.

3.4–2 IMPROVING THE IDEA SCREENING PROCESS

The process described above often strikes those trained in engineering or chemistry as too easy and too qualitative. Even when relative scores seem reasonable, they are obtained with so little effort that they are suspect. This intellectual masochism may result from the study of subjects like thermodynamics. These concept-scoring ideas are certainly simpler than standard-state chemical potentials or fugacity coefficients. As a result, many chemists and chemical engineers will concede that, "Yes, well, the concept-screening process is interesting, but we cannot really take the results seriously."

We disagree.

We believe that the result of the concept-screening matrix should be taken seriously, though the simple procedure outlined above can be improved. We suggest three specific improvements. First, we urge a careful choice of the benchmark. In many cases, this benchmark will be an existing product with the greatest market share. In some cases, it may be what we expect as a new product from competitors. In other cases, it may be what we hope we can make as the best of the existing type of product. As a check on this choice, we may sometimes try to choose a different benchmark after a first round of assessments, just to make sure our first benchmark is best.

The second way to improve this concept-screening matrix is to check the core team's scores against those of other interested experts. One obvious group are other individuals in marketing who are outside our core team. Another group are the lead users of current products. In seeking these outside opinions, we must remember that the uninformed tend to be conservative: they are those who will not buy "genetically engineered tomatoes" but will buy "improved hybrid tomatoes." While we should seek these outside opinions, we should not accept them as gospel.

The third way we should improve the screening process is to make a sensitivity analysis of the weighting factors. We will change the weighting factors within sensible limits to see if this alters our rank ordering of the ideas. In most cases,

Table 3.4–1 Concept-screening matrix for printing Chaucer's Canterbury Tales This matrix could be one developed by William Caxton, in 1476.

Selection criteria	Weighting factor	Illuminated manuscript	Printed Chaucer
Quality	0.4	5	1
Cost	0.4	5	6
Quantity	0.2	5	8
Total score		5	4.4

we will find little change. In a few cases, where the change is dramatic, we should re-examine our selection criteria, for we have probably not explicitly considered all the major issues.

While we believe that it has value, the concept-screening matrix is not a panacea, able to screen all available ideas in all situations. Some problems may come from the tendency to compress the scale of judgments. After all, if you give everything a score of 5, you are effectively abstaining from judgment. If this is the case, then that criterion has no value and should be dropped.

A much more serious problem is the assumption implicit in Equation 3.4–2 that everything can be scored and weighted linearly. This is approximately true only when the products are similar, changed in only minor ways. The assumption of linearity is generally untrue in three obvious cases:

(1) *The criterion is binary*. For a home air purifier, the product might be judged too noisy or quiet, with nothing in between.
(2) *The product will not work*. For an olefin–paraffin separation, the top-scoring hollow-fiber membrane product depends on making the membrane selective, which may not be possible.
(3) *The product changes the market*. This implies an innovation which is so good that all other criteria are irrelevant.

We can normally find ways to handle the first and second limitations. For example, the second limitation can be regulated with the ideas of risk management described in Section 4.4.

The truly innovative products, those which change the market, are the real concern. Even the vocabulary used to describe these products is different. They are "show stoppers," "game changers," or "step-out technologies." They are "out of the box," i.e. conceived beyond current thinking. They are hard to find and hard to recognize.

As one vivid example, imagine that you are William Caxton in 1476. You are trying to decide whether it makes sense to print Chaucer's "Canterbury Tales," or continue to produce illuminated manuscripts. You might develop the concept-scoring matrix shown in Table 3.4–1. Your three scoring criteria are the quality of the finished product, its cost to produce, and the quantity which you can make. The illuminated manuscript is the benchmark and so gets a score of 5 for each criterion. The manuscript is of much higher quality than the smudged and crooked printed version of Chaucer, so the printed edition gets a low score for quality. The cost of the printed Chaucer is less, but that is not a big advantage: the monks, those postgraduate students of the Middle Ages, work for nothing. What would they do if they were not copying manuscripts? Finally, you can make more printed Chaucers. But the market is limited to the literate, mostly the monks. Thus the quantity you can make is not that important. Thus, as Table 3.4–1 shows, it does not make sense to print books.

With historical hindsight, we know this conclusion is wrong: it does make sense to print books. The use of moveable type in printing was an enormous technical innovation. However, we should admit that recognizing such advances at the time will always be hard. In this sense, we should remember that when he first printed Chaucer, William Caxton was not only a printer. He was a wool merchant, and he may have been looking for an outlet for excess sheepskins, which are of course the feedstock for parchment. He may have begun printing Chaucer to use up extra sheepskins.

With this cautionary example in mind, we turn to harder, more technical examples. These examples are simpler than those encountered in practice: they only screen three to five ideas. In practice, we would normally screen around twenty ideas. If we actually had only three to five ideas, we could probably make detailed calculations of all ideas: we could jump directly to the "Selection" step of product design detailed in the next chapter. We are screening small numbers of ideas in these examples because the screening is more manageable and illustrates the method clearly. Just remember: the actual practice of idea screening will involve many more ideas, probably around twenty.

EXAMPLE 3.4–I CONCENTRATING ORANGE JUICE

Orange juice is sometimes concentrated by vacuum evaporation, frozen for storage, and then thawed and diluted for use. Its taste is inferior to untreated juice because much of the flavor evaporates during the vacuum evaporation. To replace this lost flavor, ground orange peel is sometimes added to the frozen concentrate. The peel tastes bitter.

As an alternative, we want to explore using reverse osmosis membranes to remove water and retain flavor. We are also interested in using a combination of these methods: some evaporation, some reverse osmosis, and then a blending of the two streams. Which prospects look best?

SOLUTION

Our core team decides that there are four key factors in these processes: technical maturity, engineering, cost, and product quality. After considerable discussion, the team decides on the values shown in Table 3.4–2. The low scores for the maturity of reverse osmosis reflect concern that orange pulp in the juice will tend to foul the membranes after only a short period of operation. The troubles caused

Table 3.4–2 Possible processes for better orange juice concentrate The benchmark, vacuum evaporation, could be combined with reverse osmosis in a hybrid process, which is also evaluated.

Criterion	Weighting factor	Vacuum evaporation	Reverse osmosis	Hybrid
Maturity	0.1	5	4	2
Engineering	0.2	5	2	3
Cost	0.35	5	3	3
Quality	0.35	5	8	7
Total score		5	4.7	4.3

by this fouling are serious enough to compromise the advantages of a higher quality product. This is consistent with the current market, where juice processed by reverse osmosis is not available. At the same time, the market opportunity remains: unconcentrated orange juice is outselling concentrate in spite of much higher shipping costs.

EXAMPLE 3.4–2 ZOLADEX INJECTIONS

The drug Zoladex (goserelin) is an injectable hormone used to treat breast and prostate cancer. While its sales top $800M/yr, less than 50 kg/yr are made. The drug is injected in 11 mg doses every three months, using a sterile syringe. This syringe is assembled from sterile materials. Sterilizing the loaded syringe after assembly would be an enormous cost saving. Is this practical?

SOLUTION

The core team decided to compare the current practice with two other types of sterilization, gamma radiation and high pressure. They also wanted to explore two other technologies which would avoid injection altogether: iontophoresis and microparticles. The former injects the drug by applying a large electrical potential across the skin; the latter shoots a bolus of microscopic particles at the skin at such a high velocity that they penetrate.

The results of the core team's discussions are shown in Table 3.4–3. The criteria selected are similar to those for other studies; note that ratings of "ease of use" and "comfort of use" turn out to be very similar, so these contribute little to the idea screening. In this case, the benchmark, which is the current "sterile ingredients," is not always given average scores. Two alternatives, high-pressure sterilization and iontophoresis, do not seem much better than what is done now. Two alternatives, gamma radiation and microparticles, score high enough to merit further effort.

Table 3.4–3 *Alternative sterilizations for the drug Zoladex* Gamma radiation and microparticle injection may work better than assembly using sterile ingredients.

Criterion	Weighting factor	Sterile ingredients	Gamma radiation	High pressure	Ionto-phoresis	Microparticles
Ease of sterilization	0.3	2	10	7	10	9
Complexity	0.2	6	6	6	5	6
Maturity	0.3	10	6	2	3	3
Ease of use	0.1	7	7	7	9	6
Comfort of use	0.1	7	7	7	6	7
Total score		**6.2**	**7.4**	**5.3**	**6.4**	**7.6**

Table 3.4–4 *Four methods of removing arsenic from water in Bangledesh.*

Separation	Source	Scale	Mechanism	Remarks
Rainwater	Government	1000 houses	Avoids arsenic separation	Requires government action
"SONO"	Academic	10 houses	Adsorption	Parallels existing practice
"Water for People"	Academic and industrial	300 houses	Ion exchange	Chemistry uncertain
"PUR"	Industrial	1 house	Precipitation	May be expensive

EXAMPLE 3.4–3 PURE WATER FOR BANGLADESH

Bangladesh, a country between India and Burma which is two-thirds the size of Minnesota, has a population of half the USA, around 140 million. About half of these people may drink water contaminated with arsenic. Ironically, this arsenic comes from wells promoted by the United Nations as a substitute for surface water. Thus, people have avoided cholera and dysentery in surface water only to risk poisoning by arsenic.

Removing the arsenic involves two problems. First, the arsenic concentration must be measured. Available laboratory tests, which already strain the country's infrastructure, declare that the wells are safe three times more often than they actually are. Better tests, including those based on genetically modified bacteria, are under development but not yet ready, at least at a low enough cost.

The second problem is removing the arsenic itself, which was the subject of a $1M Grainger prize offered by the US National Academy of Engineering. Four possible methods chosen from several hundred entries are shown in Table 3.4–4. The easiest collects rainwater and so avoids the arsenic altogether, but this requires coordinated government action that seems unlikely. The other three, which did receive prizes from the National Academy of Engineering, are serious contenders.

Table 3.4–5 *Screening the water-purification methods for use in Bangladesh* These three methods were winners in the Grainger prize awarded by the National Academy of Engineering (USA).

Selection criteria	Weighting factor	Rainwater (benchmark)	Adsorption ("SONO")	Ion exchange	Precipitation
Scientific maturity	0.2	5	3	9	9
Social support	0.3	5	10	5	3
Government role	0.2	5	7	5	6
Cost	0.3	5	9	4	2
Score		**5**	**7.7**	**5.3**	**4.5**

The first "SONO" method uses a series of three buckets. The first uses iron filings to remove the arsenic, the second uses carbon to remove organics, and the third uses sand to filter out particulates. The process improves a "kalski" process already used by housewives in Bangladesh. The second process uses ion exchange, possibly with alumina obtained from neighboring India. The details are not available but presumably involve measuring a breakthrough curve in packed beds. The third "PUR" process uses a sachet of water-treatment chemicals to precipitate arsenic and other pollutants.

Which method should win the top prize?

SOLUTION

While details are not public, the prize committee of the National Academy of Engineering might have used four criteria like those in Table 3.4–5. While the first, scientific maturity, is familiar, the second, social support, represents how easily the proposed purification fits into the existing social structure of the small villages. The third "government role" reflects the amount of governmental action which will be required, a real concern in a society without strong local government. The fourth item, low cost, is obviously important.

The first "SONO" method was judged strongest, and awarded the million dollar prize, even though it had been developed by one junior professor working alone. A Bangladashi himself, he carefully considered how the process would fit into village life. The ion-exchange method, an academic–industrial collaboration, may have been downgraded because the alumina needed as an ion exchanger had to be imported from India, which by Bangadeshi standards was expensive. This method was awarded a second prize. The third prize method, "PUR," was an effective purification developed by Proctor and Gamble and donated without any claim of royalty to the country of Bangladesh. The concern with this process was possible cost. Again, however, the point is that these three processes, chosen from several hundred entries, will all work well. The choice of the most suitable process depends not only on technical success, but also on social factors.

EXAMPLE 3.4–4 HIGH-LEVEL RADIOACTIVE WASTE

This complex example is unusual in that the product will be a chemical plant. The manufacturer of the product will be one of a handful of military–industrial companies; the customers will be the US Department of Energy, British Nuclear Fuels Limited, and other similar agencies. The example is included because it illustrates how the ideas of product design tend to precede those of process design, as discussed in Section 1.4.

The technical problem is as follows. In the manufacture of atomic weapons, a significant number of by-products are made. Many of these by-products are dangerously radioactive isotopes; most of these are actinides, which precipitate in basic solution. These precipitated isotopes are then separated and vitrified, i.e. made into glass. The glass is sufficiently radioactive that it can boil water. It must be safely stored for thousands of years.

However, one highly radioactive isotope of cesium, ^{137}Cs, is not precipitated in base but remains dissolved in aqueous solutions. Millions of gallons of this aqueous solution are stored in aging tanks in the locations where the atomic weapons were manufactured. If the tanks leak because of aging or earthquakes, the escaping cesium would spell disaster.

Not surprisingly, there has been a recent major effort to develop a means to make the ^{137}Cs less dangerous. Some of the 180 serious ideas suggested for this are shown in Table 3.1–4. These ideas are sorted as shown in Table 3.4–6. The organization has an unsurprising form: the first heading (I) deals with improvements to the existing process; the next four (II to V) are essentially separation processes organized as unit operations; and the last heading (VI) centers on innovations. The innovations include technically feasible ideas which are certainly politically unacceptable. In particular, idea #22 suggests setting off nuclear weapons inside the existing storage tanks, deep underground, to vitrify the entire contents *in situ*. It should work, and it would be inexpensive because the nuclear weapons are in inventory; but it is not politically feasible or morally justified.

Develop a concept-screening matrix which can choose among the ideas in Table 3.4–6 to find a small number for further development.

SOLUTION

Because this problem is complex and remains incompletely resolved, we will offer only a tentative solution here. We begin by describing our benchmark, the existing process, and by choosing screening criteria. The existing process hinges on the precipitation of the cesium cation with the tetraphenylborate anion:

$$Cs^+ + B(C_6H_5)_4^- \rightarrow CsB(C_6H_5)_4$$

This anion also precipitates potassium ions, present in non-radioactive form at much higher concentrations than cesium. However, the sodium ion, which is the chief cation present, does not precipitate. Thus we can precipitate the radioactive cesium by adding saturated solutions of sodium tetraphenylborate. The resulting process separates and concentrates the cesium about 20,000 times.

Table 3.4–6 *An outline of ideas for high-level waste*
The numbers refer to the specific ideas listed in
Table 3.1–4.

I.   *Improve current process*
   A. Increase storage capacity (14, 17)
   B. Stabilize precipitate (4, 6, 36, 39)
   C. Legalize – more benzene release (41)
   D. Vitrify without separation (38, 49)

II. *Separate by precipitation (45)*
   A. Process alternatives
      (1) Make fast (10)
      (2) Flocculate (32)
   B. Stabilize current precipitate (15)
      (1) With catalyst poison (6)
      (2) By chilling tanks (18)
   C. New selective precipitate (12, 42, 46)

III. *Separate by adsorption (44)*
   A. Process alternatives
      (1) Simulated moving bed (21)
      (2) Fluidized bed (29, 47)
   B. Regenerable ion exchange (13, 25)
      (1) Crown ethers (35)
      (2) $MnFe(CN)_6$ (31)
      (3) Electrically switched ion exchange (27, 34)
   C. Non-regenerable ion exchange
      (1) On glass
      (2) On crystalline silicon titanate (2, 24)

IV. *Separate by extraction*
   A. Process – hollow-fiber membranes (16)
   B. New chemistry
      (1) Caustic with zeolite or crown ether (3, 5)
      (2) Acidic (26)

V.   *Less conventional separations*
   A. Fractional crystallization (8, 33, 43)
   B. Decantation (40)
   C. Electrochemical methods (7, 11, 37, 48)
   D. Microbes (23)

VI. *Stabilize without separation*
   A. Into bedrock (1, 22)
   B. As concrete grout (9, 20, 30)

In the original precipitation process, the cesium precipitate is made in large quantities and then stored. It is then slowly added to the other radioactive precipitates, and eventually vitrified into glass. Unfortunately, the cesium precipitate is not stable: it decomposes to produce soluble cesium and benzene. Because benzene is volatile, the tanks where the precipitate is stored can have inflammable benzene vapor collecting in the headspace above the precipitate.

Table 3.4–7 *Sample scores for four of the processes for treating radioactive waste containing ^{137}Cs* The screening procedure described and the scores are a simplification of the actual method used.

Selection criterion	Weighting factor	Precipitation and storage (benchmark)	Precipitation and treatment (IIA.1)	Crown ether ion exchange (IIIB.1)	Crown ether extraction (IVB.1)	Concrete formation (VIB)
Mature science	0.4	5	9	3	2	6
Reliable engineering	0.4	5	7	6	9	8
Safety	0.1	5	5	4	8	10
Public response	0.1	5	5	5	5	1
Total score		**5**	**7.4**	**4.5**	**5.7**	**6.7**

As a result, this headspace is continually flushed with nitrogen. While making and storing the unstable precipitate may look foolish now, the precipitate was not expected to be unstable. It is unstable because the waste contains parts per billion of palladium, which catalyzes this decomposition. The active role of this palladium was unexpected.

The screening criteria for this waste treatment center on finding a process which works now. This means that the two most important criteria are known science and reliable engineering. Each of these was given a weighting factor of 0.4. Two other important criteria are safety and the public response, each of which was given a weighting factor of 0.1. Note that cost does not appear in these criteria, an implicit recognition that all of these choices will be expensive. Note also the relatively low weighting factor for public response. This does not imply that public response is unimportant, or that the public may not veto any one of the choices. At this point, however, those involved – the core team – felt that choosing a process which was scientifically and technically reliable was paramount. Similarly, the low weighting factor for safety does not indicate it is unimportant – clearly it is paramount. However the core team believed that if a process could work well, it could be made safe.

Each of the processes in Table 3.4–6 is then scored using these criteria. Typical scores for four of these processes are shown in Table 3.4–7. These four are all reasonable alternatives: each could work well.

The processes merit more detailed description. "Precipitation and treatment" is closest to the existing process. It begins by precipitating cesium tetraphenylborate, but under more controlled conditions than those used in the past. Once the precipitate is formed, it is separated and forwarded to vitrification before significant decomposition can occur. For this option, the science and engineering are in good shape, and the safety and public response should be similar to that for the existing process.

The second and third options both replace the tetraphenylborate precipitation with the chemistry of macrocyclic ethers, exemplified by the compound dibenzo-18-crown-6:

Compounds like this species can selectively complex specific cations like cesium. Some specific materials suitable for both ion exchange and liquid–liquid extraction are known. However, only small amounts (perhaps 100 grams) have been made. Commercial suppliers are either non-existent or underfunded academic spin-offs. This means that both these options have low scores on scientific maturity. Nonetheless, an alternative like this was developed by a laboratory at one of the sites where atomic weapons had been produced.

The final "Concrete formation" option is the most risky. In this option, we would not try to separate the cesium at all, but simply turn the entire contents of the storage tank into a concrete-like solid, called "grout." While the science for such a high-salt grout needs a little work, the engineering is straightforward. The process is safe: because we do not concentrate the cesium, the waste remains in the less dangerous "low level" form.

The risk is the public's response. Instead of concentrating the waste and shipping it to some distant desert storage, the waste is simply immobilized and left where it is. Local citizens and their elected officials will not like that, even though they have been happy to have the jobs which the making of atomic weapons has supplied. While making grout is legal within the letter of existing laws, the public's response may engender so much litigation that the solution of the problem of radioactive waste is delayed and the danger of accidental release is increased. Still, this alternative is a strong contender. In more general terms, you can now see why we wanted a complex example: it really does show how different alternatives must be carefully weighed.

3.5   Conclusions and the Second Gate

This chapter describes the generation and screening of ideas, which is the second step in our four-step template of product design. Before product ideas are developed, we identify product needs, and quantify these needs with product specifications. Identifying needs and developing specifications was the subject of the previous chapter. After these ideas are screened, we must select among the best choices, and manufacture the products. Selection and manufacturing are described in later chapters.

We normally welcome product ideas from every possible source. These sources center on the core team but also include customers, competitors, and consultants. We will depend most heavily on "brainstorming" to generate without criticism a broad selection of concepts. A useful target is often around one hundred ideas.

Because we will rarely have the resources to evaluate completely all of these, usually fragmented, ideas, we need to choose the best for further study. Such choices can be made in two stages. First, we organize the ideas, removing redundancy and pruning folly. This will normally give us around twenty remaining candidates. We will then use concept-screening methods to judge the advantages and disadvantages of these ideas. This will normally reduce the number to five or fewer. Cutting this number further hinges heavily on engineering and chemistry, as described in the next chapter.

This is also the point of the second management review, the second "gate" through which a successful product development must pass. Again, the core team will make a presentation to the same senior management group as for the first gate. Again, this presentation will include both oral and written components. As at the first "gate," management will decide whether or not to continue work.

Interestingly, business studies seem to show that management again tends to be too supportive at this stage. The management team may be charmed by suggested innovations and excited by product improvements. When the management team have non-technical backgrounds, like law or marketing, they may require help in making rational decisions if a strong chemical component is involved. As a result, the core team must be especially careful to be objective. They must make sure that management understands not only the potential rewards, but also the risks. After this second "gate," product development gets more expensive as we select the product which we hope to manufacture.

Problems

1. *Reliable birth control.* Your company manufactures birth control products. Market research has shown that rather than take a pill every day, many women would prefer a product which automatically releases hormone into their bodies in the correct dose and lasts for a period of six months. List ideas as to how this might be achieved.

2. *Bug repellent for windshields.* Bugs being squashed onto the windshield are a major cause of fouling, particularly on warm summer days, and they can be difficult to clean off. Generate as many ideas as you can for avoiding this problem.

3. *Safe smartcards.* Smartcard processors are used in applications such as direct debit cards, phone cards, and chips for decoding satellite television channels. Breaking into these chips, extracting both their hardware and software, and reproducing this on huge numbers of fakes is big business. In some countries, such as Germany and Canada, this procedure is not even illegal. Multi-million-dollar operations, complete with the most modern technology and some extremely bright scientists, exist solely for the purpose of cracking the

codes of smartcard processers. The challenge is to design a way of making it extremely difficult to break into the relevant chip without destroying the information held within it. Brainstorm to produce a list of ideas for ways to tackle this problem. Prune and sort this list.

4. *Child's heart valve.* Heart valve replacement is a relatively routine operation. However the procedure raises a particular problem when conducted on a child. Because the child is growing, the prosthetic heart valve must be replaced every year or two, so that it fits into the available cavity. Write specifications for a prosthetic heart valve suitable for implantation into children and suggest ideas for meeting these specifications.

5. *Clean feet.* Foot odor is a major cause of romantic failures. List and sort possible solutions.

6. *Salt and pepper separation.* How would you separate a mixture of salt and pepper in the kitchen? Try it.

7. *Marmite.* Marmite is a food spread made from spent brewer's yeast; a large fraction of Britain's waste brewer's yeast is recycled into Marmite. The idea for making such a yeast extract goes back to the great German chemist Liebig, and it has been manufactured in an essentially unchanged form in Burton-on-Trent since 1902. It is more or less indestructible – it is said that decades old Marmite recovered for an abandoned hut for polar explorers in Antarctica was perfectly edible. Marmite has a loyal following in Britain and beyond, Sri Lanka being one other large market. Your company wishes to expand the sales potential of the product without eroding its current customer base. Suggest options for product development to increase sales of Marmite.

8. *Solar-powered dehumidifier.* The average air humidity in the Negev desert in Israel is 64%; each cubic metre of air contains 11.5 ml of water. Could this water be extracted using only solar power? Single-person mobile units or static devices to supply whole hotels can be envisaged. Set needs for such a device and list ideas to achieve these needs.

9. *Ski lubrication.* Waxes are spread on downhill skis to lubricate them and so increase speed. The waxes are applied hot to a porous base to increase lifetime on the ski. A team at Sheffield University thought it would be advantageous to convert this batch process into a continuous one by replenishing the wax as the ski travels over the snow. This could improve speed and eliminate the tiresome ritual of hot waxing prior to racing. The Sheffield group has formulated appropriate waxes for continuous application and demonstrated that significant speed enhancements are possible. However the Federation International du Ski, which regulates ski competitions, forbids external energy sources, although it does not rule out continuous lubrication. Generate ideas for a device to apply wax continuously to the base of skis under race conditions, without using any additional energy source.

[*The Chemical Engineer*, December 2006/January 2007, 24–26.]

10. *A cholesterol-reducing drink product.* St Bartholomew Technologies Ltd has developed a range of cholesterol-reducing cheese products that are marketed in major retailers throughout the UK and are soon to be marketed in the USA. The technology involves replacing the fat phase in the cheese with plant oils that are either naturally rich in cholesterol-reducing phytosterols (EU) or are fortified with sterol esters (USA). The company's proprietry technology lies in this fat replacement process. Consumer interest in such "nutraceuticals" – foods that enhance health – is immense and St. Bartholomew Technologies has been approached by a global drinks manufacturer that is seeking concept designs for a cholesterol-reducing drink and other neutraceutical drink products. Generate ideas for such a product. Sort and screen these ideas to produce a handful of promising ideas for further research.

11. *Disposing of the sulfur mountain.* When hydrocarbon fuels containing sulfur are burned, the combustion products contain sulfur dioxide and sulfur trioxide, both of which contribute to acid rain. The sulfur trioxide also contributes to the particulates contained within the emissions. With these environmental consequences, there is considerable pressure to reduce the sulfur content of liquid and gaseous hydrocarbon fuels. Crude oil can contain significant quantities of sulfur and this sulfur is extracted in oil refineries predominantly in hydrodesulfurizing (HDS) processes and various types of cracking units. Using such technologies, refineries can now produce fuels containing less than 10 ppm sulfur. The hydrogen sulfide produced by the HDS and cracking processes is sent to the sulfur recovery units (SRUs), where technologies based on the Claus process produce elemental sulfur. The sulfur produced can be more than 99.8% pure. Over the coming decades the amount of sulfur produced by refineries will increase, since in addition to the tightening legislation for lower sulfur fuels, the crude oil supply to refineries will contain ever more sulfur as sources of low sulfur crude decline.

    In addition to the sulfur coming from oil refineries, some gas fields naturally produce gas containing a large concentration of hydrogen sulfide and this sour gas must be treated to recover the sulfur in SRUs.

    The sulfur produced by the oil and gas industries is often sold to industries manufacturing fertilizer, explosives, and sulfuric acid. However, the demand for sulfur is not as big as the supply and this oversupply is only set to increase (see Table 3.11A).

    Generate ideas for dealing with this global excess of sulfur over the coming years.

12. *Insulin aerosols.* Diabetes is a disease in which the body is unable to metabolize sugars because the pancreas does not secrete sufficient insulin, a small protein (molecular weight 5700). Patients often treat this disease by

Table 3.11A *World sulfur supply and demand.*

(kton)	1990	1995	2000	2004	2008
Sulfur supply					
Recovered from oil	9944	12 822	16 328	18 700	21 200
Recovered from gas	15 139	17 750	22 508	24 800	28 100
Other elemental	742	681	1207	2200	3300
Sulfur – other forms	9734	10 359	14 261	15 800	17 800
Pyrites	10 610	8258	5393	5100	4400
Frasch sulfur	12 355	5827	2337	800	500
Global supply	**58 524**	**55 697**	**62 034**	**67 400**	**75 300**
Sulfur demand					
Fertilizer	31 241	29 539	31 508	35 800	38 800
Non-fertilizer	27 108	24 826	27 374	28 000	30 200
Global demand	**58 349**	**54 365**	**58 882**	**63 800**	**69 000**
Global oversupply	**175**	**1332**	**3152**	**3600**	**6300**

measuring their blood glucose; when it is low, they inject themselves with shots of insulin. The injections are necessary because insulin can't survive in the stomach's acid nor cross the wall of the intestine.

Your company is interested in a product which would reduce the need for these insulin shots. As part of "idea generation" for such a product, your core team comes up with the concepts listed in Table 3.12A. (These are actually a fraction of the total.)

(1) Organize these as part of your "idea sorting." In this organization, refer to specific ideas.
(2) Choose three ideas which you feel are especially good and construct a decision matrix to illustrate your rationale for preferring one of these. Briefly justify your scoring.

13. *Manufacturing polymer composites.* Polymer composites are combinations of polymers and either fibers or particles. They have excellent strength-to-weight ratio, are not electrically conducting, and resist corrosion. They have a wide range of commercial applications. For example, they are frequently used for pipes in water distribution, for packaging film, and for household appliances. The manufacturing systems, machinery, and technology of polymer composite production have a large market potential. However, the current methods of polymer composite production have problems.

Injection molding and compression molding are the two common batch production methods. As they are batch process, their efficiency is low. Additionally, they require expensive equipment and have high labor costs. The only common continuous process of polymer composite production is pultrusion. Although pultrusion has higher production efficiency, it has limitations. For example, the shape of the plastic products must be simple.

Table 3.12A *Ideas to reduce insulin shots.*

1. Inject the insulin without needles.
2. Attach it to another molecule, so you can take it as a pill.
3. Fix this by genetic engineering.
4. Make a slow release form, so you need inject less often.
5. Build a surgically implanted pump, which releases insulin when needed.
6. Implant modified insulin, which dissolves only when glucose is present.
7. Use another drug.
8. Make the implantable pump very small.
9. Push insulin across the skin with an electric field (this is sometimes called "electroporation").
10. Modify insulin so it can diffuse across the skin.
11. Have an insulin patch, like a nicotine patch.
12. Give insulin as an aerosol, and sniff it.
13. Build a surgically implanted machine which gives off insulin pulses.
14. Put insulin in "vesicles," little microcapsules with an insulin center.
15. Build an artificial pancreas.
16. Use a mixture of different drugs.
17. Inject microspheres with attached insulin.
18. Use many very small needles.
19. Scratch the skin and rub the scratch with insulin.
20. Make an insulin reservoir and a valve which the patient can open when he needs to.

Table 3.13A *Raw ideas for manufacturing composites.*

1. Do not use polymer products.
2. Convert batch production process to continuous process.
3. Use recycled polymer.
4. Make polymer products by hand.
5. Mold polymer products by machine.
6. Compression molding.
7. Use metal instead.
8. Invent other new materials.
9. Use high temperature to reshape the plastics.
10. Modified pultrusion –stamping.
11. Modified pultrusion –knife cutting.
12. Modified pultrusion –drill or laser shaping.
13. Roll coating.
14. Make simple pieces and join them together by other means.
15. Join plastic pieces by adhesive.
16. Improve the pultrusion process.
17. Sculpture the plastic to the desired shape.
18. Join plastic pieces by welding.
19. Add more production lines.
20. Increase the number of laborers.
21. Use reaction injection molding.
22. Use modified pultrusion – floating-head method.

Table 3.14A *Raw ideas for an oxygen-impermeable film for wrapping food.*

1. Freeze the food.
2. Store food in space.
3. Canning.
4. Store food in vacuum.
5. Store food in glass bottles.
6. Store in sealed paper box.
7. Fermentation.
8. Add alcohol.
9. Store food at the north pole.
10. Use food wrap.
11. Food wrap of several layers which serve different functions.
12. Anti-bacteria treatment to food wrap.
13. Bubble carbon dioxide gas (or other inert gases) to eliminate oxygen.
14. Insert oxygen-absorbing substances.
15. Heat the food to kill the micro-organisms.
16. Store food at absolute zero.
17. Add oxygen/water absorption substances into fridge.
18. Add ultralight lamp into fridge.
19. Grind foods into dry powders.
20. Reboil the dried food.
21. Develop food substitutes.
22. Eat less food.
23. Reheat the remaining food for storage.
24. Develop non-toxic food preservatives.
25. Remove vinegar from food after pickling.
26. Eat the food immediately.
27. Store in sealed plastic bag.
28. Add vacuum pump to food storage containers.
29. Add water-proof coating to food wrap.
30. Store food in concentrated sucrose solution.
31. Store fruits in salt water and wash them with lots of water afterwards.
32. Do not open package before consumption.
33. Place the food under strong sunlight.
34. Cook the food thoroughly.
35. Change the genes of fruits and vegetables such that they are more resistant to bacteria attack.
36. Develop oxygen-, water-resistant skin of fruits and vegetables genetically.
37. Coat water-proof layer on top of the food.
38. Add water scavenger to food wrap.
39. Store at high temperature to dry up the food and kill microorganisms.

This project aims at developing new production processes for polymer composites which allow easy processing. To do so, the core team has developed the ideas shown in Table 3.13A. Sort and screen these ideas.

14. *Impermeable food wrap.* Food products spoil easily. For example, as much as one third of the fish that is caught spoils before reaching its market. This spoilage is most often due to the fish contacting oxygen. If the fish were

Table 3.15A *Raw ideas for an ultraviolet impermeable film.*

1. UV-filtering film with sticky backing.
2. UV-filtering film in a roll-on dispenser.
3. UV-filtering film with static-cling property.
4. UV-filtering film with a separate adhesive.
5. Manufacture windows out of UV-filtering material.
6. Manufacture windows with UV-filtering surfaces.
7. Bulky window filters.
8. Use no lighting and shoot down the Sun.
9. Don't use windows.
10. Use screens only.
11. UV-filtering material in a spray can.
12. UV-filtering clear lacquer formula.
13. UV-filtering dip treatment.
14. UV-filtering sprayable polish.
15. Use curtains.
16. Use a window canopy.
17. Tinted film to reduce UV.
18. Tint window surface.
19. Tinted cover to pull down over windows.
20. Don't drive during the day.
21. Make polarizer film with on/off switch.
22. Chemical additive for glass to add tint.
23. Thicker or thinner windows.
24. Smaller windows.
25. Put a filtering grid over windows.
26. Beer box panels on windows.
27. Hang blanket over window.
28. Meet with God . . . discuss replacing Sun with no-UV light source.
29. Do everything at night and sleep during the day.
30. Switch to a black-lighting system.
31. Paint windows black.
32. Liquid crystal on-off window covering.
33. Cover windows with trash bags.
34. Build homes in caves.
35. Build homes underground.
36. Build windowless houses out of adobe.
37. Use grated window surfaces.
38. Hang mini-blinds over windows.
39. Use vertical blinds over windows.
40. Remove eyes.
41. Use blindfolds.
42. Permanently cloud the atmosphere.
43. Replace windows with scenic backgrounds.
44. Board up windows.
45. Make blinds out of UV-filtering material.
46. Polarizable film with electric dimmer capability.
47. Plant trees in front of windows to block out all light.
48. Polymer contacts that filter out UV.
49. Stained-glass windows.
50. Eyeglasses that filter out UV.

wrapped in an oxygen-impermeable film, its spoilage could be dramatically reduced.

Your core team has begun working on this problem, and has come up with the ideas shown in Table 3.14A. Sort and screen these ideas, and summarize where you think the team should concentrate its future efforts.

15. *Barriers for ultraviolet (UV) light.* Your company already manufactures a wide variety of thin films using proprietary coextrusion technology. Indeed, you believe that you should be able to dominate the world's demand for such films. However, frustratingly, you have not been able to penetrate the market for coatings on glass. In what is clearly a technology push project, your core team has been urged to explore this market area.

In its initial meetings, your core team has discussed this goal, and suggested the product ideas listed in Table 3.15A. Sort and screen this list to decide which areas should receive more thought.

REFERENCES AND FURTHER READING

Appleby, **A. J.** and **F. R. Foulkes** (1989) *Fuel Cell Handbook.* Van Nostrand Reinhold, New York, NY.

Cox, **P. A.** and **Balick**, **M. J.** (1994) The ethnobotanical approach to drug discovery. *Scientific American* **270**, 82–87.

Engstrom, **J. R.** and **Weinberg**, **W. H.** (2000) Combinatorial materials science. *AIChE Journal* **46**, 2–5.

Farnsworth, **N. R.** (1990) The role of ethnopharmacology in drug development. *Ciba Foundation Symposium* **154**, 2–21.

Houghten, **R. A.**, **Pinilla**, **C.**, **Appel**, **J. R.**, *et al.* (1999) Mixture-based synthetic combinatorial libraries. *Journal of Medicinal Chemistry* **42**, 3743–3778.

Kirton, **M. J.** (2003) *Adaption – Innovation.* Psychology Press, Philadelphia, PA.

Li, **J. W. H.** and **Vederas**, **J. C.** (2009) Drug Discovery and Natural Products: End of an Era or an Endless Frontier. *Science* **25**, 161–165.

Linden, **D.** (ed.) (2001) *Handbook of Batteries*, 3rd Edition. McGraw-Hill, New York, NY.

Lowe, **G.** (1995) Combinatorial chemistry. *Chemical Society Reviews* **24**, 309–317.

Murray, **R. L.** (2008) *Nuclear Energy: An Introduction to the Concepts, Systems and Applications of Nuclear Processes*, 6th Edition. Butterworth-Heineman, Oxford.

Reddington, **E.**, **Sapienza**, **A.**, **Gurau**, **B.**, *et al.* (1998) Combinatorial electrochemistry: a highly parallel, optical screening method for discovery of better electrocatalysts. *Science* **280**, 1735–1737.

4

Selection

As explained earlier, we expect that product design will take place in four, roughly sequential steps. First, we will identify needs; second, we generate ideas to fill these needs; third, we select the best ideas; and last, we consider manufacturing. So far, we have discussed identifying needs, generating ideas to fill these needs, and choosing a shorter list for further study.

We now want to select the best ideas for further development. In some cases, we will have only one or two clear choices. In most cases, we will want to select from five or fewer possibilities, simply because the amount of work required for further development is so substantial. In selecting between these few products, we can identify two separate situations. In the first, we can compare products using only technical criteria drawn from chemistry and engineering. In the second, we must compare products not only on a technical basis, but also using more subjective criteria, like "comfort" and "safety." How we proceed in product selection depends on which of these two situations we encounter.

In the first situation involving only technical criteria, we already have the tools for selection from our technical training. In particular, we have a background in thermodynamics and so can calculate any chemical equilibria or heats of reaction suggested in our product development. We also have a background in kinetics, including an understanding of reaction rate constants and how these change with temperature. Those trained in chemical engineering will have a knowledge of heat and mass transfer.

Because of this background, we should be able to make the approximate calculations necessary for product selection. However, in our own experience we find that some aspects that are important in product selection are not stressed in conventional courses. We review these topics in this chapter. In Section 4.1, we discuss neglected thermodynamics topics. In Section 4.2, we review some kinetics topics.

The second situation, which includes both technical and subjective criteria, is more difficult than the first. Often these subjective criteria will include consumer reactions and public opinion. These criteria may change from one country to another, or evolve over time. In the vernacular, we sometimes describe this second situation as selecting between "apples and oranges." In the first situation, we would be making the easier comparison betweeen "apples and apples."

In the final sections of the chapter, we outline some ways to select among dissimilar products, i.e. among "apples and oranges." In Section 4.3, we again use the weighting factors introduced in Section 3.4 to make judgments combining subjective and objective considerations. In other words, we will try to balance factors like "comfortable" with factors like "heat retention." In Section 4.4, we explore the effect of risk. Our goal is to decide between an effective but expensive idea and a potentially superior idea which may not work. At the end of this effort, we should be prepared to select our new product, and be ready to move to the final manufacturing step.

4.1   Selection Using Thermodynamics

The selection of the best potential product out of a short list of carefully culled ideas is easiest when the proposed new products are modifications of existing products. These modifications most commonly involve ingredient substitutions or improvements in performance. They are most often based in thermodynamics.

4.1–1 INGREDIENT SUBSTITUTIONS

Changes in a product's chemical ingredients often seek to duplicate its current properties. The most common effort is the search for less volatile, less toxic solvents. For example, methylene chloride (CH_2Cl_2), one of the most useful solvents used for fine-chemical manufacture, is a carcinogen. That laboratory mainstay acetone (CH_3COCH_3) is more toxic than methanol (CH_3OH). We want to replace solvents believed to be dangerous with a more benign solvent which is less volatile, less toxic, and cheap. We want to equal product performance, but with additional benefits such as safety and cost.

Hence, our selection involves not only choice but also refinement and improvement of ideas. Our idea from the previous stage might be "use a non-chlorinated solvent" or, more generally, "use a greener solvent," that is, one with a smaller environmental impact. To compare this idea to other potential solutions, we now need to decide what replacement solvents we could use and to estimate how well they would work. As we think harder about each product idea, it evolves and improves.

The best route to discovering new solvents is by carrying out experiments. However, while we will almost always need to do some of these, we can also benefit from some kind of guide which will let us choose good candidates. This guide should be simple and quick, rather than accurate but tedious.

The best such guide uses solubility parameters. This guide, originally suggested by Hildebrand *et al.* (1970) and extended to polymers by Hansen (1999), assumes that all solutions are non-ideal and described by a relation of the following form:

$$\mu_2 = \mu_2^0 + RT \ln x_2 + \omega x_1^2 \qquad (4.1\text{–}1)$$

where μ_2 is the chemical potential of the product solute; μ_2^0 is its value in a reference state (pure 2 at the specified T and P, the "standard state"); ω is an activity parameter with the dimensions of energy per mole; and x_1 and x_2 are the mole fractions of the solvent and the product, respectively. The logarithmic term in this equation represents the free energy change of ideal mixing, and is related to entropy changes. The term with ω includes any heat of mixing. This simplest non-ideal relation is known as the "Margules equation."

The parameter ω is the key to selecting alternative solvents. It can be estimated by:

$$\omega = \overline{V}_2 \left(\delta_1 - \delta_2\right)^2 \tag{4.1-2}$$

where $\overline{V}_2$ is the molar volume of the product solute, and the δ_i values are the so-called solubility parameters. Because ω has dimensions of energy per mole and $\overline{V}_2$ is a molar volume, the δ_i values are given in the improbable dimensions of $[\text{energy/volume}]^{1/2}$. For an ideal solution, ω is zero, and the solute and solvent are miscible in all proportions. For a non-ideal solution, the solubility parameters will differ. The more different they are, the less miscible the solute and the solvent will be. In this simple model, ω cannot be negative and so the non-ideality is not allowed to favor mixing.

In almost all cases, we will not know the solubility parameter δ_2 for the product solute. We do have tables of the solubility parameters of common solvents, a selection of which is given in Table 4.1–1. To find an alternative solvent giving the same properties as the solvent we are using, we simply seek a solvent with δ_1 equal to that of the current solvent. For example, if we are currently using chloroform with δ_1 of 9.2 $(\text{cal/cm}^3)^{1/2}$, we could try substituting benzene, which has the same δ_1 value. We will use this strategy in an example later in this section.

4.1–2 SUBSTITUTIONS IN CONSUMER PRODUCTS

At this point, we interrupt outlining the methods of ingredient substitution to touch on some special characteristics of consumer products. There are two important ones. The first is that isomerically pure chemicals are the same no matter whether their origin is the farmer's field or the laboratory. We recognize that this may be anathema to those interested in "natural foods" or "natural ingredients," but it is true. If a particular chemical isomer is pure, then a sample obtained from vegetable sources is indistinguishable from a similar sample made in a laboratory.

Because of this equivalence, we should be able to substitute a "natural" chemical for a "synthetic" chemical without any difficulty whatsoever. The reverse is equally true. If we do have trouble, it is almost certainly because one of the chemicals is impure. In the vast majority of cases, the impure chemical is the "natural" one. Now this impure chemical can be superior; for example, it might contain an emulsifier or a surfactant which facilitates its mixing. Still, if there is a problem in

Table 4.1–1 *Solubility parameters* List reproduced from Hildebrand *et al.* (1970).

Formula	Substance	V (cm³/mol)	δ (cal$^{1/2}$/cm$^{3/2}$)
	HALOGENATED SOLVENTS		
C_6F_{14}	Perfluoro-*n*-hexane	205	5.9
C_7F_{16}	Perfluoro-*n*-heptane	226	6.0
C_6F_{12}	Perfluorocyclohexane	170	6.1
$(C_4F_9)_3N$	Perfluoro tributylamine	360	5.9
$C_2Cl_3F_3$	1,1,2-Trichloro,1,2,2- trifluoroethane	120	7.1
CH_2Cl_2	Methylene chloride	64	9.8
$CHCl_3$	Chloroform	81	9.2
CCl_4	Carbon tetrachloride	97	8.6
$CHBr_3$	Bromoform	88	10.5
CH_3I	Methyl iodide	63	9.9
CH_2I_2	Methylene iodide	81	11.8
C_2H_5Cl	Ethyl chloride	74	8.3
C_2H_5Br	Ethyl bromide	75	8.9
C_2H_5I	Ethyl iodide	81	9.4
$C_2H_4Cl_2$	1,2-Dichloroethane (ethylene chloride)	79	9.9
$C_2H_4Cl_2$	1,1-Dichloroethane (ethylidene chloride)	85	9.1
$C_2H_4Br_2$	1,2-Dibromoethane	90	10.2
$C_2H_3Cl_3$	1,1,1-Trichloroethane	100	8.5
	ALIPHATIC HYDROCARBONS		
C_5H_{12}	*n*-Pentane	116	7.1
	2-Methylbutane (isopentane)	117	6.8
	2,2-Dimethyl propane (neopentane)	122	6.2
C_6H_{14}	*n*-Hexane	132	7.3
C_7H_{16}	*n*-Heptane	148	7.4
C_8H_{18}	*n*-Octane	164	7.5
	2,2,4-Trimethylpentane	166	6.9
$C_{16}H_{34}$	*n*-Hexadecane	294	8.0
C_6H_{12}	Cylohexane	109	8.2
C_7H_{14}	Methylcyclohexane	128	7.8
C_6H_{12}	l-Hexene	126	7.3
C_8H_{16}	l-Octene	158	7.6
C_6H_{10}	1,5-Hexadiene	118	7.7
	AROMATIC HYDROCARBONS		
C_6H_6	Benzene	89	9.2
C_7H_8	Toluene	107	8.9
C_8H_{10}	Ethylbenzene	123	9.9
	o-Xylene	121	9.0
	m-Xylene	123	8.8
	p-Xylene	124	8.8
C_8H_8	Styrene	116	9.3
$C_{10}H_8$	Naphthalene	123	9.9
	INORGANICS		
Br_2	Bromine	51	11.5
I_2	Iodine	59	14.1
S_8	Sulfur	135	12.4
P_4	Phosphorus	70	13.1
CCl_4	Carbon tetrachloride	97	8.6
$SiCl_4$	Silicon tetrachloride	115	7.6
$SnCl_4$	Stannic chloride	118	8.7
WF_6	Tungsten hexafluoride	88	8.0
$Si(CH_3)_4$	Silicon tetramethyl	136	6.2

making a "natural" vs. "synthetic" substitution, first look for the problem in an impure natural material.

We are not arguing that "natural" ingredients are less desirable than "synthetic" ingredients. In chemical terms, we repeat that pure materials are equivalent. However, there may be major marketing advantages in using natural ingredients, for they are often perceived as superior, and command premium prices. Curiously, any chemicals made microbiologically can legally be identified as "natural." Thus, impure pepper flavoring made from genetically modified organisms which would be completely incapable of survival outside the laboratory may be labelled as "natural," while pure flavoring made in aseptic reactors cannot be. Life is strange.

In addition to this "natural" vs. "synthetic" issue, a second characteristic of consumer products which affects ingredient substitutions concerns the evaluation of consumer attributes. If we are going to make an ingredient substitution, we need a way of assessing success or failure. As an example, we consider three attributes of skin creams: "thickness," "smoothness," and "creaminess." As will be detailed in Chapter 9, "thickness" is proportional to the 0.33 power of instrumentally measured Newtonian viscosity. From other studies, we know that "smoothness" is related to the coefficient of friction. More specifically, "thickness" is proportional to the force of viscous drag; but "smoothness" is inversely proportional to the frictional force during contact lubrication. Interestingly, "creaminess" seems close to the geometric average of "thickness" and "smoothness." It is not an independent attribute.

In a case like this, we can easily substitute ingredients because we know which consumer attributes are related to which physical properties. An example illustrating such substitutions is given later in this section. Unfortunately, we usually do not know these relationships, so that ingredient substitution becomes empirical. This relationship between consumer attributes and physical properties, which currently falls between social science and engineering, would greatly benefit from more study.

4.1–3 INGREDIENT IMPROVEMENTS

We will also frequently want to improve products by using ingredients which are superior. It is often useful for ingredients to have properties which are strong functions of temperature or pH, since this allows the product's activity to be triggered. Products whose properties change dramatically with temperature are the more common case. An excellent example is a class of water-soluble absorbents, like monoethanol amines, which are used to scrub acid gases. These amines react rapidly with gases such as carbon dioxide to form water-soluble products, in this case carbonates. The carbonates are easily destroyed by gentle heating, thus reversing the reactions. However, as energy costs rise, there is a major interest in amines whose reactions are not only fast, but whose equilibria with acid gases change more radically with temperature.

To seek such amines, we again return to simple thermodynamic ideas. To focus these ideas, we consider the reaction

$$H_2S\,(g) + RNH_2\,(aq) \rightleftharpoons RNH_3^+\,(aq) + HS^-\,(aq) \qquad (4.1\text{--}3)$$

Thus,

$$\left[RNH_3^+\right]\left[HS^-\right] = K\left[RNH_2\right]\left[H_2S\right] \qquad (4.1\text{--}4)$$

where the quantities in square brackets are concentrations and K is the equilibrium constant for this reaction.

We want K to vary strongly with temperature. To seek this variation, we remember that

$$K = e^{-\frac{\Delta G^0}{RT}} = e^{-\frac{\Delta H^0}{RT} + \frac{\Delta S^0}{R}} \qquad (4.1\text{--}5)$$

where ΔG^0, ΔH^0, and ΔS^0 are the standard-state free energy, enthalpy, and entropy changes of this reaction at the temperature of interest. Thus, we seek amines which show large enthalpies of reaction when combined with H_2S.

As a case illustrating variation of a property with pH, we consider drug purification. Many drugs have either carboxylic acid (–COOH) or amine (–NH$_2$) groups which can be used to facilitate their purification by liquid–liquid extraction. As an example, we imagine an antibiotic like penicillin, which is a carboxylic acid. We expect that at equilibrium, the concentrations in an organic solvent and in water are at equilibrium:

$$\begin{bmatrix} \text{drug in} \\ \text{organic} \end{bmatrix} = K \begin{bmatrix} \text{drug in} \\ \text{water} \end{bmatrix} \qquad (4.1\text{--}6)$$
$$[RCOOH]_{org} = K[\overline{RCOOH}]_{H_2O}$$

where K is again an equilibrium constant. The drug in organic solution will normally be protonated, and hence in only one form. However, the drug in aqueous solution may either be protonated or ionized.

The ionization of the drug in aqueous solution means that the equilibrium constant K can be a strong function of pH. To see why, we first note that the concentration $\overline{RCOOH}$ actually includes different forms:

$$[\overline{RCOOH}]_{H_2O} = [RCOOH]_{H_2O} + [RCOO^-]_{H_2O} \qquad (4.1\text{--}7)$$

These forms are in equilibrium

$$[H^+]_{H_2O}\,[RCOO^-]_{H_2O} = K_a\,[RCOOH]_{H_2O} \qquad (4.1\text{--}8)$$

where K_a is the dissociation constant for this acidic drug. The true thermodynamic equilibrium across the organic–water interface is not that in Equation 4.1–6, but

$$[RCOOH]_{org} = K'\,[RCOOH]_{H_2O} \qquad (4.1\text{--}9)$$

Combining Equations 4.1–7 to 4.1–9 we find that

$$[\text{RCOOH}]_{\text{org}} = \left(\frac{K'[\text{H}^+]_{\text{H}_2\text{O}}}{K_\text{a} + [\text{H}^+]_{\text{H}_2\text{O}}} \right) \left[\overline{\text{RCOOH}} \right]_{\text{H}_2\text{O}} \tag{4.1–10}$$

Thus,

$$K = \frac{K'[\text{H}^+]_{\text{H}_2\text{O}}}{K_\text{a} + [\text{H}^+]_{\text{H}_2\text{O}}} \tag{4.1–11}$$

The apparent equilibrium constant K in Equation 4.1–6 varies dramatically with acid concentration. This variation, explored further in an example below, is useful in drug purification by extraction.

EXAMPLE 4.1–1 A BETTER SKIN LOTION

Our employer currently manufactures a variety of skin-care products. Market research shows a need for a "thinner," "creamier" skin lotion. While it is not entirely clear what these descriptors mean, they are used by consumers and market researchers to describe the desired attributes of lotions. Our idea is to develop a lotion which is twice as "thick" and twice as "creamy." How should we seek new ingredients?

SOLUTION

Experiments suggest that thickness varies with the cube root of measured Newtonian viscosity. Thus, to make the cream twice as thick, we should increase the viscosity eight times. Similar experiments suggest that thickness and creaminess are related:

$$\text{creaminess} = [(\text{thickness}) \times (\text{smoothness})]^{1/2}$$

where assessments of smoothness are inversely related to the coefficient of sliding friction. However, we increased the thickness by a factor of two, so we should look for ways to decrease the coefficient of sliding friction by a factor of two. This may be hard to do.

EXAMPLE 4.1–2 A POLLUTION-PREVENTING INK

A lithographic ink can be idealized as containing four components: a pigment, an oil, a resin, and a solvent. The pigment, frequently colloidal carbon, is important to the ink, but not a key in pollution. The oil, a mixture of natural products like castor and linseed oils, typically contains fatty acids with multiple double bonds, like linoleic acid and linolenic acid. These double bonds crosslink in the presence of oxygen, making the ink permanent. The resin is a purpose-made low-molecular-weight, highly polydisperse condensation polymer, which controls the ink's flow properties. The solvent, frequently methylene chloride (CH_2Cl_2), is used to adjust the ink's rheology to give good printing.

One of our ideas to reduce the pollution caused by printing is to replace methylene chloride with a less dangerous solvent. What new solvent should we choose?

SOLUTION

We must seek solvents whose solubility parameters are close to that of methylene chloride. From Table 4.1–1, we find that the solubility parameter of methylene chloride is 9.8 $(cal/cm^3)^{1/2}$. We then look for inexpensive solvents with similar solubility parameters. Benzene, toluene, and naphthalene are three possible choices. All are carcinogens, though toluene and naphthalene are less dangerous than methylene chloride. Benzene is excessively volatile, and naphthalene is solid at the temperatures normally used for printing. Thus, we choose toluene as a substitute for methylene chloride and start experiments. This is a conservative choice: the toluene is still toxic and will generate emissions, but the modified ink should work well.

Notice that in this example we have done two things. First, we have refined the idea by defining which replacement solvent to use. Second, we have assessed our idea to determine how well it will work, so that we can compare it to other ideas. This is typical of the selection stage, where ideas evolve and improve during the process.

EXAMPLE 4.1–3   ANTIBIOTIC PURIFICATION

We are trying to modify an existing purification for an acidic antibiotic whose pK_a is 4.52. One idea is to alter the pH and hence alter the distribution of the antibiotic between water and butyl acetate. What pH range should we use?

SOLUTION

To begin, we should return to the definition of the pK_a, which is

$$pK_a = -\log_{10} K_a = \log_{10} \frac{[H^+]_{H_2O}\, [RCOO^-]_{H_2O}}{[RCOOH]_{H_2O}}$$

where the various concentrations are defined by Equation 4.1–8. The definition of the pH is:

$$pH = -\log_{10}[H^+]$$

Thus from Equation 4.1–11, we see

$$\frac{K}{K'} = \frac{1}{1 + 10^{pH - pK_a}}$$

When the pH is much less than 4.52, K approaches K'. When the pH is much greater than 4.52, K becomes small. By adjusting the pH, we can change K and purify the drug. Knowing the pH range required to achieve this purification allows us to assess this idea accurately against others. Notice again that in the process of technical assessment the idea is refined and improved.

4.2 Selection Using Kinetics

Thermodynamics is the science of what is possible; kinetics is the science of how fast and hence how expensive. For chemical products, we normally consider three kinds of kinetics. We consider the rates of the chemical reactions by which raw materials become the desired products. Reaction rates normally cannot be predicted a priori, so this implies experimental data. However, even when data are missing, we can make estimates of the maximum rates possible by assuming that the reactions are diffusion controlled. The other two forms of kinetics are mass and heat transfer rates. Here, we are in better shape, because we can estimate heat and mass transfer coefficients. In the following paragraphs, we briefly review each of these topics to supply guidelines for product selection.

4.2–1 CHEMICAL KINETICS

Chemical reaction rates must in almost all cases be determined by experiment. Books describing these rates all describe the more common reaction mechanisms. For a first-order reaction, the rate per volume r is

$$r = kc_1 \tag{4.2–1}$$

where c_1 is the concentration of reactant species "1" and k is the reaction rate constant, with units of reciprocal time.

For a second-order reaction, the rate is

$$r = (k'c_2)\, c_1 \tag{4.2–2}$$

where c_2 is the concentration of a second reagent; and k' is a second-order reaction rate constant, now with dimensions of volume per mole per time. Note that when c_2 is present in excess, c_1 changes by a much greater percentage than c_2. Then the quantity in parentheses in Equation (4.2–2) is nearly constant, a pseudo-first order rate constant. This is one reason why so many reactions with more than one reagent can appear to be first order.

Zero-order reactions are surprisingly common, for which the rate r is:

$$r = k'' \tag{4.2–3}$$

where k'' has dimensions of moles per volume per time. At first blush, it may seem amazing that the reaction rate will not change if the amount of reagent is increased. In fact, zero-order reactions are a frequent result of chemical catalysis, for each molecule of catalyst can only react so fast. This is explored further in Section 6.3.

No reaction rate constant – k, k' or k'' – can be found without experiment. Without experiments, we are really guessing, with substantial risk. This uncertainty may cause us considerable difficulty in product selection. However, things are not as bleak as this experimental demand may seem, because we can predict the maximum reaction rate. This is because as the chemical steps in the reaction become fast, the reaction rate is controlled by mass transfer. Mass transfer is more easily estimated. If this maximum rate for a specific process is slower than we need, the process should not be selected.

The prediction of the maximum rate is easier for heterogeneous reactions. For simplicity, we will discuss only first-order, irreversible cases. For example, imagine we are interested in the combustion of solid particles in air. The apparent rate constant k of such a reaction is

$$\frac{1}{k} = \frac{1}{k_D a} + \frac{1}{k_{\text{surface}} a} \tag{4.2--4}$$

where k_D is a mass transfer coefficient, a is the particle area per volume, and k_{surface} describes the kinetics of the reaction on the particle surface. If this surface reaction is fast, k_{surface} is very large, and the process will be controlled by mass transfer. Such mass transfer can be estimated simply by assuming k_D is about 10^{-3} cm/sec in a liquid or 1 cm/sec in a gas. The rate constant k is then known if the area per volume a is known. We will return to this estimate again later in this section.

Many reactions with microorganisms, catalyst particles, or emulsion droplets are mass-transfer limited. Correlations of mass transfer coefficients exist for these systems, but they really aren't the point. If we can guess the area per volume, we can guess the maximum apparent rate constant k, and hence have a more quantitative basis for selecting between alternatives.

If the reaction is homogeneous, the estimates of the maximum rate are harder. To see why, we consider the simplest limit of irreversible reactions. If the system is initially well mixed and the reaction chemistry is instantaneous, then

$$k = 4\pi \, (D_1 + D_2) \, \sigma \, \tilde{N} c_2 \tag{4.2--5}$$

where D_1 and D_2 are the diffusion coefficients of the limiting and excess reagents, respectively; σ is a collision diameter, perhaps 5 Å; $\tilde{N}$ is Avagadro's number; and c_2 is the concentration of the excess reagent. Unfortunately, this often-quoted result is not useful in practice, because systems which are highly reactive don't start out well mixed. Instead, many fast homogeneous reactions have their reaction rates determined by the mixing. In general, the rate constant for this mixing is

$$k = \frac{4 \, (D_1 + D_2)}{l^2} \tag{4.2--6}$$

where l is the average eddy size of the mixing. In rapidly stirred liquids, l is about 30 μm and the D_i values are 10^{-5} cm^2/sec, so k is about 10/sec. If there are no experiments, this can provide a quick guess of maximum reaction speed. We must repeat that whenever available, experiments are better than estimates.

4.2–2 HEAT AND MASS TRANSFER COEFFICIENTS

Finally, we need to turn to heat and mass transfer rates. Because these are physical processes, not chemical changes, the rates vary less widely. They can be more easily estimated, at least to an order of magnitude.

Mass transfer coefficients vary least. For liquids, the most common value of mass transfer coefficient is that given above, about 10^{-3} cm/sec. One easy way to think about the physical significance of this estimate uses the film theory of mass transfer, for which

$$k_D = \frac{D}{\delta} \tag{4.2–7}$$

where D is the diffusion coefficient and δ is a "boundary layer" or a "film thickness," roughly the distance over which the concentration changes. Since liquid diffusion coefficients cluster around 10^{-5} cm^2/sec, this implies a boundary layer of 100 μm, i.e. 0.01 cm.

For gases, values of mass transfer coefficients fall in a wider range, but tend to be around 1 cm/sec. Since diffusion coefficients are about 0.1 cm^2/sec, this implies a boundary layer of perhaps 0.1 cm. The wider variation of coefficients in the gas is rarely a problem because most mass transfer is limited by the liquid.

Heat transfer coefficients cluster near these same values, but this clustering is often obscured by the units used. To see why this is so, we turn to that classic tool of chemical engineering, the Chilton–Colburn analogy (strictly, valid only for turbulent flow):

$$\frac{k_D}{v} \left(\frac{v}{D}\right)^{2/3} = \frac{h/\rho \hat{C}_p}{v} \left(\frac{v}{\alpha}\right)^{2/3} = \frac{f}{2} \tag{4.2–8}$$

where k_D and h are the mass and heat transfer coefficients, respectively; ρ is the density; $\hat{C}_p$ is the specific heat capacity; v is the average velocity of a flowing fluid; f is the friction factor with the walls; and D, α, and v are the diffusivities of mass, energy, and momentum. The quantity α is also called the thermal diffusivity, and v is also called the kinematic viscosity. The important point is that these quantities are physical properties, listed in handbooks.

We can use Equation 4.2–8 to estimate heat transfer coefficients. For gases, α and D are both around 0.1 cm^2/sec, so

$$k_D = \frac{h}{\rho \hat{C}_p} \tag{4.2–9}$$

This result is sometimes called the Reynolds analogy. Since k_D in gases is around 1 cm/sec, $(h/\rho \hat{C}_p)$ is also around 1 cm/sec. For liquids, α is around 10^{-2} cm^2/sec, while D is close to 10^{-5} cm^2/sec, so

$$k_D = \left(\frac{D}{\alpha}\right)^{2/3} \left(\frac{h}{\rho \hat{C}_p}\right) = 0.01 \left(\frac{h}{\rho \hat{C}_p}\right) \tag{4.2–10}$$

Because k_D is 10^{-3} cm/sec in liquids, $(h/\rho \hat{C}_p)$ must be around 0.1 cm/sec.

Table 4.2–1 *Rough estimates of heat and mass transfer coefficients* The values given are useful only in the order of magnitude estimates made during product selection. More accurate values should be used when planning manufacture.

Situation	k_D cm/sec	$h/\rho \hat{C}_p$ cm/sec	h, W/m² K
Flowing gases	1	1	3
Flowing water	0.001	0.1	5000
Flowing organics	0.001	0.1	1000
Condensing steam or boiling water	–	–	2000

The trouble comes because ρ and $\hat{C}_p$ differ for different chemical species. Thus while $(h/\rho \hat{C}_p)$ may be similar for many liquids, the heat transfer coefficient h will have values which vary more widely, as suggested by the estimates in Table 4.2–1. Still, our advice at this early stage is to use these estimates, and to postpone until later the more accurate but more complex dimensionless correlations for these quantities.

EXAMPLE 4.2–1   A DEVICE WHICH ALLOWS WINES TO BREATH

David Anderson, an alumnus of the University of Minnesota who says he got poor grades in chemistry, is the buyer for the best wine store in Minneapolis-St. Paul. In an interview, he described the practice of allowing wines – especially red wines – to "breathe" by exposing them to air before drinking. The following is a summary of his ideas on this subject.

Wines need to breathe to reduce the "hard" and "soft" tannins which naturally occur. These are reduced by exposing the wine to oxygen. Reactions of oxygen with polyphenols may also be involved. Such exposure enhances flavor, allowing both fruit taste and aroma to be more clearly perceived.

The amounts of oxygen required by wines vary widely. Uncorking the bottle for 15 minutes before drinking is useless; exposing wines to excessive oxygen will turn them into vinegar. Mr. Anderson suggests one good way is to pour the wine from the bottle into a decanter, leaving any residue behind. Several pours between decanters is better. Mr. Anderson even has one friend who pours the wine into a large glass, covers the glass with his hand, and gives the wine a good shake.

In thinking about this, we must remember that wine is unstable, and can be destroyed by mistreatment. Heat is a major enemy. Corks often fail, so that different bottles from the same vineyard and the same vintage may be different. Because of this, Mr. Anderson applauds the use of plastic corks. Moreover, like humans, wines can become fragile with age: a fine 25-year-old Bordeaux can degrade while it sits in the glass.

Estimate the aeration needed in a product which can let wine breathe in only a few minutes, and hence make a selection of appropriate aeration ideas.

SOLUTION

A problem like this begs for experimentation. Our experiment used four wines:

Montipulciano, *CITRA*, Italy (1996) $6/1500 ml
Cabernet Sauvignon, *Castillero de Diablo*, Chile, (1996) $8/750 ml
Gamay, *Chateau la Charge*, France (1995) $12/750 ml
Zinfandel, *Folie a Deux,* USA (1997) $18/750 ml

Aeration can make a startling difference, especially for cheap, freshly opened wine. We tried four methods:

(1)   Open the bottle for 15 minutes.
(2)   Decant into an open pitcher and let the wine sit two hours.
(3)   Decant the wine fast three times, entraining air.
(4)   Shake for ten seconds in a large glass.

Then we tested the wine.

The first method is useless, as David Anderson suggested. The other three give roughly similar taste improvements. Yet another method, putting wine in a blender, aerates it excessively, dramatically reducing flavor.

These results are consistent with estimates based on mass transfer. To see why, we make a mass balance on oxygen transferred into the wine:

$$\frac{\text{oxygen transferred } M}{\text{wine volume } V} = \{[\text{mass transfer coefficient } k_D]\ [\text{partition coefficient } H]$$
$$\times [\text{concentration in air } c_1]\}$$
$$\times [\text{area per volume } a]\ [\text{time } t]$$

In symbolic terms,

$$M/V = \{k_D H c_1\}\, at$$
$$\frac{M/V}{H c_1} = k_D at = NTU$$

where *NTU* is the number of transfer units, a dimensionless quantity commonly used in the design of absorption towers.

Estimates for our four experimental methods are summarized in Table 4.2–2. The mass transfer coefficients in this table were estimated in different ways. For the first two methods we assumed a value of 10^{-3} cm/sec, as suggested in Table 4.2–1. The aeration achieved by the third method was estimated using correlations for bubbles and by the fourth from the penetration theory of mass transfer. These methods were preferred because of the small bubbles and short contact times, which give unusually large mass transfer. The estimates in Table 4.2–2 are reassuringly consistent. Opening the bottle for 15 minutes gives less than 0.01

Table 4.2–2 *Four methods for letting red wine breathe.*

Method	Mass transfer coefficient (cm/sec)	Wine area/volume (cm²/cm³)	Time (sec)	NTU
Uncork bottle for 15 minutes	$10^{-3(a)}$	3/750	900	0.004
Decant into pitcher two hours before serving	10^{-3} (a)	100/750	7200	1
Decant three times, entraining about 5 cm³ of air	$20 \cdot 10^{-3(b)}$	$0.4^{(b)}$	120	0.5
Shake 30 cm³ wine twenty times in a 300 cm² glass	$7 \cdot 10^{-3(c)}$	300/30	$10^{(c)}$	0.7

(a) This typical value can be estimated from Table 4.2–1 or from the free convection caused by ethanol evaporation.
(b) Estimated from observed 5 cm³ of 0.1 cm bubbles entrained in each decantation.
(c) Estimated from penetration theory with a penetration time of one shake (0.5 sec).

transfer units: it does not give enough oxygen transfer. The other three methods give around one transfer unit, apparently enough to aerate the wine but not so much as to cause excessive flavor loss. These three methods supply close to one transfer unit even though they have radically different areas, times, and mass transfer coefficients.

Thus our selected design – whatever it is – should give around one transfer unit. Any engineering solution meeting this specification will give about the same performance. However, we believe that the success of a "wine breather" will depend strongly on the aesthetics of the aerator's design. It is our own confidence in our lack of ability in this aesthetic area which inhibits us in developing this product.

EXAMPLE 4.2–2   A PERFECT COFFEE CUP

We have been asked by a chain of upmarket coffee shops to develop an improved coffee cup. The current cup has a volume of about 200 cm³ and a total surface area, including the top and bottom, of 200 cm². The improved cup should keep the coffee at an optimal "drinkable" temperature, estimated to be 51 °C, for as long as possible. Data for coffee cooling in the current covered and uncovered cups are shown in Figure 4.2–1.

Generating ideas has led to three major directions for an improved cup:

(1)   a better insulated cup;
(2)   a cup with its own, self-contained heater; and
(3)   a cup with a thermal reservoir which melts around 50 °C.

Select among these ideas to see which merits further development.

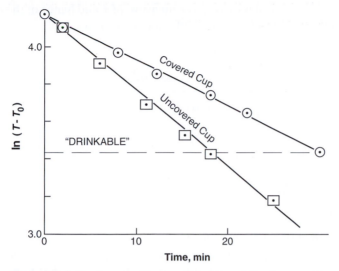

Figure 4.2–1 *Cooling a coffee cup* The slope of the data on this semilogarithmic plot is a measure of the overall heat transfer coefficient of the cup. In these experiments, the ambient temperature was 20 °C.

SOLUTION

We begin by trying to explain the data in Figure 4.2–1. From an unsteady energy balance on the coffee, we find

$$\frac{d}{dt}\left[\rho\hat{C}_v V T\right] = U A\,(T_0 - T)$$

where ρ is the coffee's density; $\hat{C}_v(\approx \hat{C}_p)$ is its specific heat capacity; V is its volume; t is the time; U is the overall heat transfer coefficient averaged over the coffee interface; A is the coffee's surface area, including the contact with the cup and with the air; and T and T_0 are the temperatures of the coffee and the surrounding air, respectively. This equation is easily integrated to give

$$\frac{T - T_0}{T(t = 0) - T_0} = e^{-\frac{t}{\tau}}$$

where the characteristic time τ is

$$\tau = \frac{\rho\hat{C}_v V}{U A}$$

This equation implies that the temperature of the cooling coffee should vary exponentially with time, which is consistent with the experiments shown in the figure.

Note that the slope of the data shown is a measure of τ and hence of the overall heat transfer coefficient. In particular, for the uncovered cup,

$$\tau = 24 \, \text{min}$$

$$\frac{\rho \hat{C}_v V}{U A} = \frac{10^3 \dfrac{\text{kg}}{\text{m}^3} \left(\dfrac{1.18 \times 10^3 \, \text{J}}{\text{kg} \, °\text{C}} \right) 2 \times 10^{-4} \text{m}^3}{U \, (0.02 \, \text{m}^2) \, \dfrac{60 \, \text{sec}}{\text{min}}}$$

$$U = 57 \, \text{W/m}^2 \, °\text{C}$$

For the covered cup, τ is 40 min and U is 17 W/m^2 °C. In this example, we have some experimental data available, and so estimates for heat transfer coefficients are not required.

We are now in a position to select among the three ideas suggested above. We first consider the idea of a better insulated cup. This idea aims at decreasing U, and hence reducing the slope of the data in Figure 4.2–1. Note that the coffee does not stay at the drinkable temperature, but rather passes that temperature more slowly. Still, there are many routes to this improved insulation, some of which are cheap. This simple, powerful idea merits further development.

The second idea, a cup with its own heater, is weaker. In this idea, we could let the coffee cool uncovered until 50 °C, and then let the heater take over. From our data, we see that the heater must provide power of about

$$Q = U A \, (T - T_0)$$
$$= \frac{57 \, \text{W}}{\text{m}^2 \, °\text{C}} \, (0.04 \, \text{m}^2) \, (51 - 20 \, °\text{C})$$
$$= 70 \, \text{W}$$

An average C-cell battery provides about a watt, so that we will need a lot of batteries for this concept. It just does not make sense.

The third idea is to build a coffee cup with a thermal reservoir. The physical form of this reservoir is not clear at this stage; it might replace the insulation with a layer of a substance which melts near 50 °C. A coffee cup with such a thermal reservoir would behave very differently to normal cups, such as that of Figure 4.2–1. When the hot coffee is added, the reservoir substance would melt, cooling the coffee. When the coffee has cooled to the compound's melting temperature, the compound would begin to freeze. The heat of fusion would keep the coffee near the melting temperature until all the compound was frozen. The coffee would then cool in the normal way.

While this could potentially work, we need more information about possible substances to make an informed judgment. Such substances are higher hydrocarbon waxes, like pentacosane ($C_{25}H_{52}$) which melts at 53 °C with a heat of fusion ($\Delta \hat{H}_{\text{fus}}$) of 220 kJ/kg, or beeswax, which melts at about the same temperature with a heat of fusion around 180 kJ/kg. We can find the mass m of the substance required to keep the coffee at the melting temperature for a time t from the relation

$$M \Delta \hat{H}_{\text{fus}} = Qt = U A \, (T - T_0) \, t$$

For example, for beeswax, we can keep the coffee at 53 °C in a covered cup for 20 minutes if

$$M \frac{180 \times 10^3 \, \text{J}}{\text{kg}} = \frac{17 \, \text{J}}{\text{sec} \, \text{m}^2 \, ^\circ\text{C}} \, (0.02 \, \text{m}^2) \, (53 - 20 \, ^\circ\text{C}) \, (1200 \, \text{sec})$$
$$M = 0.07 \, \text{kg}$$

We will need a few ounces of beeswax. While there are still major unknowns in this product, the idea merits further development. One important consideration will be how to ensure that heat is effectively transferred from the freezing wax to the hot coffee.

4.3 Less Objective Criteria

This chapter is about exploring ways of choosing between our better looking frogs, in the hope that one of them will one day grow into the prince of our dreams. In Sections 4.1 and 4.2 we saw how the traditional tools of chemistry and engineering can be used in this selection process. We made use of chemistry and engineering to compare like with like. In this effort, it is vital not to underestimate the usefulness of chemistry and engineering training in making commercially significant decisions. This is probably the most important difference between chemical product design and parallel methods of inventing new consumer products, where the technical input may be less central.

However, in making our final selection, subjective decision making will also be necessary. This may take two forms. First, we are likely to need to "compare apples with oranges," i.e. make decisions between objective but different criteria. Often this will involve balancing cost against performance. For example, we may wish to use a new form of exhaust catalyst which will improve air quality but which will also be expensive. Chemistry and engineering will allow us to estimate how well the catalyst will perform and how much it will cost, but cannot help us in balancing these two criteria.

Second, we are likely to have to evaluate genuinely subjective issues – what do people like, how much do they care, etc. For example, in the high-level waste-disposal example of Section 3.4, the public reaction to the proposed solution is a subjective criterion. The technically attractive idea of making the waste into concrete, "grout," was penalized because of a likely negative public response, although it performed well on all the objective criteria.

Introducing these additional elements into product selection is the subject of this section. The methodology we suggest is that of the concept selection matrix, already introduced in Section 3.4 for concept screening. The basic idea is the same: generate a handful of important criteria on which to judge the ideas; weight the criteria according to their perceived significance; and score each idea for each criterion, often relative to a benchmark which may be an existing product or a well-established technology. An overall score for each item is then produced by summing the products of weighting factors and scores over all criteria. The product selected is that with the best score.

The difference between the selection matrix used here and that used in the previous chapter comes in the level of detail demanded. In the Section 3.4 we were interested in reducing perhaps around twenty frogs to four or five promising candidates to be considered in greater detail. The selection phase is where that greater detail is required: all the remaining ideas are promising, so distinguishing among them becomes harder. Each number in the selection matrix (the weighting factors and the scores) must be determined with as much accuracy as possible. Where appropriate, detailed calculations should be carried out; market research may be desirable; canvassing for opinion beyond the core team is certain to be necessary. Equally, a more thorough estimate of inaccuracies is required: how much do scores vary within the core team? To what extent do external experts disagree? Once we are down to the handful of ideas left at the selection stage, we should apply a degree of effort and rigor which would be wasted on the many frogs of Section 3.4.

As chemists and engineers, we will be far less comfortable with this subjective decision making than with the quantitative analysis made possible by, for example, thermodynamics or chemical kinetics. However, we need to choose between apples and oranges and we need to worry about how people respond. This consideration of subjective criteria is necessarily woolly. The best we can hope to achieve are some useful guidelines to help us apply the selection matrix. In order to do this we first consider important aspects of selection matrices to supplement what we learned in Chapter 3.

4.3–1 WHEN TO MAKE SUBJECTIVE JUDGMENTS

Because we dislike the idea of vague subjective choices, we will postpone them as long as we can. We would like to go as far down the road of objective evaluation as possible before venturing into the quagmire of subjectivity. There is no point in worrying about whether the public will like a product if it is bound to be uneconomic anyway.

The earliest and the most important point at which subjectivity becomes unavoidable is in determining the criteria to be used in the selection matrix. Inevitably, cost and technical feasibility will be important, but what else matters? Do the customers mind noise, humidity, cold? Do environmental issues concern them? Ultimately, we will need to decide, remembering two points. First, we must not leave anything out – we must canvass opinion widely and be aware that what is dear to our hearts may be of no concern to those who will buy the product. Second, we must bear in mind that there may be more than one answer. For example, humidity in homes may be of greater concern in Kuwait than in Minnesota.

The next point at which subjectivity cannot be postponed is comparing unlike objective criteria – apples versus oranges. This element of the decision-making process will be manifested primarily in the weighting factors assigned to the various objective criteria. Again, remember that the answer depends on the audience and that your own instincts may not fit with the views of your market. A Frenchman might tell you that the most important thing about a red wine is that it is well balanced, an Australian that it is "big," and a student that it is cheap. None of these answers is more right than another. To supply the correct product for our target market, we must assign a weight to each criterion which will depend on our customer.

Finally, we will need to include the truly nebulous – what is prettier, pleasanter, smoother, more environmentally friendly? As chemists and engineers, we naturally tend to play down the importance of these issues because we have no satisfactory means of dealing with them. We can delay their consideration until last for this reason, but we make a great error to ignore the purely subjective or to underestimate its importance. The bizarre French car the Citroën 2CV sells because of its "look" as well as because of its price and reliability. Such considerations are less important in the case of chemicals, which are often somewhat divorced from the consumer, but nonetheless cannot be ignored.

4.3–2 HOW TO MAKE SUBJECTIVE JUDGMENTS

We cannot completely answer this question; all we are doing here is outlining a few basic points worthy of attention. Let us first consider how to choose the criteria for our selection matrix. As we have already mentioned, this is probably the most crucial step in subjective decision making and the easiest to overlook. Leaving out something of critical concern to the customer is likely to be disastrous. Three aspects are key:

(1) Use independent criteria. For example, crampons for mountain climbing need to be light and strong. However, these are not independent criteria: more strength implies more weight. A better single criteria would be that the crampons should be made from a material with a high strength-to-weight ratio.

(2) Avoid repetition, as this will imbalance the scores by invalidating the weighting factors you have carefully constructed. To a very large extent this boils down to ensuring that each criterion is tightly defined. We can use the high-level waste as an example again. Safety and public response will both be important. However, the public perception of each project will also depend on how safe it is. If we do not take this into account, the weighting we are giving safety will be artificially raised.

(3) Most importantly, use a complete list of criteria. The list must include all the most important factors – and remember it should be what is important to the customer, not to us. This can be difficult to achieve. Canvass opinion as widely as possible before finalizing the criteria to be used. Information from market research and independent "experts" is particularly valuable at this point. The list of needs is a good starting point for these criteria.

Having produced a list of criteria, we must weight each and then score the competing ideas. Inevitably this is a somewhat arbitrary process, so absolute significance is not assigned to the results. At the point of selection it is usually worth putting more effort into canvassing opinion, firming up the scores and performing sensitivity tests. Only a few, reasonably attractive ideas should remain, because product design now becomes substantially more expensive. Backtracking is easy, and often valuable, up to the point of selection – mistakes then become costly.

4.3–3 WHY WE USE SELECTION MATRICES

There are clearly many other possible way of making the decision we require. We could get someone else to choose – a manager or a customer. We could simply go with our intuition. We could prototype and test each of our ideas. This will be very reliable, but costly and, crucially, slow – time to market is often the prime determinant of a product's success. Nurofen and Adoil dominate their market, though they are more expensive than generic ibuprofen, which is chemically equivalent but arrived later in the shops.

Although selection matrices are far from perfect, we do believe that there are several advantages in using the systematic approach they imply.

(1) The selection stage is the point of no return. Costs escalate rapidly once the best one or two ideas are chosen, as an extensive program of testing, prototyping, and market research becomes necessary. Simultaneously, this will be the point of fierce management review, so we must be able to justify our choice to those outside the core team. The decision matrix forces you to stop and consider seriously each stage of your decision-making process – and this at a point when there will be considerable pressure from managers keen to minimize time to market. It also ensures that you thoroughly document your deliberations.

(2) The need to weight and score each idea forces the core team both to efficiently pool its resources and to search for external input. The need for justification of criteria, sensitivity tests, etc. will make the consulting of experts and customers inevitable. The use of numerical scoring tends to make it harder for a single personality to dominate the core team's deliberations.

(3) The selection stage is the last point at which we can sensibly combine aspects of different ideas to produce an improved product. The separate scoring of the different criteria ensures that the strengths and weaknesses of each idea will be very obvious and opportunities for enhancements by combination should stand out. Good but imperfect frogs may be combined into improved models.

Despite the systematic approach offered by the concept matrix, there is no need to ignore intuition. It is always possible to override the conclusions of the matrix – but it is better to realize you are doing it. In reality, many business decisions are made on gut feeling and products are launched more on a hunch than on sound reasoning. For example, throughout the nineteenth century, railway investments almost always lost money. However, the investors continued to be attracted by the romance of travel and progress – entrepreneurs and engineers continued to

Table 4.3–1 *Selecting a Prince of Wales* The incumbant and two alternatives have similar scores but for very different reasons.

	Weighting factor	Prince Charles	Hapsburg Prince	Romanov Prince
Health	0.5	5.5	10	1
Looks	0.5	5.5	1	10
Total		5.5	5.5	5.5

make large fortunes by building railways. Using a concept matrix gives one the opportunity to exercise "informed intuition."

EXAMPLE 4.3–1 MONARCHY SUBSTITUTION

Since we are describing the search for princes, we take as an example a replacement of the British Prince of Wales. How could we go about doing this?

SOLUTION

This problem is a summary of the design template of needs, ideas, and selection. First, we must establish our selection criteria. There are really only two things required of a prince. The first is good health: monarchy is a symbol of stability and this is badly undermined if royalty keep dying off. The second is good looks: a prince must meet the demands of television appearances, damsels in distress, and so on. We will weight these criteria equally. Next we need some ideas. We could do worse than look at the unemployed royal families of Europe; the most noble are the Hapsburgs and the Romanovs. How do they match up to our requirements? We will use the current market leader, Prince Charles, as our benchmark, giving him the average score of 5.5 on each criterion. On looks, the Romanovs score big. The Hapsburgs sadly suffered from a notorious chin, perhaps a result of the severely limited breeding opportunities available to European royalty. On health, the Hapsburgs perform excellently. The penultimate Emperor, Francis Joseph, died in 1916 at the age of 87, having ruled for 68 years. Sadly, the Romanovs score poorly on health. Their particular genetic defect was hemophilia. The Tsarevich Alexis was so infirm that he often had to be carried at state functions by the sailor Darevenko. We might produce the selection matrix in Table 4.3–1. Interestingly, all three options have the same scores, but for widely different reasons. Changing these scores implies changing the weighting factors, or introducing new criteria for a successful prince. An alternative would be a genetically engineered Hapsburg–Romanov "super-prince," combining the health of the Habsburgs and the good looks of the Romanovs. We should always be looking for ways to combine the best aspects of our different product ideas.

EXAMPLE 4.3–2   THE HOME VENTILATOR

In older houses, the mean residence time of air is often around 40 minutes. Demands for draft reduction and energy efficiency mean that houses are increasingly well insulated and sealed, so the mean residence time can be one day in modern houses. This is good for comfort and efficiency, but is unhealthy. The air inside well-sealed houses contains excess CO_2, CO, radon, and formaldehyde. (This last evaporates from carpets and drapery, particularly when they are new.) The formaldehyde is often the first to reach dangerous levels. However, cigarette smoke is a very efficient scavenger of radon, so in the houses of smokers radon may be an even greater problem. Currently, there is no legislation regulating the turn-over time of air in homes. Anomalously, fresh-air replacement is required every three hours for laboratory animals in the USA. Fresh air is recommended every six hours for humans.

Assuming that complete air replacement is required every three hours, is it possible to combine the economics and comfort of a tight house with the health advantages of a leaky one?

SOLUTION

The simplest answer is to open a window! This of course destroys much of the benefit of carefully sealing the house, but it is easily achieved and will have the desired health benefit. We will take as our benchmark product an automatic window opening device which monitors air flow into the house and keeps the opening of a window at a level such as to maintain the flow necessary for three-hour exchange of air. This device will be cheap and easy to retrofit into "tight" houses.

Can we do better than this? Another approach is to exchange the air through a heat exchanger to minimize energy loss. This will be more comfortable, reducing the drafts to be expected as a result of an open window. A disadvantage, particularly in very cold climates, is that the incoming air will be very dry while that leaving the house will have a significant water content. Thus, the house will dry out and become uncomfortable. One possible solution, to evaporate water, will be expensive due to the latent heat required. An alternative is to devise a ventilator which exchanges both heat and water vapor from the outgoing air stream into the incoming one. Although such units have not been commercially constructed, available literature suggests that polyimide membranes could be used for this application.

We have two alternative products to compare to our benchmark: heat exchange with water evaporation and simultaneous heat and mass exchange. We will start by using an engineering analysis to estimate the feasibility, cost, and energy saving for each product. We will then use a decision matrix in order to assess the products' value relative to the benchmark, including some subjective criteria. We perform our analysis for the cases of Minnesota and Cambridge. In what follows, the calculation is followed in detail for the case of Minnesota, with notes where the Cambridge case differs significantly.

Setting up the problem. We assume an average house in Minnesota has a floor area of 250 m². In Cambridge houses are smaller, perhaps half the size. Thus we find:

$$\text{Volume of air in house} = 250\,\text{m}^2 \times 3\,\text{m ceilings} = 750\,\text{m}^3$$

Number of moles of air in house, n, is given by:

$$n = 750\,\text{m}^3 \times \frac{1}{22.4 \times 10^{-3}}\frac{\text{mol}}{\text{m}^3} = 33{,}500\,\text{mol}$$

We now assume an exterior winter temperature of $-10\,^\circ$C ($5\,^\circ$C in Cambridge) and an interior temperature of $20\,^\circ$C.

The enthalpy, H, required to heat the amount of air contained in the house from the exterior temperature to the interior temperature is

$$H = n \underbrace{\left(\frac{7}{2}R\right)}_{\text{gas heat capacity}} \Delta T = 29{,}000\,\text{kJ}$$

This is our energy cost per complete air exchange in terms of specific heat. But what about humidification and latent heat?

Now we assume the external relative humidity is 0% (60% in Cambridge). At typical room temperature, humans are comfortable at relative humidity between 20% and 60%. However, to avoid condensation it is best to operate at the lower end of this range, particularly for low external temperatures. Therefore we will assume 25% relative humidity inside. The energy required to evaporate water to reach this humidity is

$$\Delta H\,(\text{vap}) = \left(y^{\text{in}}_{\text{H}_2\text{O}} - y^{\text{ex}}_{\text{H}_2\text{O}}\right) n \Delta \tilde{H}_{\text{vap}} \approx 9000\,\text{kJ}$$

where $\Delta \tilde{H}_{\text{vap}}$ is the molar enthalpy of evaporation of water (approx $45\,\text{kJ mol}^{-1}$); and $y^{\text{in}}_{\text{H}_2\text{O}}$ and $y^{\text{ex}}_{\text{H}_2\text{O}}$ are the mole fraction of water vapor inside and outside the house, respectively. In Minnesota

$$y^{\text{in}}_{\text{H}_2\text{O}} = (0.25 \times 0.023) = 0.0058 \qquad \text{and} \quad y^{\text{ex}}_{\text{H}_2\text{O}} = 0$$

where 1.0 bar is the total pressure and 0.023 bar is the saturated vapor pressure of water at $20\,^\circ$C.

In Cambridge, the external mole fraction of water is

$$y^{\text{ex}}_{\text{H}_2\text{O}} = (0.6 \times 0.0087) = 0.0052$$

where 0.0087 bar is the saturated vapor pressure of water at $5\,^\circ$C. This is close to the internal mole fraction. Therefore, we expect on a "normal" day in Cambridge that there will be no need to replenish the house's water content. Of course, the weather fluctuates a lot. On some days it may be desirable, but on average the cost will be low.

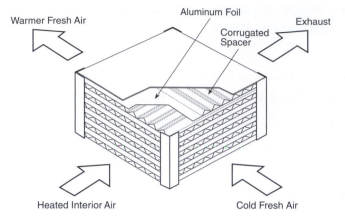

Figure 4.3–1 *Schematic of a domestic crossflow heat exchanger*
Warm, stale inside air is used to heat cold, fresh outside air. This
heating, which reduces domestic energy use, takes place across
sheets of aluminum foil.

Back in Minnesota, the heating costs for the tight house, with air turnover
every 12 hours, are

$$\frac{(29{,}000 + 9000)\,\text{kJ}}{12\,\text{hr}\left(3600\dfrac{\text{sec}}{\text{hr}}\right)} \times (24 \times 30)\ \frac{\text{hrs}}{\text{month}} \times \frac{\text{£0.03}}{\text{kW hr}} = \text{£19 per month}$$

The heating costs for a house with a window opener with air turnover every three
hours are

$$\frac{(29{,}000 + 9000)\,\text{kJ}}{3\text{hr}\left(3600\dfrac{\text{sec}}{\text{hr}}\right)} \times (24 \times 30)\ \frac{\text{hrs}}{\text{month}} \times \frac{\text{£0.03}}{\text{kW hr}} = \text{£76 per month}$$

We expect that a heat exchanger recovers 70% of the specific heat. The heating
for a house with such a heat exchanger plus humidification and a three-hour air
turnover costs

$$\frac{(0.3 \times 29{,}000) + 9000}{3 \times 3600} \times (24 \times 30) \times 0.03 = \text{£35 per month}$$

Finally, we expect that the heat and water vapor exchanger recovers 70% of
both heat and water vapor. The cost of heating a house with both heat and mass
exchanger and a three-hour air turnover is

$$0.3 \times 76 = \text{£23 per month}$$

Making the heat exchanger. We aim to build a unit with 70% efficiency for heat
recovery. Typically such units use cross flow geometry as shown in Figure 4.3–1.
However for simplicity we will assume a countercurrent unit with parallel plate
geometry. The molar flow rate of air F is

$$F = \frac{33{,}500\,\text{mol}}{3\,\text{hr} \times 3600\,\dfrac{\text{sec}}{\text{hr}}} = 3.1\,\text{mol/sec}$$

The total heat transfer rate across the whole exchange surface,

$$Q = 0.7 \times \frac{29{,}000 \times 1000 \, \text{J/air change}}{3 \, \text{hr} \times 3600 \dfrac{\text{sec}}{\text{hr}}} = 1900 \, \text{W}$$

and

$$Q = UA\Delta T$$

where A is the heat exchange area, and ΔT is the temperature drop across the heat exchanger surface, a constant 9 °C (5.5 °C) throughout the unit. The overall heat transfer coefficient U is given by:

$$U = \frac{1}{\dfrac{1}{h_{\text{ex}}} + \dfrac{1}{h_{\text{wall}}} + \dfrac{1}{h_{\text{in}}}}$$

where h_{ex}, h_{wall}, and h_{in} are the heat transfer coefficients of the exhaust air, the wall, and the entering air, respectively.

The exchange surface can be constructed of a 1 mm thick aluminum sheet: this will have a large h_{wall} and a negligible thermal resistance. Thus the h_{wall}^{-1} term can be ignored. The heat transfer coefficient for the incoming air, h_{in}, will equal that for the exhausting air, h_{ex}. Thus

$$U = \tfrac{1}{2} h_{\text{ex}}$$
$$h_{\text{ex}} = \frac{k_{\text{T}}}{\delta}$$

where k_{T} is the thermal conductivity and δ is the film thickness over which heat transfer is occurring.

If we assume that the flow is laminar and that we have parallel plate geometry, then δ is approximately one quarter of the channel dimension in the heat exchanger (i.e. half way to the middle). Let us assume a channel dimension of 3 mm, so that

$$U = \frac{1}{2} h_{\text{ex}} = \frac{1}{2} \left(\frac{k_{\text{T}}}{\frac{1}{4} \, \text{channel}} \right) = \frac{1}{2} \frac{0.026 \, \text{W/m k}}{\frac{1}{4} (0.003) \, \text{m}} = 17 \, \text{W/m}^2 \, \text{K}$$

The area A required for the exchanger is now

$$A = \frac{1900}{17 \times 9} = 12 \, \text{m}^2$$

(The analogous calculation for Cambridge gives an area of heat exchanger of 5 m^2). This type of simple parallel-plate heat exchanger, a well-established technology, can be built for around £6 per m^2. The production cost will be about £72 (£30 for Cambridge) and we might expect to sell the units for perhaps eight times this, a few hundred pounds.

Before we proceed, we should check that the flow is indeed laminar. To do so, we recognize that the volume of heat exchanger is $12 \times 0.003 = 0.036 \text{m}^3$. If the exchanger is cubic, the cross-sectional area available for flow in each direction is

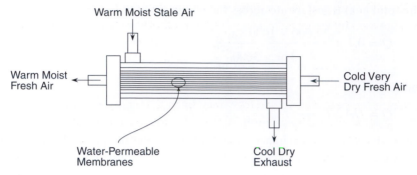

Figure 4.3–2 *Schematic of a countercurrent heat and mass exchanger* Warm, conditioned, but stale inside air heats and humidifies cold, dry, fresh air. A water-permeable membrane replaces the aluminum foil of the previous figure.

about

$$= 0.5\,(0.036)^{2/3}$$
$$= 0.055\,\text{m}^2$$

The superficial velocity of air v is given by

$$v = \frac{750\,\text{m}^3}{3\,\text{hr} \times 3600\dfrac{\text{sec}}{\text{hr}}} \times \frac{1}{0.055\,\text{m}^2} = 1.27\,\text{m/s}$$

Thus, the Reynolds number is

$$\text{Re} = \frac{dv}{\nu} = \frac{1.27 \times 0.003}{1.3 \times 10^{-5}} \approx 300$$

i.e. the flow is laminar. Remember that d is the spacing and ν is the kinematic viscosity.

Making the heat and mass exchanger. The equipment is shown in Figure 4.3–2. It is similar to the heat exchanger just considered. However the problem is now more complex because we have simultaneous heat and mass transfer.

We will assume the same geometry as for the heat exchanger, but replace the aluminum exchange surface with a 30 μm thick polyimide membrane. Such a membrane is selective for water transport and has a negligible resistance to water exchange. It is also sufficiently thin that its thermal resistance can be ignored (as was done for the aluminum in the normal heat exchanger). We can therefore assume that its performance as a heat exchanger will be the same as we have just calculated, i.e. a 12 m² (or 5 m² for Cambridge) unit will give 70% efficiency for the conditions described. We now need to calculate the efficiency of this exchanger for mass transfer.

We will first look at heat transfer and consider a point on the exchange surface.

The heat flux q at such a point is given by a combination of an enthalpy term (heat carried by water moving across the membrane) and a conductive term:

$$q = \underbrace{C_p N (T - T_0)}_{\text{enthalpy term}} - \underbrace{k_{\text{T}} \frac{\partial T}{\partial z}}_{\text{conductive term}}$$

In taking the heat transfer performance to be that calculated previously, we have assumed that the enthalpy term in the heat transfer equation is small compared to the convective term – it is easy to check that this is indeed the case, for the small temperature changes involved, once we know the mass flux. Thus,

$$q = -k_{\text{T}} \frac{\partial T}{\partial z} \approx -k_{\text{T}} \frac{\Delta T}{2\delta}$$

where ΔT is the total temperature drop across the exchanger; δ is the film thickness over which heat transfer occurs, as defined before. The factor of two comes from there being one such film on each side of the polyimide membrane.

The mass flux N at the same point is given by an analogous equation:

$$N = -Dc \frac{\partial y}{\partial z} \approx -Dc \frac{\Delta y}{2\delta}$$

where D is the diffusion coefficient and Δy is the water mole fraction drop across the exchanger. We have neglected the convective term in this mass transfer equation, a valid assumption since water is dilute.

Consider the ratio of the total molar flow of water W to the total heat transfer Q in the device:

$$\frac{W}{Q} = \frac{N}{q} = \frac{Dc \Delta y}{k_{\text{T}} \Delta T}$$

Also,

$$W = F \left(y_{\text{H}_2\text{O}}^{\text{in}} - y_{\text{H}_2\text{O}}^{\text{ex}} - \Delta y \right)$$

For an N_2–H_2O mixture at 308 K, $D = 0.256$ cm^2/s.

$$\text{and } c = \frac{P}{RT}$$

Thus

$$3.1 \, (0.0058 - 0 - \Delta y) = \frac{1900 \times 0.256 \times 10^{-4} \times 101\,325 \Delta y}{0.026 \times 9 \times 8.314 \times 298}$$

$$\Delta y = 0.0015$$

The efficiency of water exchange in our 12 m^2 polyimide exchanger is

$$\frac{0.0058 - 0.0015}{0.0058} \times 100\% \approx 74\%$$

Thus both heat and mass exchange efficiency targets are met simultaneously for this 12 m^2 polyimide exchanger.

We summarize the results of these calculations in Table 4.3–2. This summary recognizes that the cost of polyimide membranes is likely to be greater than the

---

Table 4.3–2 *Summary of house ventilation* costs.

	Minnesota	Cambridge
House with three-hour air turnover	£76 per month	£14.60 per month
+ Heat exchange (and humification)	£35 per month	£4.40 per month
+ Heat and mass exchange	£23 per month	£4.40 per month
Unit price for heat exchange unit	£500 + installation	£250 + installation
Unit price for heat + mass exchange unit	£500 + installation	£250 + installation
Pay-back time: heat exchange unit	12 months	25 months
Pay-back time: heat + mass exchange unit	9 months	25 months

Note: pay-back times are months of *winter*

Table 4.3–3 *Decision matrix for house ventilation* (bracketed values are for Cambridge).

	Weight	Three-hour exchange (benchmark)	Heat transfer + humidification	Heat + mass transfer
Heating cost	0.3	5	8 (7)	10 (7)
Capital cost	0.2	5	3 (3)	3 (3)
Health benefit	0.2	5	5 (5)	5 (5)
Comfort	0.2	5	8 (6)	10 (7)
Noise	0.1	5	2 (2)	2 (2)
Total		5.0	5.8 (5.1)	6.8 (5.3)

price of aluminum sheet. However, most of the cost of exchanger units can be expected to be in the construction, so we can assume that the cost of production for the heat and mass exchanger is similar to that for the heat only exchanger already estimated. On the purely objective criterion of economic benefit, both heat and heat/mass exchangers look like winners in Minnesota but not in Cambridge.

However, this analysis is not sufficient to make a decision. There are other, subjective factors to consider. For simplicity, we restrict ourselves to three additional ones: health (as air quality), comfort, and noise. We now have five criteria: heating cost, capital cost, health, comfort, and noise. We have separated heating cost and capital cost rather than include a single economic criteria because we might wish to give slightly higher weight to the reduction of heating bills on environmental grounds. We will now use these five criteria to illustrate how a decision matrix can help in the final selection process.

Table 4.3–3 exemplifies the decision matrix we can draw up. The values in this table include both the engineering results of Table 4.3–2 and the subjective factors. These merit discussion. "Health" scores identically for all products: it is redundant. We have included it for completeness and it would certainly be needed if a larger range of products was being considered. It probably should be dropped.

The results for comfort are more interesting. Both exchangers give humidity control when the outside is dry. However, the heat and mass exchanger also gives the potential for humidity reduction (in summer). Comfort is not such a large problem in Cambridge, where conditions are less extreme and so the scores are more moderate. Both exchangers need a pumping system, which may result in irritating noise. Notice that we have scored the silent benchmark as 5: this inevitably leads to scale compression. We could avoid this by scoring the benchmark differently. The conclusions from this decision matrix are pretty much what we already know. In Cambridge the viability of these exchangers is marginal. In Minnesota, their viability looks good. The heat and mass transfer unit looks somewhat better than heat transfer with humidification.

4.4   Risk in Product Selection

In our focus on product selection, we must normally make comparisons between very different product options. In some cases, the products try to accomplish the same objective in slightly different ways. In these cases, our decisions are straightforward. For example, we may be making ingredient substitutions, as described in Section 4.1. We may be simply comparing performance, as in the wine aeration example in Section 4.2.

In other cases, our product selection will be complicated by a combination of objective and subjective factors. In such cases we must balance considerations like heat transfer with more subjective questions like health and comfort. We must then make estimates of the relative importance of these factors, and decide on ratings for each product alternative. This process, which is similar to that used for idea screening in Chapter 3, often makes technically trained persons uncomfortable. It makes us as authors uncomfortable. Still, those making these estimates, including us, are reassured that the results so obtained often seem reasonable. The basis for our product selection is exposed.

In some cases, however, we may not be sure that all aspects of all product options will work. For example, we may be uncertain of the details of a chemical synthesis, or unclear as to how the synthesis can be accomplished at greater than laboratory scale. We may not know if the mixing can be as quick as we need, or if the heat transfer to a fluid of unusual rheology will be as fast as predicted from standard correlations.

In these cases, we are selecting between products with varying degrees of risk. We consider this risk in our selection in two ways. First, we must judge how serious a particular risk is, and how much this risk will affect our product. Often, this risk assessment means that we are making estimates of extra product-development money and of additional product-development time. Second, we will want to reduce the risk as much as we can, perhaps by some quick experiments. Such risk reduction seeks to manage our chances of success. These two considerations, risk assessment and risk management, are the subject of the following paragraphs.

4.4–1 RISK ASSESSMENT

The assessment of risk most commonly involves three steps. First, we must identify and catalog all risks. Second, we must decide if these risks can be estimated with engineering tools, or if they simply generate uncertainty. Third, we must compare our possible products in terms of both cost and time.

The identification of risk begins by making a list of any possible difficulties. Making this list includes the same techniques used in the generation of ideas. We must discuss the risks with our core team and with others in our organization, especially those in manufacturing, who up to now may have been less involved with product design than other groups. We must again contact our customers, especially the lead users of our product. We may again check with consultants, looking for problems outside our own organization's experience.

Armed with this list of risks, we on the core team must then choose a probability and a consequence of each risk. We can think of the probability and consequence of a risk as being like a man jumping over a chasm. The probability of something going wrong depends on the width of the chasm, whereas the consequence if it does go wrong depends on the depth of the chasm. The probabilities should range from zero to one. If the probability of the risk happening is negligible, it should have a score of less than 0.3. If the probability of the risk is significant, it should have a score around 0.5. If the probability is likely, it should be scored above 0.9.

Those in nuclear engineering have much more exact definitions for assigning these probabilities. While we do not think these are required for most chemical products, we do think that these definitions are worth mentioning. For example, a probability less than 0.01 is defined as happening once in 10,000 years. A probability of 0.7 will happen once in 10 years, that is, within the normal lifetime of chemical-process equipment.

In addition to the risk's probability, we must assign consequences of the risks. Again, we choose a scale which varies from zero to one. Defining this scale is important: a low score means that the consequence is minor, and a high score means the consequence is major. Thus, if the consequences of a risk are small, it should have a score of less than 0.3. If the consequence is significant, it should have a score around 0.5. If the consequence is severe enough to kill the project, it should have a score greater than 0.9.

The evaluation of the probability and the consequences should largely be the responsibility of the core team. As before, we suggest that each individual member assign values separately, and that the team reach its joint assessments by consensus. Assigning values as individuals forces every member of the core team to think about each risk, and to consider what he himself knows and what he does not. That prepares each member for the next core team meeting. In the meeting itself, core team members can reassess their scores, often by seeking information from those with different expertise. Again, we stress our belief that evaluation by consensus is better than evaluation by simple averaging. Consensus demands discussion.

Once the core team have agreed on the probabilities and the consequences, we must define the level of each risk:

$$[\text{risk level}] = [\text{risk probability}] \times [\text{risk consequence}] \qquad (4.4\text{--}1)$$

We will normally decide to analyze in detail only those risks which are above a specified level, perhaps 0.5. The risk level of such "significant" risks is clearly arbitrary, and again must be decided by consensus.

This combination of risk and probability concerns us deeply. If we are making a comparison of similar chemical products which have relatively little risk, we find this "risk-level" concept reasonable. However, if we are comparing secure technologies with those which may fail completely, then the simple combination suggested above seems questionable. We draw no conclusions from this concern; we only urge caution when one product idea really has a large chance of failure, or the consequences of failure are very high (for example, a serious explosion).

The risk levels defined above do focus our attention on the key concerns with each possible product. We must now separate these into categories, those which can be clarified by chemical or engineering analysis, and those which cannot. Examples of the risk levels which can be analyzed are a heat transfer correlation which might be inappropriate, or a chemical reaction which could be slower than expected, or a product which could be unexpectedly viscoelastic. In these cases, we can judge how serious the risk level is using chemical and engineering analyses like those in the first three sections of this chapter. Often, these risks will be evaluated not by the core team, but by those specialists who are supporting the core team.

The high-level risks which cannot be clarified by chemical or engineering analysis must be evaluated by the core team itself. These risks will include the consequences of the market place and of politics. For example, we will be at risk if our chemical raw materials are only available from one supplier. We will be at risk when our manufacturing requires new licenses, or when we expect litigation from the local community to delay our facilities' expansion.

For these risks, the core team must guess the extra time and money which such risks imply. Doing so requires our judgment, based on our collective experience. For example, if we suspect heat transfer correlations are in error, we may judge that it will take one engineer six months to develop corrected correlations. If we expect a public hearing about our request for a building permit, we may anticipate a three month delay from experience of earlier struggles.

4.4–2 RISK MANAGEMENT

Having identified and tried to quantify both the risks and their potential consequences, we must decide on the appropriate response. We have two choices:

(1)   Reduce the risk before proceeding with product development
(2)   Accept the risk and proceed immediately.

The first strategy is the more traditional method of risk management in product development; it is outlined below. It will often require developing several different ideas in order to establish which is the most effective. This method works well, but it is slow and can be expensive.

The second strategy is also frequently adopted. Speed and especially time to market are important considerations in chemical product development. The emphasis on speed is why many companies are moving towards a project-based organization. Risk reduction involves research, experiments, process design, and market testing, all of which are time consuming. Delaying is in itself a commercial risk; a competitor may reach the market first and even if their product is inferior, they are then likely to gain most of the market share.

There is an analogy here with mountain climbing on glaciers. Climbing courses will correctly put a strong emphasis on safety: extensive use of ropes, ice screws, and snow stakes will be encouraged. However, using these safety measures takes time, and the safety gear is heavy, slowing progress further. The longer one spends on a glacier, the more chance there is of being caught in an avalanche; this is increasingly true later in the day as the sun softens the snow. To a considerable extent, safety is speed and delay can be fatal. Similarly, delay in product design, even for good reasons of risk management, can kill our product ideas. For this reason, having evaluated the risks inherent in our different product ideas, we may wish to proceed directly to the manufacturing stage.

This approach often makes good commercial sense. We may risk losing less money by proceeding with a product idea which might not work than by delaying and risking loss of market share. If it is a simple question of balancing financial considerations, we are happy with this approach. We are more uneasy when we are balancing the risk of delay against considerations such as health, environment, or safety. This is perhaps why these areas are heavily regulated to ensure companies are not tempted to cut corners to speed product development.

The final aspect of risk concerns our efforts to resolve the most major pitfalls before we become overcommitted. This is sometimes phrased as two guidelines:

(1)   If the risk is high, keep the investment low. As the risk decreases, raise the investment.
(2)   Break the risk into increments, deciding where you will stop work if unsuccessful.

For example, if we are developing a new device which carries out a chemical change, we could proceed in three steps. First, we repeat our engineering estimates, removing the most major simplifications, and using physical property estimates which are pessimistic. If the idea still looks good, we build a realistic model at a scale convenient for the laboratory. We use experiments with the model to see if our estimates are reasonable. If things look good even now, then we build a pilot scale model with which we can get the data for a final selection.

Table 4.4–1 *Risk during new drug development.*

Status	Chemical status	Quality status	Risk
Preclinical efforts	Major process work needed	Few methods available	High
Phase I clinical trials	Laboratory procedures available	Analytical development necessary	High
Phase II to Phase III clinical trials	Pilot plant production	Analytical methods in place	Moderate
Late-stage clinical trials	Production process fixed	Methods validated	Low
Mature product	Plant process available	Quality control key	Low
Generic drug	Patents available	Methods sometimes available	Moderate

Source: Charles M. Boland, Cedarburg Laboratories, quoted in *Chemical and Engineering News* Feb. 14, 2000.

As a second example, consider how risk wanes during development of a new drug, as suggested schematically in Table 4.4–1. Risk is highest during the initial development, when the efficacy of the drug is uncertain. If clinical trials continue to be positive, risk drops as the drug's synthesis is better established and the drug's value become more certain. Risk is smallest for a mature drug, produced at carefully monitored quality. Ironically, risk is increased for a generic drug, because a new manufacturer will know the patents, but not the trade secrets, for the drug's manufacture.

EXAMPLE 4.4–1   POWER FOR ISOLATED HOMES

In many European countries, electricity companies are required to provide power to homes at a fixed connection fee and standard cost per unit consumed, regardless of their remoteness. Laying many kilometres of cable to connect a single house to the national grid is clearly uneconomic.

Investigate alternative sources of electric power for isolated homes.

SOLUTION

We briefly review how one might follow the design template suggested in this book to reach the stage at which risk should be considered.

Needs. We will not attempt to provide electric heating, but will aim to fulfill all other normal domestic requirements, such as cooking, lighting, cooling, etc. A little research indicates typical power requirements to average 3 kW, with a peak loading of 15 kW (mainly a result of cooking). This provides our specification.

Ideas. There are a very large number of ways of generating electricity, some obvious (such as hydroelectric power), others more bizarre (natural gas from manure). Idea generation and initial screening might lead one to consider four leading contenders: a diesel generator, wind power, solar power, and a fuel cell.

Selection. For us, as the electricity provider, the primary selection criterion is going to be cost, both in terms of capital and the running cost of providing the specified power. (Remember we can only charge the standard, national rate.) Clearly our solution must also be acceptable to the consumer; aesthetic and environmental considerations will also be significant.

Diesel generator. A 15 kW generator costs around $6500. Operating costs can easily be estimated from the price of gasoline and an assumed efficiency of 30%. The combustion of gasoline is

$$\tfrac{1}{8}C_8H_8 + \tfrac{25}{16}O_2 \rightarrow CO_2 + \tfrac{9}{8}H_2O \quad \Delta H = -733.8\,\text{kJ/mol}$$

At the efficiency given, this suggests that

$$\frac{733.8\,\text{kJ/mol}}{\tfrac{1}{8}\,(104\,\text{g/mol})} \times 800\,\text{g/L} \times 3.76\,\text{L/gallon} \times 0.3/\$3\,\text{per gallon}$$
$$\approx 17{,}000\,\text{kJ per}\$$$

In a year, we need

$$3\,\text{kW}(3600\,\text{s/hr})24\,\text{hr/day}\,(365\,\text{days/year})$$
$$= 95 \times 10^7\,\text{kJ per year}$$

Thus our fuel cost for the diesel generator is

$$\frac{95 \times 10^7}{17{,}000} = \$5600\,\text{per year}$$

While noise might be an issue, we can expect a diesel generator to work well. Since it is very well established technology, we will use it as our benchmark.

Wind power. A 3 kW generator has 3 m diameter blades and costs at least $5000. We must make use of a battery to provide the peak power load. (This will probably be necessary anyway to smooth the uneven wind energy.) We need two such generators. Once installation and battery costs are included, we are unlikely to escape with capital costs under $20,000. Running costs will be negligible. Aesthetics could be an important issue.

Solar power. Even the most efficient solar cells only manage to convert around 10% of the absorbed solar energy. Furthermore, we can only use the cells on average 12 hours per day, less in winter when power requirements will be high. Incoming solar energy is about 100 W/m^2 in much of Europe. To supply our 3 kW average power, we will need:

$$\frac{\left(\dfrac{24\,\text{hr}}{12\,\text{hr}}\right)\,3\,\text{kW}}{\left(0.1\,\dfrac{\text{kW}}{\text{m}^2}\right)\,0.1\,\text{efficiency}} = 600\,\text{m}^2\,\text{of solar panels!}$$

Again, we will require a battery, the efficiency of which we have ignored. In winter, the power available will be lower. Solar panels cost around $100 per m^2, giving a capital cost of at least $60,000. We reject this idea on economic grounds.

Table 4.4–2 *Risk assessment for wind power* The most serious risks are regulatory and reliability.

Risk	Probability	Consequence	Risk level
Customer acceptability	0.5	0.5	0.25
Regulatory acceptability	0.5	0.7	0.35
Maturity of technology	0.1	0.3	0.03
Reliability	0.7	0.3	0.35

Fuel cell. While the running cost will parallel that of the diesel generator, we can now avoid the limitations of Carnot efficiency and hope to reach 70% efficiency. Therefore running cost is

$$\$5600 \times 0.3/0.7 = \$2400 \text{ per year.}$$

Fuel cells are still substantially more expensive than the equivalent conventional generator: we can expect to pay at least $15,000.

A decision matrix based on the considerations just outlined might leave three contenders: the diesel generator, wind power, and the fuel cell. Before immediately proceeding with the idea which seems most attractive at this stage we must consider risk.

In this case, the major risks are:

(1)   customer acceptability, including noise and environmental considerations;
(2)   regulatory acceptability, including pollution and planning permission;
(3)   maturity of technology; and
(4)   reliability.

The third of these takes account of the effort likely to be required to bring the technology to the level where it can be slotted into our application. This risk applies primarily to the fuel cell, a relatively unestablished technology. Fuel cells operate reliably on a large scale with hydrogen fuel, but considerable uncertainty remains about running them at small scale in remote locations with gasoline. The final risk listed reflects the likely costs of repairs and maintenance. Here the wind generator, vulnerable to storms, is likely to be the biggest problem.

Our risk assessment for wind power is given in Table 4.4–2. We are unsure about reducing these risks. Planning restrictions are severe in most of Europe. We cannot change the chances of wind turbines being rejected on aesthetic grounds. Similarly it is hard to alter negative consumer reactions to these unsightly and mesmeric objects. Research may help in improving reliability, but on the whole the wind power option looks risky. We are unlikely to pursue this option.

This example illustrates that the quantification procedure is a good way of identifying which risks need careful consideration. Deciding what to do about them will depend more on our assessment of how they can be mitigated and what the alternatives are than the numerical assessment itself. In one case a risk factor

> **Table 4.4–3 *Risk assessment for the fuel cell*** These risks assume
> hydrogen can be handled safely.
>
> | Risk | Probability | Consequence | Risk level |
> |---|---|---|---|
> | Customer acceptability | 0.3 | 0.5 | 0.15 |
> | Regulatory acceptability | 0.1 | 0.3 | 0.03 |
> | Maturity of technology | 0.5 | 0.7 | 0.35 |
> | Reliability | 0.5 | 0.5 | 0.25 |

of 0.4 might result in the project being stopped, while in another, we might decide
to continue with the same calculated risk factor. The numbers are there to help
us – in particular to make sure important risks are not ignored – not to constrain
our decisions.

Our risk assessment for fuel cells is in Table 4.4–3. Maturity of the technology
and reliability are significant risks. While both of these could be mitigated by
further research, we should proceed by installing reliable diesel generators in the
short term. At the same time, we may decide to pursue research on the fuel-cell
option.

EXAMPLE 4.4–2 TAKING WATER OUT OF MILK AT THE FARM

Remote dairies in New Zealand can face major expenses in shipping their milk
to a central processing facility, where the milk is largely made into cheese. These
dairies would benefit from a method of concentrating the milk on the farm,
removing only water. For a typical farm, this means reducing 4000 kg/day of raw
milk to about 1000 kg/day of concentrate. How we might do so is the focus of this
problem.

Our efforts to resolve this problem have focused on four unit operations: evaporation, absorption, spray drying, and reverse osmosis. Evaporation is the best
established, and is used for products like "evaporated milk" and "condensed
milk." It requires careful energy integration.

Absorption of water in inorganic or organic gels has significant problems. The
inorganic gels which are selective require a lot of energy for the regeneration
required for reuse. The organic gels – like polyisopropylacrylimide – are easily
regenerated but are not sufficiently selective. Spray drying works well only with
a feed of 50% solids, much more than that in raw milk. Reverse osmosis membranes foul too easily.

Thus, our best idea is evaporation. From an extensive energy analysis, not
included here, we decide to run the evaporator at 60 °C, using 64 °C steam. The
steam is produced by sending the 60 °C evaporated water through an electrically
driven heat pump. (We should remember that a heat pump is approximately a
Carnot engine run backwards, using work to move heat up a temperature gradi-
ent.) The use of a heat pump reflects the fact that hydroelectric power generation
is common in New Zealand, and so electricity is relatively cheap.

Within our choice of evaporation, we have three possible forms of evaporators. The first is the conventional, falling-film unit, whose performance is well established and which is the sensible benchmark. The second is the centrifugal evaporator, which uses centrifugal force to stabilize thin milk films and hence improve evaporation efficiency. This method works well but the equipment is expensive. The third is a membrane evaporator, where the milk films are stabilized between membranes, which can impede evaporation. This membrane method has not been carefully explored and so has considerable risk.

Select which of these ideas is best.

SOLUTION

The solution to this problem implies a total of five steps. The first step is to determine the general specifications which any evaporator must meet. The next three steps are to find the size and cost for each of the three evaporators. The final step is to consider the risk, which in this case is largely associated with the membrane evaporator.

General specifications. We must first specify the general heat transfer characteristics of any successful evaporator. Doing so depends on choosing values for the physical properties of milk. Because the evaporation increases the concentrations of milk solids and non-volatiles, the viscosity increases from 0.9 cp to around 10 cp during evaporation. We will include this change in our calculations, but will assume that other properties of the milk remain close to those of pure water. Thus, the milk's density is taken as 1000 kg/m^3 and its thermal conductivity is about 0.60 W/mK.

The total heat transferred Q is proportional to the mass evaporated N_1:

$$Q = UA\Delta T = \Delta \hat{H}_{vap} N_1$$

where U is the overall heat transfer coefficient; A is the evaporator area; ΔT is the temperature difference, in this case 4 °C; and $\Delta \hat{H}_{vap}$ is the specific heat of vaporization at 60 °C, here about 2430 kJ/kg. Because N_1 is 3000 kg/day or 0.035 kg/sec,

$$UA = 21\,\text{kW/K}$$

But

$$\frac{1}{U} = \frac{1}{h_{steam}} + \frac{1}{h_{wall}} + \frac{1}{h_{milk}}$$

where h_{steam} is the individual heat transfer coefficient of the condensing steam, around 5000 W/mK; h_{wall} is that of the evaporator surface, typically 20,000 W/mK; and h_{milk} is that of the milk itself. We assume that this is given by

$$h_{milk} = \frac{k_T}{\delta}$$

where k_T is the thermal conductivity of the milk, and δ is the milk film thickness. Thus if we can estimate δ, we know h_{milk} and hence U, and so can find the area of a particular evaporator. This will be the key parameter in our selection.

Falling-film evaporator. The first unit we consider is the conventional falling film evaporator. In this unit, the film of milk must spread smoothly over the evaporator surface in order to use all of the surface efficiently. Such a smooth film means that the Weber number *We* must be greater than a critical value of 2:

$$We = \frac{\rho v^2 \delta}{\sigma} \geq 2$$

where ρ is the milk's density, v is its velocity, and σ is its surface tension. For a falling film,

$$v = \frac{\rho g \delta^2}{3\mu}$$

where g is the acceleration due to gravity and μ is the viscosity. Combining gives

$$\delta = \left(\frac{18\mu^2\sigma}{\rho^3 g^2}\right)^{1/5}$$

$$= \left(\frac{18\,(0.1\,\text{g/cm sec})^2\,30\text{g/ sec}^2}{(1\text{g/cm}^3)^3\,(980\,\text{cm/ sec}^2)^2}\right)^{1/5}$$

$$= 0.09\,\text{cm}$$

To make sure we have a stable film, we assume we want about twice this value, or

$$\delta = 0.2\,\text{cm}$$

From the above, we then find that h_{milk} equals [(0.60 W/mK)/0.002 m], U is about 280 W/m²K, and the evaporator area A is

$$A = 75\,\text{m}^2$$

This evaporator area, the benchmark for our selection, is large because the temperature difference is small (4 °C).

Centrifugal evaporator. The centrifugal evaporator uses centrifugal force to keep the milk film smooth, thin, and stable. As the milk film moves outwards on the centrifuge discs, its higher viscosity caused by evaporation is more than balanced by the increased centrifugal force. While the details of the fluid mechanics are beyond the scope of this book, the result is that the average film thickness is about

$$\delta = 25\,\mu\text{m}$$

Parallel with our earlier arguments, we now find that h_{milk} equals [(0.60 W/mK)/ 25×10^{-6} m], U is about 3400 W/m² K, and the evaporator area A is

$$A = 6\,\text{m}^2$$

Using a centrifugal evaporator cuts the surface area required for evaporation by over ten times.

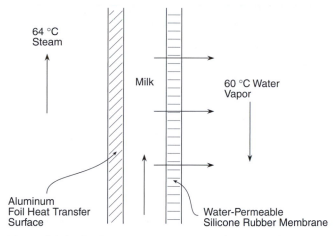

Figure 4.4–1 *Membrane evaporator* Water in warm milk evaporates across the thin membrane shown. Because the membrane is selective, volatile flavors are not lost.

However, this dramatically reduced area is dearly purchased. The only serious estimate which we could obtain for building a centrifuge like this was over $50,000. This seems too expensive for most farmers. As a result, we turn to the third method for evaporation.

Membrane evaporator. Like the centrifugal evaporator, the membrane evaporator can sustain very thin, stable milk films during evaporation. As shown in Figure 4.4–1, the thin films are now sustained not by centrifugal force but between two thin membranes. One of the membranes is a metal foil which transfers heat from the 64 °C steam to the 60 °C milk. This membrane has a heat transfer coefficient around 20,000 W/m^2 K.

The other membrane, which separates the 60 °C milk from the 60 °C steam produced by the evaporation, is the barrier for the evaporation. Interestingly, its heat transfer resistance and mass transfer resistance are predicted to be negligible under these conditions. If this is true, then the significant resistance to heat transfer must be the film of milk itself. In many membrane devices like this, the two membranes are held apart by a spacer, which fixes the thickness of the milk film. Typically, the thickness δ of this spacer is

$$\delta = 600\,\mu m$$

By the same arguments as before, h_{milk} is [(0.60 W/m^2 K)/6 × 10^{-4} m], U is about 800 W/m^2 K, and the evaporator area A is

$$A = 26\,m^2$$

This is one third the area of the falling-film evaporator, but four times the area of the centrifugal evaporator. Significantly, membrane experts agree that membrane modules like this can be built for about $10/m^2, independent of the chemical structure of the membrane used. Thus, we should be able to build a membrane evaporator for less than $1000. Such a system is attractive commercially.

Table 4.4–4 *Risk assessment for the membrane evaporator.*

Risk	Probability	Consequence	Risk level	Mitigation
1. Difficult to make heat transfer membrane	0.1	0.5	0.05	Use parallel heat exchange technology
2. Difficult to make evaporation membrane	0.3	0.5	0.15	Existing data suggest, at most, required membrane area doubles.
3. Cannot easily manifold the module	0.5	0.2	0.10	Can mitigate with larger steam channel.
4. Evaporation flow is slow	0.5	0.2	0.10	Use larger membrane spacer in steam channel.
5. Cannot sterilize effectively	0.3	0.9	0.27	Chemical cleaning preferred, but requires no dead spots.

Risk assessment. The three evaporators discussed above show a vivid contrast of advantages and disadvantages. The traditional thin-film evaporator has the largest area because it operates with the thickest milk film. The centrifugal evaporator has a very small area but a very high price. The membrane evaporator has a moderate area and a very low price, but it may not work. Using the membrane evaporator is risky.

Five of the major risks of the membrane evaporator are shown in Table 4.4–4. The first, that we have trouble making the heat transfer membrane, is unlikely because there are already foil-based heat exchangers on the market. The obvious strategy is to use the manufacturing procedures of these foil exchangers as a guide. The second risk, that the membrane across which evaporation occurs offers a major mass transfer resistance, is more serious. While such trouble would be inconsistent with earlier studies of membranes with high permeability, we suspect that the membranes used in those studies may be difficult to make in the large, flat sheets needed here. However, even if the water permeability is only 20% of that reported earlier, the membrane area required increases only slightly.

The other risks depend on the design of the evaporation module. The third risk concerns the design of the inlets and outlets, and should not be especially difficult to resolve. The fourth risk reflects the concern that the evaporated water will not easily flow out of the module. This is easily mitigated by using a larger membrane spacer in the steam channel. Sterilization of the milk channel is the most severe risk. While the membranes may not be able to stand high temperature, most farms use chemical cleaning anyway. We must ensure that sterilization is complete, without any dead spots. Even this risk, scored as the most serious, does not seem crippling. We should build a prototype and show by experiments if this new but risky idea merits selection.

4.5 Conclusions and the Third Gate

This chapter aims at selecting the best one or two product ideas to prepare for product manufacture. Before this point, we have identified product needs and ideas which might fill these needs. We have qualitatively screened these ideas until we are left with five or so leading possible products. We want to choose the best one or two of these products.

How we do so has been described in this chapter. The methods used for product selection often use quick estimates of thermodynamics and rate processes. These estimates are especially useful when we are trying to improve existing products, either with new ingredients or with more responsive systems.

The challenge comes when we compare new, very different products with improved but familiar ones. In this case, we return to concept selection matrices, which attempt to weigh the relative importance of different attributes. We must also consider the importance of risk, for a new proposed product may not work as well as we hope. We must consider what new engineering and chemistry are needed to mitigate the risk.

After these efforts, we will have chosen our best potential product, and be ready to consider how it can be manufactured. But first, we must face a third management review.

This third management review, or "gate," will be by far the hardest and most critical which we encounter. Remember that the first gate was early in the process, when we had first developed our specifications. At that point, management will not be especially critical, often because they had a role in identifying the original need. The second gate came after idea generation and sorting, when we had produced five or fewer good ideas. Again management will remain relatively uncritical, partly because they will see merit in a few of the ideas, and partly because they may not understand the chemical details of the new product ideas. Remember that management consultants feel that these two first gates are often too casual, and that many product design projects are allowed to continue for too long.

The third "gate" will be the hardest because we on the core team will be asking for a lot of money. Management may not understand chemistry, but they do understand money. As before, we should write a report and prepare an oral presentation. Now, however, we are going to be critically examined, even grilled. This is the stage at which the product development is most likely to be cancelled. But if our product still looks good, we will be ready to consider how it can be manufactured. This is the subject of the next chapter.

Problems

1. *Comparing fruit.* Construct a selection matrix to compare the merits of apples and oranges.

2. *Offshore wind energy.* This millennium has seen an explosion of interest in renewable energy resources and the construction of many wind-powered

turbines in offshore sites. A fundamental problem of windpower is its variability, which has led people to consider battery systems connected to wind farms in order to deliver a steady supply of electricity. Batteries are simply a way of storing electrical energy as chemical energy, to be released as needed. The task for this project is to consider the scope for windfarms to use their variable electricity supply to generate chemicals in stand-alone units offshore, which can then be pumped out or collected as needed. Set specifications which must be met by any such chemical energy storage ideas. Use these to assess the following ideas for storing intermittent wind energy:

(1)   Electrochemical generation of hydrogen from seawater.

(2)   A water tower for each wind turbine.

3.   *Warm seats.* You want to test two possible ideas for a cushion to keep your seat warm during a three-hour sports event. The first idea is a pad containing rechargeable batteries; the second is for a pad filled with an inorganic hydrate which freezes at a comfortable temperature, releasing its heat of fusion. Quantitatively assess these two ideas.

4.   *Concentrating cork liquids.* Cork boiling wastewaters contain phenolic compounds which have potential value in the tanning industry. However, these liquids are too dilute to be useful without concentration. Concentration by evaporation is too expensive and tends to degrade the chemicals which are important to tanning. Concentration by reverse osmosis, which forces the solution through a non-porous membrane, is more attractive. Table 4 of the article "Sustainable membrane-based process for valorization of cork boiling wastewaters" [A. R. S. Teixeira, J. L. C. Santos, and J. G. Crespo, *Separation and Purification Technology* **66** (2009), 35–44] gives two membranes which have been studied for this process. Which is better?

5.   *Batteries vs. fuel cells.* A company has asked for proposals to develop a cordless, wall-mounted, power supply, giving six volts presumably for security cameras, smoke detectors, and other equipment. The new power supply should replace four D-cells, which last about six months. The company states that the benchmark gives 0.55 mA continuous current and 15 1.2 second, 800 mA bursts per hour. The new power supply should last twice as long; if it needs to be replaced, the procedure should take only a couple of minutes. The company seeking this technology has focused on three ideas:

(1)   a better battery;

(2)   a small fuel cell;

(3)   a battery supplemented by a solar cell running off building lights, which recharges the battery.

Compare the three possible technologies including the benchmark. Choose the candidate that you think comes closest to meeting the company's criteria. Discuss the first experiments you would make to mitigate the risk of your alternatives. To aid in this comparison, a table of different power sources is given in Table 4.5A.

Table 4.5A *Summary of alternative power sources.*

	Voltage	Energy/volume	Charge time	Cycles	Cost
D-cells	1.5 V	60 W hr/L	2 hr	1000	$15
Lead acid	2 V	40 W hr/L	8 hr	1000	$10
Lithium	3.6 V	100 W hr/L	1 hr	1000	$50
Hydrogen fuel cell	0.8 V	10 W hr/L	2 min	Large	$500
Methanol fuel cell	0.5 V	80 W hr/L	2 min	Uncertain	$500

6. *Snacks in aeroplanes.* A popular brand of potato snack is packaged in a rigid cardboard tube. An airline wishes to serve it as a snack on its flights. Unfortunately, due to the reduced cabin pressure at altitude, the lids of the snack can blow off prior to serving, resulting in significant wastage. Four solutions are under consideration:
   (1) Pack the tubes at reduced pressure, by including carbon dioxide in the gas mixture and subsequently absorbing it using a sachet of sodium carbonate included in the tube.
   (2) Include a pressure relief valve in the lid.
   (3) Make the lid into a screw cap so it will not blow off.
   (4) Serve a different snack.
   Select between these options.

7. *Disposal of spent coffee grounds.* A large plant for making instant coffee produces around 20,000 tons (dry weight) per year of waste. The waste is spent ground-up coffee beans and consists mainly of woody material (cellulose, lignin, etc.) and insoluble oils. The coffee manufacturer wishes to evaluate environmentally benign alternatives to the current means of disposal which is landfill. The alternatives have been whittled down to three:
   (1) Incineration for energy generation (the plant uses a large amount of steam).
   (2) Anaerobic digestion to produce methane.
   (3) Gasification followed by Fischer–Tropsch to produce hydrocarbons.
   Select the best solution.

8. *Solar panel for new builds.* The UK government has just introduced legislation which guarantees a price of £0.41 per kW hr for electricity generated domestically by solar power. Similar schemes exist or are proposed in other countries. This is resulting in increasing viability for solar generation of electricity at home. Most solar-panel offerings for domestic situations are aimed at retro-fitting existing houses, but your company feels that there is a space in the market for a solar-electricity generating system designed specifically for new buildings. Three devices are under consideration:
   (1) The solar brick. This will have the size, shape and structural properties of a standard brick or breeze block, but will include a polyvinyl

coating externally and all the required electronics within the brick to allow connection directly to the mains system.

(2)   The solar window. This device will absorb some light to generate electricity, but will also transmit light into the house. It will be particularly useful in buildings suffering from excessive solar thermal gain but requiring natural illumination. Such devices already exist for the south-facing side of greenhouse roofs where a reduction in solar intensity is required.

(3)   The solar roof. A solar panel would become part or all of the roof. Insulation, integration into standard roofing materials, sealing, and easy replacement are key issues.

Using a standard bolt-on polyvinyl panel system as a benchmark, select between these alternatives for your company's first commercial offering.

9.   *Extending applications for plastic money.* In 1988, Australia issued its first polymer note and subsequently became the first nation in the world to convert entirely from paper to polymer bills. The substrate of Australian currency is Guardian®, a biaxially oriented polypropylene, produced by Securency Ltd. This material, which is not available commercially, has a distinctive feel, so counterfeits cannot be made with the use of other polymers and color copiers. Australian currency also uses unprinted transparent windows where the substrate shows through the bill. The bills also use micro printing, raised printing, symbols that appear only when held to light, and serial numbers that fluoresce under UV light. While the polymer notes cost twice as much to manufacture as paper bills, they can remain in circulation four times longer.

Your core team is interested in extending the benchmark co-extrusion technology used for Australian currency to other product areas. The core team has decided to concentrate on five areas:

(1)   A multilayer adhesive to apply to documents for authentication.

(2)   An optical currency substrate that uses multilayer technology.

(3)   A polymer currency substrate similar to the polypropylene Australian currency currently uses.

(4)   A passport identification page made of polymer.

(5)   A surface treatment.

Select the area which you think has the greatest promise, and describe the information that you will need to evaluate this promise more completely.

10.   *Mining the tire mountain.* Dealing with waste tires has become a major problem at local, regional, national, and international levels. As a whole, industrialized countries dispose of about 6.5 million tons of waste tires per year: most of this waste goes directly to landfill. Landfilling is not only a harmful, unsustainable method of disposal but it is also wasteful. The chemicals and the energy stored in the tires are permanently lost when these are buried. Moreover, other components of the tires (flame retardants, stabilizers, colorants etc.) can leach from the bulk and be released to the environment

causing further problems of pollution in water and soil. It is clear that better ways to dispose of used tires are needed for environmental, political, and economic reasons. Your core team is considering four possibilities:

(1)   Granulation followed by road surfacing.
(2)   Use as thermal insulation.
(3)   Thermal pyrolysis to useful products.
(4)   Microwave pyrolysis to useful products.

Develop these ideas in more detail and select the best.

11.   *Coolers for oil platforms*. The size of oil and gas exploration platforms is extremely sensitive to the size and weight of the equipment to be supported above sea level. Key components of gas production platforms are the gas compressors used to transport the product to shore; their efficiency and size is optimized by the use of intercooled multistage units. The coolers, however, are often larger than the compressors – one prevalent existing technology is to use banks of fin-fan coolers (air/gas or air/liquid coolant), which are limited by the maximum obtainable rates of heat transfer to the surrounding air. More efficient cooling solutions are under investigation. The following are the most promising three alternatives to fin-fan coolers:

(1)   Packed bed cooling. The hot gas is countercurrently contacted with pure, cool water. The pure water is then cooled by seawater.
(2)   Direct injection of liquid coolant. Latent heat and sensible heat are transferred to the added liquid, thus cooling the natural gas.
(3)   Heat pipe. Heat is transferred from the evaporation section of a heat pipe to the condenser section using the equilibrium between liquid and vapor water under vacuum conditions.
(4)   No cooling. Just lag the pipeline and let it cool as it is piped away from the platform under the sea.

Assess these ideas relative to the benchmark of fin-fan cooling.

Data:

Natural gas enters the cooler at 170 °C and leaves at 90 °C, 50–150 bar. 100–250 million ft^3 of gas must be cooled per day.

12.   *A mini Haber process*. Ammonia is largely made by the Haber process, an early triumph of industrial chemistry. This process burns natural gas to produce a mixture of nitrogen, hydrogen, and carbon dioxide. After the carbon dioxide is removed by amine scrubbing, the nitrogen and hydrogen are reacted at 450 °C and 200 atm over an iron catalyst to produce ammonia. Argon in the product must be purged; unreacted N_2 and H_2 are recycled.

While the Haber process has been optimized by almost a century of effort, the extreme reaction conditions require large capital investment in large plants, and a complex distribution system to deliver the ammonia to farmers. As a result, your company is interested in developing small ammonia plants for individual farms. Your development team has decided to focus on two technologies. The first is the Haber process itself, scaled down and carried out in high-temperature, high-pressure microreactors. The second is a high-temperature, atmospheric pressure electrochemical process [*Science* **282** (1998), 98–100].

In this process, hydrogen passes across a nickel anode to make protons. These are electrochemically driven across an inorganic membrane separator at 570 °C to react at a palladium cathode to produce ammonia. The reaction is 78% efficient; 22% of the protons are lost as heat. Still, a 78% conversion is much better than the Haber process, where conversion is typically around 20%.

Set up a selection matrix to help decide if you want to pursue development of either of these processes.

13. *Pipeline plugging.* Wax deposition in sub-sea pipelines which are used for the transport of crude oil can cause flow reductions or even blockages. This problem has been estimated to cost oil companies $100 million annually. The main challenge in clearing such pipeline blockages is in supplying heat to downstream regions of the pipeline in order to melt the wax, which is then dispersed using injected surfactants. In sub-sea pipelines the oil is continually cooled by the sea at 4 °C and blockages may be tens of kilometers downstream of the point of injection. It is therefore necessary to provide the heat selectively at the point of blockage. It is proposed to achieve this by using a fused exotheric chemical reaction; that is one with a built-in, controlled delay before commencing. The reaction proposed is:

$$NH_4Cl(aq) + NaNO_2(aq) \rightarrow 2H_2O + NaCl(aq) + N_2(g)$$
$$\Delta H = -334.5 \, kJ/mol.$$

 The reaction is catalyzed by acid.

Two methods of achieving and controlling the time delay are being suggested:
(1) Encapsulation of the reactants and catalyst in separate capsules which then disintegrate over a controlled time. The reaction is initiated when the catalyst mixes with the reactants.
(2) Pulses of reactants and catalyst are separated by an inert spacer – contacting then occurs as a consequence of dispersion as the material travels down the tube.

Consider criteria on which to judge the methods and so justify selection of one of them. Suggest also information which would be needed before manufacture of the product could proceed – what experiments might you wish to do?

[P. Singh and H. S. Fogler, *Industrial Engineering and Chemical Research* **37** (1998), 2203–2207 and

K. Subramanian and H. S. Fogler, Proceedings of the 4th Italian Conference on Process Engineering, Florence, 1999, pp. 43–46.]

14. *Stuck on gum.* Chewing gum goes back at least to the ancient Greeks, who chewed mastic gum, a resin extracted from the bark of the mastic tree, which

grows on the island of Chios. It was particularly popular with ancient Greek women as a tooth cleaner and breath freshener.

Modern chewing gum production dates back to February 14 1871, when US Patent # 111,798 was issued to Thomas Adams for a process to manufacture gum. His gum was made from chicle, a natural latex, chewed for centuries by Mexicans. Adams sold the first flavored gum, Black Jack (liquorice flavored), in New York in the same year.

Chewing gum has been popular the world over ever since. Clearly people like to chew.

Approximately 50% of chewing gum ends up on pavements or other public surfaces. At present most gums are based on polypropylene, a material so stable that it will last for 25 years. Removal is costly and unpleasant. The UK chewing-gum market is worth about £280 million per year and the cost of clean up is estimated to be £150 million. The government is considering legislation to pass this clean-up cost on to the gum-manufacturing companies, and your company is looking for ways to reduce these costs. Four ideas are under consideration:

(1) Provide paper "gum wraps" packaged with the gum itself, so that the gum could be wrapped before disposal. This would be combined with an awareness campaign to encourage consumers to dispose of their gum responsibly.

(2) A gum additive to reduce adhesion.

(3) Digestible gum, e.g. corn zein gum.

(4) Freezing gum – use a polymer with a T_g just below body temperature, so that it "freezes" when removed from the mouth, making cleaning much easier. One current method of removal from pavements is cryogenic freezing; the principle would be the same, but at ambient temperatures.

Select the best idea for further development by a major gum-manufacturing company.

15. *High-temperature oxygen separation.* Oxygen is one of the most important commodity chemicals made. It has three main uses. About 55% is used to make steel: a typical steel mill uses 2000 tons/day. Another quarter is used by the chemical industry to make ethylene oxide and ethylene glycol, intermediates used in polyesters. The last 20% is used in a spectrum of operations, including water treatment, medicine, and welding.

The demand for oxygen is expected to grow, driven especially by the goal of reducing carbon dioxide emissions. For example, if coal is burned in oxygen, the resulting flue gas is largely CO_2, so that capture and sequestration is greatly facilitated. Converting coal to liquid fuels also requires large amounts of oxygen. Oxygen producers are attracted by these new markets, but understand they will require lower-cost oxygen. For many of these new applications the oxygen need not be very pure, just concentrated. These new applications have made oxygen producers look beyond their traditional

Select the best idea – if any – for further development. Your selection should include calculations supporting your choice, summarized in tables and detailed in an appendix. Be careful to identify all assumptions, especially how long you expect the barrier to remain intact. In addition, identify the risks of your choice, and explain how you will mitigate them.

REFERENCES AND FURTHER READING

Crouhy, M., Galai, D., and **Mark, R.** (2006) *Essentials of Risk Management.* McGraw Hill, New York, NY.

Cussler, E. L. (2009) *Diffusion.* Cambridge University Press, Cambridge.

Hansen, C. M. (1999) *Hansen Solubility Parameters.* CRC Press, Boca Raton, FL.

Hildebrand, J. H., Prausnitz, J. M., and **Scott, R. L.** (1970) *Regular and Related Solutions.* Van Nostrand Reinhold, New York, NY.

Louvar, J. F. and **Louvar, B. D.** (1997) *Health and Environmental Risk Analysis.* Prentice-Hall, New York, NY.

McMillan, J. (1996) *Games, Strategies, and Managers.* Oxford University Press, Oxford.

Murray, R. L. (2000) *Nuclear Energy: An Introduction to the Concepts, Systems and Applications of Nuclear Processes,* 5th Edition. Butterworth-Heineman, Oxford.

Poling, B. E., Prausnitz, J. M., and **O'Connell, J. P.** (2001) *Properties of Gases and Liquids,* 5th Edition. McGraw Hill, New York, NY.

Ulrich, K. T. and **Eppinger, S. D.** (2007) *Product Design and Development,* 4th Edition. Irwin McGraw-Hill, New York, NY.

5

Product Manufacture

By this point, we are close to a decision on what product we will make and sell. We have identified a customer need, and we have quantified the need in terms of product specifications. We have sought a large number of ideas which could meet this need, and we have organized and edited these ideas until we have a manageable number. We have selected the best one or two ideas. Now we are close to deciding what we will manufacture.

The initial section of this chapter explores three aspects leading to product manufacture. The first, discussed in Section 5.1–1, concerns intellectual property. Often, our new product will include some aspects of invention. In these cases, we will want to consider whether or not to seek patent protection. Patents can give us an exclusive licence to market our invention, and hence command higher prices which let us more quickly recover our development cost. In return for this exclusive licence, we must make a full disclosure of what our product is, and how it works. Sometimes we will decide to seek patent protection, but sometimes we will choose to keep trade secrets.

In Section 5.1–2, we turn to assembling missing information required to realize our product. Sometimes, this information will be necessary to make sure our selected product will function as we expect. In other cases, it may be part of what is needed for any patent applications. Usually, the information must be obtained from actual chemical and physical experiments, which are almost always tedious. As a result, we are concerned in this section with planning experiments which are as efficient as possible.

The third aspect we must consider, discussed in Section 5.1–3, is the environmental impact of our product. This involves making a cradle-to-grave analysis of how the manufacture, use, and disposal of the product will impact our environment. This is complex, because there are many different criteria for environmental damage, including water use, pollutant discharge, and landfill requirements; and many different scales on which we might worry about these, from contamination of a local river to global CO_2 output. In considering the environmental consequences of our product, we must judge which of these patent application. Thus, any product developer who knows that the new product is useful and novel should move promptly to make the patent application.

As well as being useful and novel, a patentable product should be "non-obvious." This requirement means that the differences between the new product and earlier products – "the prior art" – must be sufficient that they are not obvious to one having "ordinary skill in the art to which the invention pertains." This non-obvious requirement introduces substantial uncertainty into the assessment of whether the new product merits a patent. The United States Supreme Court tried to sharpen this "non-obvious" requirement by urging inquiry into three areas:

(1) the scope and content of the prior art;
(2) the differences between the art and the claimed invention; and
(3) the level of ordinary skill in the field of the invention.

The Court also listed several secondary considerations. Clearly, we must supply information to patent attorneys, but we normally must depend on their judgment.

We can aid the process of patent application by keeping careful records of our development of the product. The standard is a hand-written, bound laboratory notebook, kept in ink and witnessed weekly by a supervisor or a knowledgeable peer who is not an inventor. This witness is asserting that the notebook has been "read and understood." However, from our own experience with patent litigation, we believe that such careful records are the exception, rather than the rule. All too many times, the records include incomplete descriptions of what was done and why, as well as scraps of paper which appear to have been added later. Such poor records are incompetent.

Moreover, the explosion of electronic files and computer-generated printouts means that most actual data are no longer carefully crafted columns of numbers written in blue ink. We recognize that computer files are more efficient and do not for a moment suggest that they be abandoned. However, for the foreseeable future, we still think that it makes sense to collect printouts on a weekly basis, to glue them into a laboratory notebook, and to have them read and witnessed by a knowledgeable peer. With these precautions, establishing the basis for a patent should be much easier.

Finally, we should stress that US patent laws divide the inventive process into two steps: conception, and reduction to practice. Conception is the formation by the inventor of a definite idea of the complete invention, including every feature sought to be patented. Conception is complete when one of ordinary skill in the art could practice the invention without extensive research or experimentation. Posing a problem is not conception.

Because conception is mental, the courts also require reduction to practice, which is evidence that the invention works. Reduction to practice takes two form, actual and constructive. Actual reduction to practice requires construction of a device or preparing a composition. Then the inventor must demonstrate that the invention fulfills its intended purpose. This actual reduction to practice often includes getting missing information, using methods like those in the next section. Constructive reduction to practice is filing a patent application. This is normally the lawyers' responsibility.

There are two crucial points to remember about intellectual property law. First, it is complex: when we need to get involved in patents, it is imperative that we do so with the assistance of a specialized lawyer. Second, intellectual property law has little to do with justice as it is normally understood. When asked to sum up patent law as concisely as possible, one patent lawyer described it simply as an opportunity to employ lawyers. While we would not wish to encourage quite this level of cynicism, we must remember that patent law is a set of rules, not a means of enforcing truth, justice, or a better way of life.

EXAMPLE 5.1–1 THE INVENTION OF THE WINDSURFER

The windsurfer was first commercialized by Hoyle and Diana Schweitzer in California, in the late 1960s and early 1970s. They set up a company, Windsurfing International Inc., which did much of the early development of windsurfers, and trademarked the name "Windsurfer" in 1973. Together with an aerospace engineer, Jim Drake, they filed the first patent relating to windsurfers in 1968, and it was granted by the US Patent Office in 1970. The Schweitzer–Drake creation incorporated all the essential elements of the modern windsurfer, including a triangular sail, "wishbone" booms, and a universal joint. For many years this company produced its own windsurfers and received license fees from other manufacturers. It was not until the late 1970s that windsurfing really took off, when the craze hit Europe; its greater popularity there is sometimes attributed to the longer holidays common in Europe. Were the Schweitzer–Drake patents effective?

SOLUTION

For around 20 years, the original inventors' patents were unchallenged. However, in the late 1970s, windsurfer manufacture became highly profitable, and large-scale manufacturers such as Bicsport, F2, and Mistral became heavily involved. These companies found it irksome to continue to have to pay a royalty to Windsurfing International every time they sold a board and began to search for ways to circumvent the patent.

In 1983, Windsurfing International Inc. sued the Swiss company, Mistral, for patent infringement. Mistral defended their position by referring to the work of Newman Darby, who produced the "Darby Sailboard" in Wilkes-Barre, Pennsylvania, in 1964. Darby published his design in 1965 in "Popular Science" magazine, including pictures of his sailboard being ridden by his wife. The Darby board included a hand-held diamond "kite" sail and a curved boom, on a floating platform, for recreational use. The rig was not attached to the board: the mast was held by the rider in a depression in the board. However, the published version includes reference to a "more complex swivel step for advanced riders not shown," and this was sufficient for the US courts to eventually rule that the Schweitzer–Drake board was an obvious development of Darby's published invention. Windsurfer International's patent was worthless in the USA.

Also in 1983, an Australian court attributed the first legally recognized use of a "windsurfer" to Richard Eastaugh. Aged between 10 and 13, from 1946 to 1949, with the help of his younger brothers, Eastaugh equipped a series of canoes with sails and split bamboo booms and sailed them on the Swan River near Perth.

In 1985, Windsurfing International Inc. sued Tabur Marine Ltd. (the precursor to Bicsport) for infringement of their patent, in a celebrated case which was to have important implications for the interpretation of the "inventive step" and "non-obviousness" in British law. Tabur Marine defended their position by claiming to have discovered prior art, in the form of material published prior to the patent application date. In 1958, a British boy, Paul Chilvers, then 12 years old, experimented with something that looked a bit like a windsurfer on Hayling Island. Tabur Marine were able to exhibit in court a homemade film of Chilvers sailing his invention. Chilvers' "windsurfer" included a board and mast with a universal joint, but was steered by a rudder and used a "straight split boom," rather than the "wishbone" type used on commercial boards and patented by Schweitzer and Drake. However, the British court ruled against Windsurfing International, judging that the "wishbone" boom was merely an obvious development once the principle of the windsurfer was established.

After these hotly contested legal battles, the patent was undermined by prior art in the USA, Australia, and Europe. It is interesting to note that separate legal cases were required in each jurisdiction and in each case the relevant prior art came from within that jurisdiction. Mistral and Bicsport no longer pay license fees (and nor does anyone else). By the late 1980s, Windsurfer International had ceased trading. We do not wish to suggest any ethical conclusion as to who deserve financial benefit. We use this example to draw attention to the uncertainties and vicissitudes associated with the ownership of a patent.

5.1–2 SUPPLYING MISSING INFORMATION

In Chapter 4 we saw how to make the final selection of our most promising idea. At this stage we had sufficient information to convince ourselves that the idea we chose was a winner, a prince among frogs. This information may have come from the available literature, from external experts, or from back-of-the-envelope calculations. However, the information is unlikely to be complete and rigorous. Because we are about to embark on an expensive program of product development, we had better be sure exactly how well our product is going to work.

Discovering these product details requires further research and experimentation. Up to this point, we tried to minimize the work at each stage: simplified calculations have always been employed, experimentation kept to a minimum, and literature research used only to establish if something is possible, with little attention to the details of how it might be achieved. This streamlines product design, allows easy comparisons between ideas, and minimizes time to market. Now, however, detailed information is indispensable: we must confirm experimentally any information used already and fill in the many gaps in our knowledge. Our prince must be clothed and educated to become a working monarch.

It is of course difficult to generalize about what missing information will be necessary and how best it can be obtained. Every project will have its own problems and the information available will vary enormously, depending on both the level of literature interest and a company's prior activity in this area. The minimum requirement will be experimental verification of relevant reported data. At the other extreme, a full experimental program may be necessary to demonstrate the viability of a new and untested idea.

To illustrate the type of approach required to fill in the gaps in our information, we give two examples.

EXAMPLE 5.1–2   STERICALLY HINDERED AMINES

Acid gas removal from gas streams (sweetening) is a very common process in the chemical and refining industries. For example, in a hydrogen plant, methane is converted by steam reforming to hydrogen and CO_2. The CO_2 must then be removed to leave a pure product. In an existing plant, this CO_2 removal is often the bottleneck for capacity expansion. Our company would like to improve CO_2 removal from gas streams. How can we do so?

SOLUTION

Conventionally, gas sweetening is achieved using amines by the following reaction:

$$2R-NH_2 + CO_2 \rightarrow R-NH_3^+ + R-NH-COO^-$$

Reaction occurs in a gas–liquid column at low temperature (40–80 °C) followed by amine regeneration at higher temperature (120 °C) and low CO_2 partial pressure. This reaction requires two moles of amine per mole of CO_2 removal. In 1974, Sartori, at Exxon's research laboratories, realized that by reducing the stability of the carbamate ion ($RNHCOO^-$), the stoichiometry of the reaction could be changed:

$$R-NH_2 + CO_2 + H_2O \rightarrow R-NH_3^+ + HCO_3^-$$

Only one mole of amine is now required per mole of CO_2 absorbed. This is potentially a great improvement in efficiency. The carbamate ion can be destabilized by using a hindered amine, such as diisopropylamine:

This species is destabilized both by the high electron-pushing power of the side groups and the fact that the bulky side chains prevent free rotation of the acid group.

However, if the amine is highly hindered, the rate of reaction becomes so slow that it is useless for CO_2 removal. What is required is a moderately hindered amine, such that the second reaction above, which makes the bicarbonate and dominates the first reaction, which makes the carbamate. At the same time, a reasonable reaction rate must still be achieved. This is possible with amines similar to that shown above. However, we do not know exactly which hindered amine to use. Our specifications are likely to go as follows:

(1)   We require a new product which will double the capacity of the old plant or reduce the size of absorption columns in a new plant. This means that we need to achieve an increase in the CO_2-carrying capacity of the absorbing liquid via the second mechanism above.
(2)   The rate of reaction must be at least as high as that for the conventional amines, or our capacity gain will be offset by a poor rate of absorption.
(3)   We want to retrofit the old plant with our new product. Therefore, operating conditions must be similar to those used currently, i.e. absorption at 40–80°C, and regeneration at 120 °C.
(4)   In an operating plant, a corrosion inhibitor containing V^{5+} is present in the absorbing liquid. The hindered amine must be stable in the presence of this inhibitor.

In order to develop the final product, Sartori and co-workers (1987) screened a wide range of possible hindered amines for their performance on each of these four criteria. It is likely that they tested hundreds of possible amines to establish the optimum product. Data on around a dozen hindered amines are published in the open literature, showing the trends established as a function of the size and the chemical nature of the hindering groups.

First, Sartori established that hindered amines do indeed react via the second mechanism rather than the first. Next, he investigated the rate constants for CO_2 absorption. He found that moderately hindered amines showed almost an order of magnitude drop in rate constant relative to unhindered ones and that highly hindered amines were over an order of magnitude worse again. This led to rejection of highly hindered amines in favor of moderately hindered ones. Although the rate constant is reduced by moving to hindered amines, the rate of CO_2 absorption is given by an expression of the form:

$$\text{rate} = k[CO_2][\text{amine}]$$

Since the stoichiometry is 1:1 for hindered CO_2 absorption rather than 2:1 for conventional amines, there will be certain operating conditions under which the actual rate of CO_2 absorption is higher in the hindered amine case than in the conventional situation. Sartori and his collaborators were able to show that for some moderately hindered amines this was the case for typical plant operating conditions. Indeed, they found that under these conditions the rate of CO_2

absorption was limited only by CO_2 diffusion into the liquid. Thus, by choosing from a range of moderately hindered amines, the requirement of increased capacity without loss of rate can be met.

Having established the required degree of steric hindering for the amines, the requirements of solubility and thermal stability were satisfied by altering the chemical nature (but not size) of the side groups. For example, the solubility in aqueous solution is usually enhanced by using alcohol side chains. For this reason, the standard unhindered amine used in conventional CO_2 absorption is diethanolamine (DEA). Undoubtedly Sartori incorporated similar chemical features into the hindered amine in the final product.

The final stage in the experimental program was to test the stability of the possible amines in the presence of the V^{5+} inhibitor. It turns out that the same chemical features which make the carbamate anion unstable also make the amine stable to V^{5+}. Thus, the new product is better than conventional amines in this respect also. The new amines are now produced commercially.

EXAMPLE 5.1–3   SILVER BULLETS FOR ZEBRA MUSSELS

Zebra mussels are a freshwater mussel native to Europe which reached North America via bilgewater in cargo ships. They have become prolific, out-competing native bivalve species and aggressively colonizing freshwater habitats. They present a particular problem for industry because of their propensity to block the intake pipes of cooling water systems, such as those for power-station heat exchangers.

Zebra mussels feed by filtering nutritious particles, typically algae, from the water. It is proposed to control them by feeding them poisoned capsules, "silver bullets," thus utilizing their own filtering activity as a means of concentrating the poison. This will allow bulk poison concentrations in the water to be many times lower than that which would be required if the poison were placed directly in the cooling water system. Hence, cost and environmental damage should simultaneously be minimized. What further information would be required before this product could be commercially developed?

We must answer two questions:

(1)   What should the size and composition of our capsules be?
(2)   What concentration of capsules do we need in order to achieve a given kill rate, say 90%? How does this compare with the bulk-water concentration of toxin we would need to poison the mussels?

SOLUTION

Question (1) is relatively simple to answer. There is a considerable amount of literature available on toxins for bivalves in general and zebra mussels in particular. One simple but promising candidate is KCl, which induces heart attacks in mussels. Another is the Ethiopian soapberry, which is rich in surfactants and is used extensively for washing clothes in Ethiopia. Waterways used for such washing are

Goal definition. The system under consideration is our plant plus our suppliers. The environment, as in the thermodynamic sense, is more vague, usually described as "that which surrounds the system." The goal is achieving minimum impact of the system on its environment. Unfortunately, going beyond this platitude is almost always controversial; for example, would closing your plant to reduce emissions and importing from China constitute a reduction of environmental impact? We must choose the factors which we are trying to minimize. For example, a common goal is to minimize CO_2 output globally; another might be to minimize NO_x emissions in an urban environment; and a third might be to minimize the biochemical oxygen demand (BOD) in a local waterway.

Inventory analysis. This more familiar territory is essentially a series of mass and energy balances, taken over the product life. Environmental burdens, which can be resource depletion or waste emissions, must be analyzed. The analysis must include steady-state production, start up, and shut down. In doing this analysis, we may need to reconsider our goals; for example, it may turn out that the CO_2 footprint is low, but that we may have a major impact on local water supplies.

In producing an inventory of a product's environmental impact, we need to include factors like the following:

the raw-materials suppliers' processes;
our manufacturing process, including energy use and by-products;
waste and its treatment;
packaging, including cost, impact, and function; and
final disposal.

Impact assessment. We now need to establish a way of comparing the competing demands we have established. While this is rarely possible quantitatively, we must make the effort, or the hard work of inventory analysis will be wasted. Engineers again tend to become nervous about the qualitative aspects inherent in this step. First, we need to classify the inventories in groups, such as "resources," "emissions, "goods made." Next, we need to characterize the impact of each group, for example the value of making CO_2-free electricity and the burden of radioactive waste.

Improvement evaluation. We seek a single figure of merit, an "environmental impact function," by attaching weightings to the impacts assessed in the previous step. We can then design the product to minimize this environmental-impact function within the other constraints of the design. Inevitably this will be controversial, since it is based not only on science but also on the preferences of the stakeholders. Even determining who should be consulted as stakeholders is difficult. Our choice may include decision makers (chief executive officers, politicians, etc.), experts (e.g. public-health practitioners, process engineers), and environmental advocates (the Sierra Club, the Council for the Preservation of Rural England, local activists). While we have not made cost explicit in this analysis, determining the best way forward will inevitably balance cost against perceived benefit. These ideas are illustrated in the following examples.

EXAMPLE 5.1–4 REMOVAL OF SO_2 AT THE DRAX POWER STATION

The DRAX power station in northern Yorkshire is the largest electrical-power-generating plant in Europe, producing 4000 MW, which is about 7% of the power generated in Britain. The plant burns 430 kg of coal per second, an astonishing amount, making the plant the largest emitter of carbon dioxide in the country. Because the coal is two percent sulfur, the plant produces 17 kg of SO_2 per second. To meet environmental regulations, 90% of the sulfur must be removed.

Three chief technologies to capture the sulfur oxides are available: wet limestone treatment, the double alkali process, and dry sodium carbonate processing. In the wet limestone process, a calcium carbonate slurry reacts with the flue gas to make calcium sulfate, i.e. gypsum. Every second, this process uses 27 kg of limestone to make 46 kg of gypsum. The gypsum is then used to make sheetrock or wallboard.

The second, double alkali process mixes calcium carbonate and sodium carbonate in water and uses this mixture to capture the sulfur oxides. The higher concentration of carbonate in solution means that the reaction kinetics are faster, and the equipment required is smaller. Thus, the removal of the SO_2 is more efficient. The third process injects dry particles of solid sodium carbonate, rather than using an aqueous slurry. The solid–gas reaction can be carried out at higher temperatures, and so is still fast. The particles of sodium sulfate that are produced are captured in an electrostatic precipitator.

Which process has the least environmental impact?

SOLUTION

Each of these three processes is producing carbon dioxide in order to capture sulfur dioxide (Azapagic *et al.* (2004)). In other words, each is trading more global warming for less acid rain. The three processes are evaluated on the basis of the tons of CO_2 produced per ton of SO_2 recovered. Small values of this parameter are good.

Estimates for the three processes are shown in Figure 5.1–1 on the basis of three criteria: resource depletion, global warming, and photochemical smog. The wet limestone process is the winner on each criterion. However, if every large power plant adopted this technology, the entire world would be awash in excess wallboard. Wet limestone was the technology used by DRAX, but it should not be universally adopted.

EXAMPLE 5.1–5 DECIDING WHICH NAPPIES (DIAPERS) TO USE

Mr. and Mrs. Verdi are about to have their first child and are considering what type of nappies to use. They identify three options: disposable nappies,

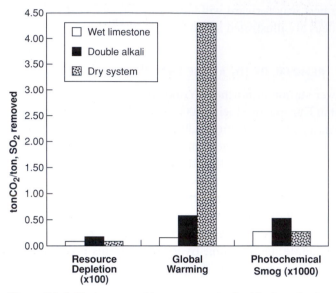

Figure 5.1–1 *Environmental impact per unit of acid rain reduction*
The graph compares three technologies that increase global warming (as additional CO_2) in order to reduce acid rain (as captured sulfur).

commercially laundered cloth nappies (including a delivery service), and home-laundered cloth nappies. While cost and convenience are of course important considerations, the Verdis are also concerned that their decision is environmentally responsible.

What should they do?

<u>SOLUTION</u>

This choice depends on cost, convenience, and environmental issues. Each is discussed below.

Cost. Based on 5000 changes over a two-and-a-half-year period, total nappy costs per child are as follows:

> disposable nappies: £700–1100 depending on size and quality
> washing service: £750–1050 including delivery, liners, and accessories
> home-laundered   £300–900 including detergents, energy, and a washing
>   cloth nappies: machine

No allowance is made for the cost of disposal of solid waste, which costs about 10% of the nappies' cost.

Convenience. Disposable nappies provide the ultimate in convenience. However, commercially laundered cloth nappies, with a collection service, do not represent a significant additional burden, except when travelling. Home laundering requires substantially more time and effort. The development of shaped cotton

Table 5.1–1 *Environmental consequences per 1000 nappies.*

	Disposable	Commercially laundered	Home laundered
Energy requirement, GJ	2.0	2.2	4.0
Solid waste, m^3	0.46	0.06	0.06
Atmospheric emissions, kg	3.8	2.0	4.4
Waterborne wastes, kg	0.7	2.6	2.8
Water requirement, L	4900	12 800	10 200

nappies and flushable liners have improved both the performance and convenience of cloth nappies. This variety of available quality also explains the wide range of prices for home-laundered nappies shown above.

Considering only convenience and cost, disposables would come out the clear winner.

Environmental issues. Eight million nappies per day are thrown away in the UK. On the face of it, the environmental question would therefore seem simple. However, in order to make a good comparison, we must consider energy requirements, solid waste, waterborne and atmospheric emissions, and water usage, using a full life-cycle analysis, including manufacture, washing, and disposal. This yields the estimates given in Table 5.1–1.

As is often the case with environmental issues, what seems a simple question turns out to be complex, requiring the balancing of competing demands. The key question is whether you consider energy use, water consumption, or landfill requirement to be more important.

Interestingly, the case against disposable nappies is not clear cut – the conclusion will depend on the importance ascribed to the different categories of waste. Surprisingly, commercial laundering is significantly less environmentally damaging than home laundering. In the UK, landfill sites are at a premium, tipping the balance against disposables. In other countries, another element of the environmental consequences, such as water use, might dominate.

In making their decision, Mr. and Mrs. Verdi draw up the decision matrix in Table 5.1–2 (scores are again out of 10 with high scores being good). Because both Verdis work, and they are reasonably affluent, they weight convenience above cost. They decide to go for commercially laundered cloth nappies with a home-delivery service. For other parents, the conclusion might be different.

EXAMPLE 5.1–6 THE CASE OF THE SWEDISH MEATBALLS

There is an increasing trend towards eating pre-prepared meals, rather than cooking at home. In public debates, this is often claimed to increase the environmental impact from foods, because food manufacture and preparation accounts for a significant fraction of the total environmental impact of humans. The major

Table 5.1–2 *Decision matrix for nappies.*

Criterion	Weight	Disposables	Commercially laundered cloth	Home-aundered cloth
Cost	0.2	5	5	9
Convenience	0.5	10	9	2
Environment	0.3	5	8	5
TOTAL		7.5	7.9	4.3

flat-pack furniture retailer, IKEA, best known for the claim that 10% of Europeans are conceived in IKEA beds, is also famous for selling delicious meatballs. Being a Swedish company, IKEA are concerned to minimize their environmental impact, and they wish to analyze the impact of using pre-prepared meatballs compared to making them fresh in their shops.

What is the best environmental choice for Swedish meatball meals?

SOLUTION

Sonesson *et al.* (2005) have thoroughly analyzed a similar problem by performing a rigorous life-cycle analysis on three scenarios, each delivering one meal, consisting of meatballs with potatoes, bread, carrots, and milk, onto the table of a Swedish household. All three meals include the same ingredients at the point of delivery (the mouth). The three scenarios are:

- A homemade meal: each dish prepared from raw ingredients at home, meatballs and bread homemade.
- A semi-prepared meal: meatballs prepared in a central kitchen, bread from an industrial bakery, vegetables prepared at home.
- Ready-to-eat meal: meatballs and vegetables microwaved from a pre-prepared package, bread from an industrial bakery.

Each meal was analyzed for air emissions (CO_2, CO, CH_4, volatile organic compounds (VOCs), NH_3, NO_x, SO_x, N_2O, HCl), water emissions (chemical oxygen demand (COD)/BOD, NH_4^+, NO_3^-, P), and energy use. Even for a simple case like this, a good life-cycle analysis is laborious – the results occupy eight closely spaced pages in Sonesson's report. Sonneson concludes that all preparation methods show similar environmental impact. The pre-prepared meal scored slightly worse than the others on energy use, the homemade meal was marginally the worst for eutrophication and global warming, and the semi-prepared meal was worst for photochemical substance emission. All differences, however, were small, primarily because the environmental impact is dominated by raw-material production, notably beef. A cow can be thought of as a chemical reactor for turning grain or grass into protein, with an efficiency of around 6%; 70% of the grain produced in the USA goes for animal feed. In this context, it is not surprising that the details of cooking, storage, and transport have a small impact on the overall environmental cost of eating meatballs. The way to an environmentally

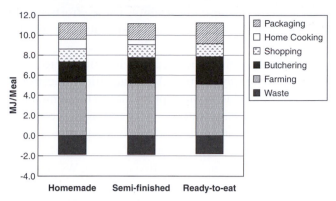

Figure 5.1–2 *Energy Required for a Meal of Meatballs* The energy used for home cooking is about equal to that for a meal prepared commercially and reheated from frozen.

clean conscience is to eat less grain-fed meat. Additional details are given in Figure 5.1–2.

This example is instructive for three reasons. First, doing a good life-cycle analysis is hard work. Second, the conclusions are often not clear, and in particular depend on the environmental impact parameter which is considered most important. Third, popular rhetoric and careful analysis rarely converge. Life-cycle analysis is an important tool and can be useful, but needs to be carefully carried out and interpreted.

5.2   Final Specifications

5.2–1 SETTING A SPECIFICATION

At this point, we are considering making only a small number of products for manufacture. Normally, this will be one or two. The products may be a chemical, like a new drug for counteracting depression, a mixture of chemicals like chocolate, or a device, like a new catalytic convertor for reducing nitrogen oxide (NO_x) emissions.

We want to choose final specifications for this product. To begin this task, we are wise to carefully review where our work so far has led us. We should describe the chemical product we want to make. If it is a chemical, we need to specify its molecular structure, its final form, and its required purity. If the product is a device for chemical change, we must specify its physical size and shape, and its expected mode of operation. Again, we suggest that each member of the core team briefly write out these specifications, and that the core team resolve any differences by consensus.

Our final specifications should also re-examine our competition. We want to compare our new product with the best existing product. We want to identify improvements we expect, and to state how large these improvements can be. We want to re-state all the assumptions which we are making, and to decide which of these assumptions involves the most uncertainty. These efforts should circumscribe our problem, and identify the technical constraints with which we must deal.

Our thinking in setting these final specifications is different to that required earlier. Before, we were especially interested in innovation, because we hoped that such "out-of-the-box" thinking would supply clues to big commercial advantages. We urged setting general specifications; we encouraged eccentric ideas; we sought surprising selections. We understood that we would reject most of these ideas. We understood that we needed to kiss a lot of frogs to find a prince.

We now want adaptive thinking, not innovative thinking. We want to improve our carefully selected ideas by slight modification, not by random invention. In many ways, we are best guided by a maxim taught as a diagnostic guide to second-year medical students:

When you hear hoof-beats in the street, think horses not zebras.

For the medical students, this means that sick patients are more likely to have common diseases than exotic ones. For us, it means that our products can most likely be improved by careful, evolutionary changes, not by enormous mutations.

Setting these final specifications can often be aided by a three-step strategy. First, we define the product structure, a task which is relatively straightforward. Second, we rank the product's most important attributes, an effort which forces a review of how the product will be used. Third, we review any chemical triggers, i.e. chemical stimuli which cause major changes in product properties. These three strategic steps are detailed in the following paragraphs.

Product structure. Specifying the product structure usually involves considering the four items listed below:

(1) *Chemical composition.* What is the planned product made of? If it is chemically pure, what is its chemical structure? If it is a device, how much can its composition be changed without affecting its performance?

(2) *Physical geometry.* What product characteristics are fixed? Are there fixed macroscopic dimensions? Is there any unusual physics?

(3) *Chemical reactions.* Does the product change chemically during use? Do additives like acids, bases, and salts affect these changes?

(4) *Product thermodynamics.* What is the product's phase? Is this phase thermodynamically stable or metastable? What is the characteristic size scale in any mixed phases?

We must emphasize that these concerns with product structure will apply differently to different products. Drugs will be different to shut-down battery separators. Nonetheless, we recommend this checklist, even if it only stresses the product's uniqueness.

Central product attributes. We next turn to re-examining the most important attributes of the product we are almost ready to manufacture. Most often in preparation for manufacture, we will be willing to make a long list. That is not the goal here. We want to choose three or fewer important attributes; we would prefer identifying only one as most important.

The long list of product attributes can often be organized under three headings:

(1) *Structural attributes*. These include the product's physical properties, like its strength and elasticity. These attributes are most important for devices.

(2) *Equilibrium changes*. Many chemical products, particularly microstructured ones, will show major changes in equilibrium as a consequence of altered temperature, pH, or some other process variable.

(3) *Key rate processes*. The most obvious is the rate of any important chemical reaction. Less obvious but often important are rates of heat transfer, fluid flow, or diffusion, which are often manipulated by changes in interfacial area.

In our search for final specifications, we should use this organization to find the most important attributes.

Chemical triggers. The final strategic step seeks to identify any chemistry which makes the product become active. What frees the product from its original thermodynamic bondage? This step, most important for molecules and microstructured products, usually involves a variable such as one of the following:

(1) *Solvents*. These dissolve or disperse the product so it becomes useful.

(2) *Temperature changes*. The most common example is regenerating a product – like an adsorbent – by heating or cooling.

(3) *Chemical reactions*. The most common occur because of pH changes or hydrolysis.

(4) *Other physical changes*. These may include pressure, detergency, and electric field.

When these three steps are complete, we should be in a position to imagine how manufacturing can occur. Before outlining this next step, we turn to examples which illustrate how this strategy can improve product design.

EXAMPLE 5.2–1   FREON-FREE FOAM

Refrigerators are normally insulated with polyurethane foam. The foam is made by injecting reactive monomers into the space between the inner and outer walls of the refrigerator. Traditionally, freon was injected along with the reagents. As the reaction proceeded, the freon evaporated, producing a foam with about 95% bubbles containing freon.

The result is a very effective insulator. The properties of this insulation have been used to establish standards for home refrigerators. The external dimensions of refrigerators have become standard, so a new refrigerator will fit into the space

> **Table 5.2–1 *Properties of gases used in insulating foam*** Freon's large diameter and
> high molecular weight give it the lowest thermal conductivity.
>
> | | Molecular weight, daltons | Molecular diameter, Å | Boiling point, °C |
> |---|---|---|---|
> | Nitrogen (N_2) | 28 | 3.8 | –196 |
> | Carbon dioxide (CO_2) | 44 | 3.9 | –79[a] |
> | Freon12 (CCl_2F_2) | 121 | 5.3 | –30 |
>
> [a] Sublimes.

occupied by the old one. The internal dimensions have also become standard, so that milk bottles fit conveniently inside. The insulation required for energy efficiency is legally restricted, with laws based on the properties of freon-containing foam.

However, when freon is released to the environment, it destroys the layer of ozone which protects the Earth from excess ultraviolet radiation. As a result, an international agreement has banned the production and use of freon. To be sure, freon in insulating foam seems less abusive than freon in single-use products like hair sprays. Nonetheless, the freon in foam will eventually leak out, perhaps long after the refrigerator has been scrapped. The manufacture of polyurethane foam blown with freon is being phased out, as a consequence of increasing regulatory restrictions.

We need to build refrigerators with the same dimensions and the same degree of insulation as those with freon-containing foam. The degree of insulation achieved in any refrigerator depends most dramatically on the thermal conductivity of the gas in the foam's bubbles. This thermal conductivity k_T is given in W/mK by

$$k_T = \frac{0.08}{\sigma^2 \Omega} \sqrt{\frac{T}{\widetilde{M}}}$$

where σ is the molecular diameter, in Å; Ω is dimensionless and of order one, a weak function of temperature; T is the absolute temperature, in K; and $\widetilde{M}$ is the molecular weight in daltons. Thus if we replace the freon with CO_2, we find from Table 5.2–1

$$\frac{k_T(CO_2)}{k_T(CCl_2F_2)} = \left(\frac{\sigma_{CCl_2F_2}}{\sigma_{CO_2}}\right)^2 \left(\frac{\widetilde{M}_{CCl_2F_2}}{\widetilde{M}_{CO_2}}\right)^{\frac{1}{2}}$$

$$= \left(\frac{5.3}{3.9}\right)^2 \left(\frac{121}{44}\right)^{\frac{1}{2}} \approx 3$$

The foam blown with carbon dioxide will provide only one third the insulation of the same thickness of foam blown with freon. A foam blown with nitrogen is even worse, with only one fourth the insulation.

We need a better foam. A careful search for ideas has produced many interesting alternatives, including materials made of many layers of aluminum foil. After careful analysis, however, we decide that our best choice is polyurethane foam modified in some way to reduce its thermal conductivity.

Use the strategy given above to suggest final product specifications.

SOLUTION

The three-step strategy given above suggests defining the product's structure, specifying its chief attributes, and identifying any chemical triggers which make the product active. In this case, there is no chemical trigger, but the other steps are important.

Product structure. Defining the structure is easy. We want a polyurethane foam containing 95% gas bubbles. The bubbles should be small to avoid free convection: free convection in any larger bubbles will compromise insulation. The idea that the bubbles could be much smaller than in the present foam is interesting, but we defer discussing this until later. There are no chemical interactions in the present foam. Again, the interesting idea of such interactions is deferred until later. In general, foams are metastable, especially if the bubbles are very small, but this should not be a major problem in this case.

The central product attribute. The foam is a good internal insulator. This key attribute is directly a result of the thermal conductivity in the foam's gas-filled bubbles. As a result, we can benefit from a review of this transport property. For a monoatomic dilute gas, the thermal conductivity k_T is given by

$$k_T = \frac{1}{3}\left[\begin{array}{l}\text{distance between}\\ \text{collisions}\end{array}\right]\left[\frac{\text{energy}}{\text{volume}}\right]\left[\frac{\text{volume}}{\text{area}\times\text{time}}\right]$$

The volume per area per time is nothing more than the average molecular velocity v. For a monoatomic gas, this velocity depends on temperature T via the kinetic energy:

$$\frac{1}{2}mv^2 = k_B T$$

where m is the molecular mass and k_B is Boltzmann's constant. The energy per volume, the product of the molecular concentration c and the molar heat capacity $\tilde{C}_V$, is given by

$$c[\tilde{C}_V] = \frac{p}{k_B T}\left[\frac{3}{2}k_B\right]$$

where p is the pressure. We only need to estimate the distance between collisions.

There are two limiting cases of this collision distance, valid for large bubbles and for small bubbles. For larger bubbles, the distance is the mean free path λ that a gas molecule travels before it collides with a second gas molecule. This mean free path is related to the volume per molecule:

$$\frac{\pi}{4}\sigma^2\lambda = \frac{V}{n}$$
$$= \frac{k_B T}{p}$$

where σ is again the molecular diameter, V is the bubble volume, and n is the number of gas molecules in the bubble. Solving for λ and combining with the above, we find that

$$k_T \propto \frac{1}{\sigma^2}\sqrt{\frac{T}{m}}$$

This variation of the thermal conductivity with molecular size and weight is equivalent to that presumed in the problem statement. It is the variation which let us estimate how much poorer CO_2-blown foam would be compared with freon-blown foam.

This large-bubble result is dramatically different to that for small bubbles. For small bubbles, the gas velocity and the gas energy per volume are unchanged, but the distance between collisions is different. For small bubbles, this distance is proportional to the bubble diameter. Unlike in large bubbles, where gas molecules collide with each other, a molecule in a small bubble bangs from one point on the wall to another. As a result, the thermal conductivity is now

$$k_T = dp\sqrt{\frac{k_B}{2mT}}$$

where d is the bubble diameter. Note how different this result is from the previous equation. While k_T varies with the inverse square root of molecular weight for both large and small bubbles, k_T increases with temperature in large bubbles but decreases with temperature in small bubbles. More importantly, the thermal conductivity is independent of pressure and bubble size in large bubbles, but is proportional to the product (dp) in small bubbles. Thus, we can make a better freon-free insulating foam by having small bubbles or a low gas pressure.

Setting final specifications. To complete our product specifications, we must decide what is a large bubble and what is a small bubble. From the above, we see that this difference depends on the Knudsen number Kn, the ratio of the mean free path λ and the bubble diameter d:

$$Kn = \frac{\lambda}{d}$$
$$= \left(\frac{4}{\pi}\right)\frac{k_B T}{p\sigma^2 d}$$

When $Kn \ll 1$, we have intermolecular collisions, and hence large bubbles. When $Kn \gg 1$, we have molecule–wall collisions and hence small bubbles.

We want small bubbles. While we can try to make these mechanically, we will find it difficult to get bubbles smaller than 1 µm, not small enough to be "small."

The reason is that the surface energy of such bubbles is high, so that some of the bubbles tend to grow at the expense of others. This process is sometimes called "Ostwald ripening."

However, we could make the bubbles behave as if they were small by reducing the gas pressure, and hence raising the Knudsen number. The product designers who were involved in making a better foam did just this by a very clever invention. They blew polyurethane foam with carbon dioxide in the normal way, under established reaction conditions, but they blew it into a bag made of metal foil. The bag is essentially completely impermeable to all gases. Just before the bag was sealed, the designers added a spoonful of sodium hydroxide to the bag. The sodium hydroxide reacted with any CO_2 that slowly diffused through the foam to react. It turns out that a chemical trigger is involved in our product manufacture after all.

The result is a foam initially the same as any other CO_2-blown foam, but which gets to be a better insulator with time. Eventually, as the gas pressure gets lower and lower, the foam conducts even less than the original freon-blown foam which it replaced. The final product specification that the new foam must have a thermal conductivity no higher than freon-blown foam will be exceeded.

EXAMPLE 5.2–2   BETTER BLOOD OXYGENATORS

For open heart surgery, we must use a machine to bypass the patient's heart and lungs while the heart is being repaired. The machine must move the blood at roughly the normal rate, which is relatively easily accomplished with a pump. The machine must also add oxygen and remove carbon dioxide. It must perform the same function as the lungs, a much more difficult task. In almost all cases, oxygen addition is more difficult to accomplish than carbon dioxide removal, so that blood oxygenators are normally designed using oxygen transfer as the benchmark.

We can oxygenate blood using many familiar chemical engineering operations. For example, we could oxygenate blood in a packed tower, letting blood trickle downwards over Rashig rings while air flows upwards, countercurrently to the blood. This type of operation is not attractive for two reasons. First, any free interface between air and blood tends to cause clots, just as any open cut on our hands tends to clot. Blood clots can cause strokes. As a result, past designs of blood oxygenators tend to carry out the mass transfer across a membrane. Originally, silicone rubber membranes were used, which offered significant resistance to mass transfer. More modern designs use microporous hydrophobic membranes which offer no significant resistance to mass transfer.

The second reason that blood oxygenators cannot use conventional equipment like packed towers is that the volume of blood required to start up such a packed tower would quite literally drain the patient white. The tower could be started with blood transfusions. Unfortunately, these carry the ever-present risk of infecting the patient with HIV or hepatitis. We must use the smallest blood oxygenator which gives enough oxygen transfer.

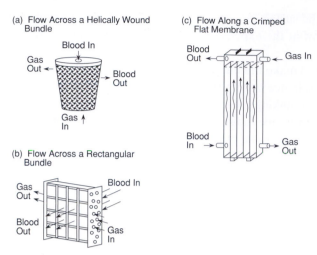

(a) Flow Across a Helically Wound Bundle

(b) Flow Across a Rectangular Bundle

(c) Flow Along a Crimped Flat Membrane

Figure 5.2–1 *Commercial blood oxygenators* Each design adds oxygen and removes carbon dioxide from blood.

Thus, we need a mass transfer device which offers the greatest amount of mass transfer per volume across a microporous membrane. Some designs are shown in Figure 5.2–1. Originally, the membranes used were flat. Later, to get more area, the membranes were corrugated, like furnace filters or the air filters in automobiles. Now, blood oxygenators usually use hollow-fiber membranes, which give the best performance yet achieved.

Imagine that we want to build a new hollow-fiber blood oxygenator which outperforms other models. To do so, we want to maximize the oxygen transferred per blood volume. We know that the oxygen flux per volume J_1 is given by

$$J_1 = Ka(c_1^* - c_1)$$

where K is the overall oxygen mass transfer coefficient, based on a liquid side resistance; a is the membrane area per volume; c_1^* is the blood oxygen concentration at saturation, kept constant by using excess air; and c_1 is the actual oxygen concentration in the blood. In this case, K is dominated by the individual mass transfer coefficient in the blood k. Thus our problem is simple: we must select the oxygenator design which maximizes ka.

SOLUTION

In this example, the key is the product's structure, epitomized by the product ka. The central product attributes duplicate those of competing oxygenators. There are no real chemical triggers. We just need a big ka. We begin this selection by considering the area per volume a. For the hollow fibers, we expect

$$a = \frac{\text{fiber area}}{\text{oxygenator volume}}$$
$$= \frac{\text{fiber area}}{\text{fiber volume}} \times \frac{\text{fiber volume}}{\text{oxygenator volume}}$$

> **Table 5.2–2** *Mass transfer correlations across hollow-fiber membranes.*[a,b]
>
> | Flow geometry | Flow range | Correlation |
> |---|---|---|
> | Within fibers | $Sh > 4$ | $Sh = 1.62Gr^{1/3}$ |
> | Outside and parallel to fibers | $Gr < 60$ | $Sh = 1.3\left(\dfrac{d_e^2 v}{vl}\right)^{0.9} Sc^{1/3}$ |
> | Outside and across fibers | $Re > 2$ | $Sh = 0.4Re^{0.8}Sc^{0.33}$ |
> | Outside and across fiber fabric | $Re > 2$ | $Sh = 0.8Re^{0.49}Sc^{0.33}$ |
>
> [a] Dimensionless groups defined as follows:
> Graetz number $Gr = d^2v/Dl$; Sherwood number $Sh = kd/D$; Reynolds number $Re = dv/v$; and Schmidt number $Sc = v/D$
> [b] Variables defined as follows: d is fiber diameter; d_e is the equivalent diameter in the shell outside the fibers; v is average blood velocity; D is oxygen diffusion coefficient; l is hollow fiber length; k is mass transfer coefficient in blood; and v is the kinematic viscosity in blood.

$$= \left(\frac{\pi dl}{\frac{\pi}{4}d^2 l}\right)\phi$$

$$= \frac{4\phi}{d}$$

where d is the fiber diameter, l is the fiber length, and ϕ is the volume fraction of fibers in the module, normally around 0.5. Commercially available microporous hollow fibers typically have a diameter of about 300 μm. Thus, a is reasonably circumscribed, and any advantages will come from the mass transfer coefficient k.

Some of the correlations which are reported for the hollow-fiber mass transfer coefficient are given in Table 5.2–2. In these correlations, the mass transfer coefficient k is given as a function of many variables, in particular the fluid velocity, v. While this velocity can vary dramatically with the geometry of the hollow fibers, the velocity per length in blood oxygenators is normally fixed:

$$\frac{v}{l} \approx \frac{1}{\text{sec}}$$

Higher velocities usually imply higher shear, which can damage the blood.

We can now look at three special geometries of hollow-fiber oxygenators. In every case, we will look at the Sherwood number, for the largest Sherwood number means the largest k, and hence the fastest oxygenation. For flow inside the fibers, no matter how the fibers are arranged, we have

$$Sh = 1.62\left(\frac{d^2 v}{Dl}\right)^{1/3}$$

$$= 1.62\left(\frac{(300 \cdot 10^{-4}\text{cm})1/\text{sec}}{10^{-5}\text{cm}^2/\text{sec}}\right)^{1/3}$$

$$= 4.5$$

homogenization. "Cholesterol reducing" cheddar can then be produced by essentially the same process as traditional cheddar, using a carefully selected bacterial culture.

Prior to developing a full-scale manufacturing facility, we decide that we need to test a prototype for two reasons. First, we want to launch our product as cholesterol lowering. This claim will be considerably strengthened if we have experimental evidence that it is so, even though phytosterol alone is known to have this effect. We will need material for these experiments. Second, there will inevitably be some differences in flavor, texture, and odor between our product and conventional cheddar. While we believe these to be unimportant, we want to test the new product on prospective customers before investing heavily in a manufacturing facility. These tests are demanding, since we seek a prototype which is not only chemically similar to the final product, but also has the same physical structure and properties.

With the help of a specialist cheese manufacturer, we are able to produce enough material of good quality to test our prototype. Whom should we test it on? To investigate the cholesterol-lowering properties of our cheddar, we want to focus on a group who tend to have elevated cholesterol levels, such as middle-aged men. Double blind trials indicate almost 6% reduction in cholesterol over a three-week period of consuming 65 g per day of cheese, giving a 95% confidence level that the product is effective.

When it comes to testing the consumer acceptability of the product, we want to focus on those who buy cheese, mainly women. Market researchers subdivide consumers by age, aspirations, wealth, etc. into categories such as "Ms. Healthy," "Mrs. Homemaker," and "Mrs. Livelonger." While we may be appalled by such crass categorizations, we must cover the spectrum of consumers, which will require some such simplification. Our market testing is generally positive – all groups of consumers recognize some loss of quality relative to what they describe as "real" cheddar, but large numbers are willing to accept this and a price premium to achieve health. There seems a good chance that we can sell our healthy cheddar at a profit.

Our research does identify some causes for concern. In traditional cheddar, the fat droplets are solid at room temperature; but the fat in the cholesterol-reducing cheese is liquid. Although the encapsulation by protein is good, there is inevitably some leakage. Once removed from the packaging, this is barely noticeable, forming a thin oily film on the surface, but in plastic packaging it is a visible oily deposit. The group described by marketing as "Ms. Healthy" finds this unappealing and, more seriously, identifies the oil as unhealthy. Both "Mrs. Homemaker" and "Mrs Livelonger" use a significant fraction of the cheddar they purchase in cooking and report that our product does not melt like "real" cheese, but forms a greasy sponge when heated, which is also unacceptable. This is a consequence of the tougher protein matrix required to encapsulate the liquid fat, which is why a carefully selected bacterial strain was required.

We decide to proceed with the launch of our product in opaque packaging, labelled as unsuitable for use in cooking. We also decide to pursue research to improve the cheese's performance on heating, experimenting with the addition of different fat mixtures. Had we felt that the negative aspects revealed in the testing were too great, we might either have abandoned the project or returned to the selection stage to look for an alternative product, such as a yoghurt-based drink, which is unaffected by the liquid phytosterol fat.

This example illustrates how prototyping and testing is often beneficial before the investment of large sums into manufacturing plant, particularly when the product's success is dependent on the subjective judgment of consumers. It also shows that, even during prototyping, details of the product continue to evolve.

5.3 Scale-up/Scale-down

When we come to manufacture, we will often want to make more of our desired product, be it a commodity chemical or an expensive drug. If we are building a device, like a small stove for developing countries, then we will want to make more devices. Occasionally, we will want to make less of our product, as in a nanosphere adsorbed to a tumor that is releasing a chemotoxic agent. In every case, we will want to make a very different amount to our prior experience.

5.3–1 GENERAL ISSUES OF SCALE

This problem of scale is discussed frequently, not only in the scientific literature, but also in more popular books, ranging from the Bible to the present. For example, when Gulliver arrives on the island of Lilliput, he reports that (Swift, 1726):

> The Reader may please to observe, that in the last Article for the Recovery of my Liberty, the Emperor stipulates to allow me a Quantity of Meat and Drink, sufficient for the support of 1728 Lilliputians. Some time after, asking a Friend at Court how they came to fix on that determinate Number, he told me, that his Majesty's Mathematicians, having taken the Height of my Body by the Help of a Quadrant, and finding it exceeded theirs in the Proportion of Twelve to One, they concluded from the Similarity of their Bodies, that mine must contain at least 1728 of theirs $[12^3 = 1728]$, and consequently would require as much Food as was necessary to support that Number of Lilliputians. By which the Reader may conceive an Idea of the Ingenuity of that People, as well as the prudent and exact Economy of so great a Prince.

Is the Emperor's estimate correct?

Scaling laws, notably for metabolic rate, have been a hot topic in biology for over a century. The first theoretical speculation argued that heat loss from warm-blooded animals should be in proportion to surface area. This led in 1847 to Bergmann's rule stating that closely related species of animal should get larger as their habitat approaches one of the poles – hence polar bears are the largest of the bears, elk are the largest of the deer, etc. The first experimental data were gathered by Rubner in 1883 on dogs ranging from 3 to 30 kg, and confirmed the "surface law" that metabolic rate scales as surface area, not as body mass. In 1888, von Hoesselin showed that oxygen consumption in fish also scaled approximately in proportion to their surface area. Fish are cold blooded and so this cannot be heat loss. If we think of an animal as a chemical reactor, heat or mass transfer limited in one way or another, it should come as no surprise to chemical engineers to find that this "surface law" actually holds remarkably well for classes of organisms from bacteria to mammals. In 1932, Kleiber showed that the scaling is closer to the three quarters power of body mass. Whether the data are sufficiently different from the two-thirds power to be significant and, if so, what is the cause has been enthusiastically debated ever since. For example, West *et al.* (1997) ingeniously argue that the scaling of metabolic rate to the three quarters power of body mass can be justified on the basis of a distribution system (for example blood vessels) with a minimum unit size (for example capillary diameter) which is invariant with size of the organism, resulting in a fractal analysis of the problem. In the case of Gulliver, this would indicate around 200 Lilliputian portions is enough. The Lilliputian emperor got the wrong answer. Without careful analysis, we would, too.

To make this more careful analysis, we must consider the rate at which we want to make our product. Normally, we will already have made the product under known conditions. For example, if we are making a chemical synthesis, we will commonly know the percent conversion as a function of time. For an irreversible first-order chemical reaction, we may have

$$X = 1 - \frac{c_1}{c_{10}} = 1 - e^{-k\tau} \tag{5.3-1}$$

where c_{10} and c_1 are the initial and final concentrations of reagents, $(1/k)$ is a time characteristic of the chemistry, and τ is a time characteristic of the reactor. For a batch reactor, c_{10} and c_1 are the concentrations at the actual times zero and τ. For a steady-state plug flow reactor, c_{10} and c_1 are the inlet and outlet concentrations, and the time τ is the "space time," the quotient of the reactor volume V and the volumetric flow Q (or of the reactor length l and the average velocity through the reactor v).

We want the chemical time $(1/k)$ to be short. In more conventional terms, we want the reaction rate constant to be large. We can often get large rate constants by increasing the temperature. However, these gains taper off at higher temperatures because the process becomes diffusion controlled, and diffusion varies less with temperature than chemical rate constants do. The reaction rate constant now becomes $k_D a$, where k_D is a mass transfer coefficient, and a is an interfacial area per volume. This area per volume depends on the particle (or droplet or eddy) size and concentration, but k_D varies with stirring and fluid properties. To scale up a reactor we need to ensure that the residence time in the reactor τ remains the same and that the characteristic size of segregated material (particles, droplets, or eddies) is also constant. Often, this last condition can be achieved by keeping the same stirring power per unit volume on scaling up.

The scale-up of separation processes is a parallel to that of chemical reactors, though this is often obscured by details of the separation. Just as before, we consider the rate at which we want to separate our product. Again, the simplest case is an irreversible process. For example, for irreversible gas absorption, the result is

$$\frac{c_1}{c_{10}} = e^{-k_D a t} \tag{5.3-2}$$

where c_{10} and c_1 are the starting and ending concentrations of the gas being absorbed, $1/k_D a$ is a time characteristic of the absorption, and t is the time for absorption. The dimensionless group in the exponent is important enough to be given a specific name: the "number of transfer units," NTU.

Of course, Equation 5.3–2 has exactly the same mathematical form as the result in Equation 5.3–1 for a chemical reactor. This is a consequence of the fact that separations are first-order processes. Thus for a steady-state absorption, c_{10} and c_1 are the inlet and outlet concentrations, and t is again a space time, equal to V/Q.

The dimensionless group in the exponent of Equation 5.3–2 is easier to analyze than the corresponding group (kt) for reactors. The current group is the result of physical processes, not chemical ones. These are easily tabulated: just think of the values of viscosity, density, and diffusion coefficient in handbooks. There are no similar tables of chemical reaction rate constants. The smaller range of physical properties facilitates the organization of the $k_D a$, using dimensional analysis. Before developing this analysis, we consider three examples.

EXAMPLE 5.3–1 HIGH-FRUCTOSE CORN SYRUP

We have been running a 300 cm^3 batch experiment which gives 90% conversion of 0.05 g/cm^3 glucose to fructose in 6 minutes. The reaction, limited by chemical kinetics, uses an enzyme supported on 400 μm beads. Using the same reaction conditions, we want to use a plug flow reactor to make 1 kg fructose per minute. How big should the reactor be?

SOLUTION

For the same conditions, we want the same value of $(k\tau)$, which in the new reactor is (kV/Q). Because k is unchanged in the new reactor, V/Q equals 6 min. The amount made in the old reactor is $0.9 c_{10} V$, 13.5 g. We need (1000/13.5) times as much, requiring a reactor volume of 22 L. This suggests a flow rate of (22 L/6 min) for the larger reactor. Remember that this assumes that the process is chemically limited, unaffected by mass transfer. This may not be a good assumption in this case, since the flow pattern over the beads supporting the enzyme will be completely different in the two cases; some experiments will be needed to check this.

EXAMPLE 5.3–2 REACTING SUSPENDED STEROIDS

You are reacting a suspension of steroid particles, about 2.60 mm in diameter, with butyl lithium in tetrahydrofuran at 253 K. You believe that the reaction is mass transfer controlled, with the mass transfer coefficient k_D given by the correlation (Boon-Long *et al.*, 1978):

$$\frac{k_D d}{D} = 0.46 \left(\frac{dd'\omega}{\upsilon}\right)^{0.28} \left(\frac{gd^3}{\upsilon^2}\right)^{0.17} \left(\frac{M_T}{\rho d^3}\right)^{-0.011} \left(\frac{d'}{d}\right)^{0.02} \left(\frac{\upsilon}{D}\right)^{0.46}$$

where d is the particle diameter; D is the reagent diffusion coefficient in the liquid; d' and ω are impeller diameter and speed, respectively; υ and ρ are the kinematic viscosity and density of the liquid, respectively; g is the acceleration due to gravity; and M_T is the particle mass.

You want to scale up the reaction 1000 times using the same size particles. How should you proceed?

SOLUTION

This is a good example of a problem with extraneous information. If you look at the correlation, you see that the only variables you can control are d' and ω. But

$$k_D \propto (d'\omega)^{0.28}(d')^{0.02}$$

If we are scaling up 1000 times with a geometrically similar reactor, then d' increases ten times. Thus to keep k_D the same, we should decrease ω about twelve times. This is close to scaling at constant Reynolds number ($dd'\omega/\upsilon$).

EXAMPLE 5.3–3   PACEMAKERS FOR PETS

Heart implants and organ transplants are becoming increasingly routine surgery for humans. Our healthcare company proposes to diversify its market by launching a range of pacemakers for pets. Management argues that mammalian hearts are all similar, that our company's current design of pacemaker is a small implantable electronic device, and so the current model designed for humans should be suitable for pets almost without modification.

Should we expect this to be correct?

SOLUTION

Pacemakers stimulate the heart at a specified rate. We will need to estimate the rate at which we set the pacemaker for different animals. The blood volume required for metabolism by any mammal is a product of this rate and the heart volume. The blood volume for metabolism is proportional to the body mass to the three quarters power. The volume of the heart per body mass is remarkably similar for all mammals: between 5 and 6 g of heart per kg of body. (The exceptions, racehorses and greyhounds, have unusually large hearts. The legendary horse Eclipse was found to have a heart weighing 7 kg at his death in 1789, twice the weight for the average horse of the period; Secretariat, a descendent of Eclipse, was estimated to have a 10 kg heart.) For normal mammals, the rate for the pacemaker, proportional to the blood volume needed divided by the heart volume,

Table 5.3–1 *Basal heart rates of various mammals* The values shown are consistent with our expectation that heart rate should be proportional to the mammal's mass to the (–0.25) power.

Mammal	Basal heart rate/ beats per minute
Whale	20
Horse	45
Human	70
Cat	150
Hamster	330
Shrew	600

should be proportional to the body mass to the minus one fourth power of the body. This seems to be true, as shown by the data in Table 5.3–1. Our management is correct: this is a possible market.

5.3–2 DIMENSIONAL ANALYSIS

The principle tool used by chemical engineers to achieve scale up or scale down of processes is dimensional analysis, especially the use of dimensionless groups. Such groups are a powerful idea. Indeed, dimensional analysis is possibly the greatest intellectual contribution of engineering to science. Whatever their merits in analyzing physical problems, dimensionless groups are undoubtedly useful for intimidating chemists, who feel that they should know what the groups mean, but do not. Groups with Germanic names – Schmidt, Prandlt, Thiele, Damköhler – are especially good intimidators, because they echo vanished heroes of German chemical science.

Dimensional analysis is based on the principle of dimensional homogeneity, which states that both sides of an equation describing physical quantities must be dimensionally equivalent, something which will seem close to a tautology to anyone schooled in engineering or science. Newton understood this idea, which he called the "great principle of similitude"; Rayleigh formalized its presentation. To begin this analysis, we must accept that all physical quantities can be described by a limited set of fundamental dimensions. Exactly how many of these there are is a matter of some debate, but for our purposes length [L], mass [M], time [t], temperature [T], and amount of material [n] suffice. Details of these are given in Table 5.3–2. It is occasionally convenient to add a sixth fundamental dimension, heat. While heat is interchangeable with work and has dimensions [mass × length^2/time^2], problems where there is negligible interchange between thermal and mechanical energy are simplified by including heat as an additional fundamental dimension.

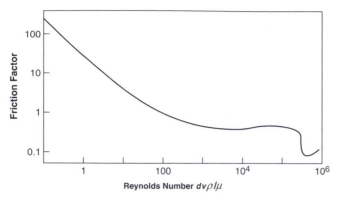

Figure 5.3–1 *Drag Coefficient vs. Reynolds Number for Smooth Spheres* Many sports are played with balls moving at a similar Reynolds number. Hence, during play, the balls have about the same drag coefficient.

polygons which make up their surfaces, cricket balls have raised seams, golf balls have dimples, baseballs have stitching. Why are sports balls the sizes they are and why are they not smooth?

SOLUTION

We can start to understand this by using the dimensional analysis given above for the force on a sphere moving through a static medium. Figure 5.3–1 shows a plot of Reynolds number against friction factor. Our dimensional analysis suggested that all spheres should show the same behavior and so this should be a single curve, and this is confirmed by the figure which is made up of data from smooth spheres of many different sizes and speeds. However if we made the same plot for spheres with rough surfaces, we would get a somewhat different result, approximately the same curve but shifted left a bit. This illustrates a limitation of the analysis. We did not include surface roughness in our list of variables which could affect force on the ball, and so it inevitably did not get included in our dimensionless groups. If we start with an incomplete list of relevant variables, we will inevitably end up with an incomplete analysis. Nonetheless the Reynolds number plot is very useful – any smooth ball will fall on the same curve, regardless of its size; any rough ball will fall on a different but qualitatively similar curve.

Table 5.3–3 shows the sizes of balls in some sports, the top end of ball speeds at which the games operate, and the resulting Reynolds number. All these sports operate at approximately the same Reynolds number, 2×10^5. Volleyball and soccer operate at slightly higher Reynolds numbers, possibly because these balls are smoother and lighter than those of other sports. This Reynolds number corresponds to the dip in the curve for a rough sphere – this is the point at which turbulent separation of the boundary layer occurs, reducing the form drag. A rough surface encourages local turbulence and so reduces the Reynolds number at which this transition occurs. Sports balls are the size they are so that they

> **Table 5.3–3 Reynolds numbers for various sports balls**
> Note that the kinematic viscosity of air (the ratio of viscosity to density) is 15.11×10^{-6} m^2 s^{-1}. It varies slightly with temperature, pressure, and humidity.
>
> | | Typical top speed (mph) | Ball diameter (mm) | Reynolds number |
> |---|---|---|---|
> | Golf | 160 | 43 | 2.04×10^5 |
> | Baseball | 95 | 75 | 2.11×10^5 |
> | Tennis | 115 | 64 | 2.18×10^5 |
> | Cricket | 90 | 68 | 1.81×10^5 |

operate in the window of minimum friction, allowing the games to be played at maximum speed. Moreover, if the ball slows, it can transfer into a different flow regime, causing discontinuous changes of behavior with speed. This makes for more interesting games – the bend on Beckham's free kicks, the break on a baseball curve, and the late swing of a cricket ball are in part due to transitions of flow regime as the ball slows down over its trajectory. In some sports, such as squash or ping pong, the balls are smooth because they are too small to allow them to get anywhere close to the turbulent transition. These games are played in a different flow regime, without a discontinuity of behavior, and so there is no point in roughening the surface of the balls.

EXAMPLE 5.3–4 MASS TRANSFER CORRELATIONS

For the diffusion-controlled reaction on the surface of catalytic particles, we expect the mass transfer coefficient k_D to vary with the flow velocity v, the diffusion coefficient D, the kinematic viscosity υ, and the diameter of the catalyst particles d. Use dimensional analysis to develop a possible correlation for k_D.

SOLUTION

We have five quantities involved: k_D, v, D, υ, and d. Because these are functions of two dimensions (L and t), there will be $(5-2)$ or three dimensionless groups. If we choose d and D as the primary variables, we get

$$\frac{k_D d}{D} = f\left(\frac{dv}{D}, \frac{\upsilon}{D}\right)$$

The group $(k_D d/D)$ is called the Sherwood number, (dv/D) is the Péclet number, and (υ/D) is the Schmidt number. Alternatively, if we choose υ and v as the primary variables, we obtain

$$\frac{k_D}{v} = f\left(\frac{dv}{\upsilon}, \frac{\upsilon}{D}\right)$$

where (k_D/v) is the Stanton number, and (dv/υ) is the Reynolds number. Neither of these results is that most often used, which is

$$\frac{k_D d}{D} = f\left(\frac{dv}{\upsilon}, \frac{\upsilon}{D}\right)$$

All three results are equally correct. This example illustrates both how dimensional analysis helps to organize our data, and that dimensional analysis may not give us an organization that we want to use.

EXAMPLE 5.3–5   CHICKEN ROASTING TIME

How long should you roast a chicken?

SOLUTION

We must first determine what are the variables controlling how long it takes to cook a chicken. As a simplification we will consider the chicken to be immersed in a fluid of constant and unchanging temperature – if we have a good oven this will be a reasonable approximation. A chicken may be considered cooked when all points within it reach a certain minimum temperature. Hence, to determine cooking time we need to be able to calculate temperature at an arbitrary point within the chicken as a function of time. Immediately we have five variables: three position coordinates (x, y, and z), time (t), and temperature at the defined position (T). We will choose to use increase in temperature from the initial temperature of the chicken (T_0) rather than actual temperature, since we have no control over the starting temperature (normally room temperature). Another obvious variable is the oven temperature (T_1). Again, we will choose to express this as a difference to the initial temperature of the bird. In expressing our temperature variables as these differences we are implicitly assuming that the important mechanism of heat transfer is conduction (proportional only to temperature difference) and not, radiation (which is also a function of absolute temperature). We see how using dimensional analysis relies on physical insight as well as application of a set of rules. The remaining variables are the thermal diffusivity of the chicken α, assumed to be constant over time and space, and the size of the chicken. If we assume chickens of all sizes to be geometrically similar, then we can express this as a single length dimension, L; this might for example be the distance from the parson's nose to the neck. Thus we have:

$$\{T - T_0\} = f(x, y, z, t, \{T_1 - T_0\}, \alpha, L)$$

We have eight variables in all. There are three dimensions involved: distance, time, and temperature. Thus we are looking for five dimensionless groups. The positional coordinates are easy: we will divide each by the characteristic length, L. A fourth obvious dimensional group is the ratio of the two temperatures. This

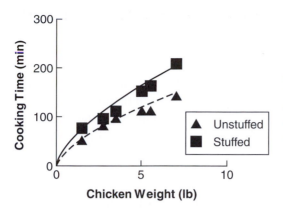

Figure 5.3–2 *Chicken roasting times* The solid curves, obtained from dimensional analysis, agree with the data points, which are experimentally determined roasting times.

leaves t and α (dimensions $[L^2 t^{-1}]$, which can be formed into the dimensionless group $\alpha t / L^2$, giving:

$$\frac{\{T - T_0\}}{\{T_1 - T_0\}} = f\left(\frac{x}{L}, \frac{y}{L}, \frac{z}{L}, \frac{\alpha t}{L^2}\right)$$

For cooking to be complete, we want the equivalent geometric position (the middle of the chicken) to have just reached the threshold temperature for meat to be considered cooked, whatever the size of the chicken. We expect to achieve this if the group $\alpha t / L^2$ is a constant independent of chicken size. As a result,

$$t \propto L^2 \propto (\text{mass})^{2/3}$$

We must compare this with experiments. Figure 5.3–2 shows a plot of recommended cooking times versus weight for chickens, either stuffed or unstuffed. The solid lines are consistent with the predictions of dimensional analyis. Clearly our analysis has worked. Note that we get different straight lines for stuffed and unstuffed birds, because the two cases are not geometrically similar.

5.4 Economic Considerations

Although it has been implicit in much that has gone before, we have not yet talked explicitly about money. To be sure, we have repeatedly mentioned that costs and prices are important, but we have not included these in any systematic way. Those with business training may find this omission overwhelming. These business persons may argue that chemical product design must include detailed discussions of financial issues. They may point to the large and detailed business literature, which includes examples aimed both at beginners and at accomplished professionals. They are correct to stress the value of these resources.

At the same time, our goal in this book is to focus on the chemistry and engineering which are central to chemical product design. Making detailed financial projections for a product which implies scientific nonsense is clearly folly. If the product does not work, what good is it? Moreover, in our judgment, the resources, including both people and literature, are much better developed for the financial aspects of product design than for the technical aspects of this same design, particularly in the area of the chemical industry. This is part of the reason why we have written this book.

We do want to include the briefest outline of the finances of chemical product design. We do so partly so that chemists and engineers will understand how the financial arguments are likely to be made. After all, we need the financial people to be our allies, so we need to know at least part of their language.

We also need to discuss the economics of chemical products because they are phrased in different terms to the economics of chemical processes. At present, most training in industrial chemistry describes the production of commodity chemicals, those made in optimized, dedicated equipment in huge amounts. Much less describes non-commodity chemical products, produced in whatever equipment is available often in quantities of a few handfulls. Many chemical engineers will have worked on chemical process design, where they will always know what they want to make. These engineers may be starting to work on chemical product design, trying to decide what to make. For example, many with years of industrial experience in petrochemicals may have taken new jobs in areas like consumer products or pharmaceuticals, where they are suddenly involved in deciding what to make.

For these groups, we want to supply a précis of economic issues. We begin in Section 5.4–1 with a description of the differences between products and processes. This distinction centers on how much of the desired product is made. In Section 5.4–2, we describe the economics of chemical processes. This branch of economics implies designing and building a chemical plant dedicated to the large-scale production of one product. In Section 5.4–3, we discuss product economics, especially in terms like net present value. This allows us to quickly estimate how much money we can hope to make from a particular product. We outline the business plan, which summarizes economic concerns, in Section 5.4–4. This entire section is intended only as the briefest of outlines, but one which stresses differences between the economics of non-commodity chemical products and commodity chemical processes. These differences are important in accurately assessing the economic potential of the possible chemical products.

5.4–1 PRODUCT VS. PROCESS DESIGN

The rationale for this book is that growth in the chemical industry will depend on new products as well as new processes. This is consistent with current corporate strategies. With a few exceptions, the large chemical companies are tending to de-emphasize their commodity chemical business to focus on non-commodity chemical products. In some cases, these companies have left the commodity chemical business altogether. In other cases, companies planning to make commodities have become private, presumably because they feel more confident of withstanding market cycles. In still other cases, especially in continental Europe, public companies plan to continue manufacturing commodities in parallel with an increased commitment to other products.

In this situation, we do well to ask what the difference is between commodity chemical processes and non-commodity chemical products. We are especially interested in how chemical products and commodities are judged financially. In this subsection, we explore these differences.

Commodity Products

To begin, we suggest differentiating between chemical products and commodities on the basis of three criteria: how much product is made, what equipment is used, and which producer makes the most money. We begin with commodity chemicals:

(1) *How much is made?* Commodities are normally made in quantities greater than 10,000 tons per year.

(2) *What equipment is used?* Commodities are normally manufactured in dedicated equipment which is operated continuously.

(3) *Which producer makes the most money?* As a general rule, the one with the lowest manufacturing cost will be the most profitable.

These generalizations deserve discussion.

The choice of 10,000 tons per year is the rough consensus of those in the chemical industry. Many of these chemicals are made from petroleum. They provide most of the examples in the chemical engineering undergraduate curriculum. While some inorganics are mentioned, usually in discussions of stoichiometry, the great majority are petrochemicals. Ethylene, polypropylene, and vinyl chloride are good examples.

These commodity organics are almost always made in very large chemical plants focused on making a single product. This has been the case since about 1970. Vinyl chloride is a good example: between 1964 and 1972, the number of US producers of vinyl chloride shrank by 70%, but the median plant size went up by more than a factor of ten. The reason is that the cost of a chemical plant is roughly proportional to the two thirds power of its capacity. While the reasons for this are complex, we can rationalize it by saying that the plant cost is proportional to the amount of steel needed, which is roughly proportional to the equipment's surface area, which is proportional to the two thirds power of the equipment's volume.

Thus, if we want to make commodity chemicals, we must be prepared for a huge capital investment. This is why the capital investment per employee is larger in commodity chemicals than in any other industry. This is why we are forced to operate continuously. We cannot afford to ever have so much expensive equipment sitting idle. We will normally be most profitable if we operate all day, every day of the year.

Most chemical commodities have been made for decades, using technology which does not change much from one year to the next. All commodities are sold into competitive markets. Moreover, the commodities are chemically well defined. For example, there is no difference whatsoever between ethylene made

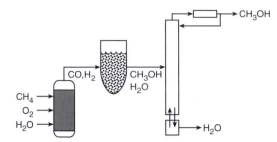

Figure 5.4–1 *Simplified methanol synthesis* Natural gas is burned to make carbon monoxide and hydrogen. These gases react to produce water and methanol, which are separated by distillation.

10,000 tons per year, which almost always implies a continuous process with dedicated equipment. We will never normally even consider batch equipment which would be used for several different processes. We will go straight to the continuous process with dedicated equipment.

Our second step will be the input–output structure, most commonly in the form of a block diagram. One example is for the manufacture of methanol, a liquid fuel. Methanol is made from natural gas. In a first reactor, methane is reacted with steam over a nickel catalyst at 850 °C and 40 bar:

$$CH_4 + H_2O \rightarrow CO + 3H_2 \tag{5.4–1}$$

Because this reaction is endothermic, some oxygen may also be fed to produce more carbon monoxide:

$$CH_4 + \tfrac{1}{2}O_2 \rightarrow CO + 2H_2 \tag{5.4–2}$$

This second reaction is exothermic, and can balance the heat needed. Third, the carbon monoxide–hydrogen ratio can be adjusted using the water gas shift reaction:

$$CO + H_2O \rightarrow CO_2 + H_2 \tag{5.4–3}$$

Next, in a second reactor, the gases are reacted over a copper–zinc catalyst, supported on alumina, to make the methanol product:

$$CO + 2H_2 \rightarrow CH_3OH \tag{5.4–4}$$

This reaction, carried out at 70 bar and 250 °C, only goes to about 40% completion. A block diagram of this simple process is shown in Figure 5.4–1.

The third step will involve identifying recycles. In this the most important will be to re-use the unreacted carbon monoxide and hydrogen.

Finally, in the fourth step of the design hierarchy, we need to specify the appropriate separations and the energy requirements. It is at this point that we will begin to use most aspects of chemical engineering. For example, we must decide how to separate the carbon monoxide, the hydrogen, and the methanol. In this case, we will use a combination of liquefaction and distillation.

Economic Potential

While we are designing the process using the hierarchy described above, we will make three sequential tests on the economic potential of the process. The first is to estimate the potential based only on the current prices of the product and the raw materials.

$$\begin{bmatrix} \text{economic potential} \\ \text{(first estimate)} \end{bmatrix} = \begin{bmatrix} \text{revenue from product} \\ \text{sales per year} \end{bmatrix}$$
$$- \begin{bmatrix} \text{raw material} \\ \text{cost per year} \end{bmatrix} \qquad (5.4\text{--}5)$$

We would expect this potential to be positive for an attractive chemical process.

Three implications of this simple definition of economic potential are worth mentioning. First, the stoichiometry of the process will be important. For example, in the case of methanol, the first synthesis gas reaction makes three moles of hydrogen per mole of carbon monoxide, but the methanol synthesis requires only two moles of hydrogen per mole of carbon monoxide.

The second implication follows from the first: we must have finished steps "i" and "ii" of the design hierarchy. In other words, we must have a reasonably exact idea of the process streams. Our estimate of economic potential will be dramatically altered by whether or not we run the water-gas shift reaction. If we do not, we get two moles of hydrogen from each mole of methane burned with oxygen. If we do run this extra reaction, we can approach three moles of hydrogen from each mole of methane.

However, the truly dramatic implication of this simple definition of economic potential occurs when we try to apply it to both commodity chemicals and to non-commodity chemical products. For commodity products, it is easy: we can look up the current prices of ammonia and methane. Imagine that we try to do this for the shut-down battery separator, or the pollution-preventing lithographic ink, or a new drug for treating depression. The battery separator sells for dollars per kilo, but the polyolefin from which it is made sells for a few cents per kilo. The new ink costs less than the old one because it no longer uses a solvent; it is made from the same raw materials but with altered reaction conditions. The drug may be made from expensive raw materials, but its selling price is not really known. The price will depend not only on the drug's effects, but also on an elaborate cultural situation. Just think of Viagra, the drug which enhances penile erections.

If the economic potential is positive after this first simple test, we can then apply a second criterion:

$$\begin{bmatrix} \text{economic potential} \\ \text{(second estimate)} \end{bmatrix} = \begin{bmatrix} \text{revenue from} \\ \text{product per year} \end{bmatrix} - \begin{bmatrix} \text{raw material} \\ \text{cost per year} \end{bmatrix}$$
$$- \begin{bmatrix} \text{utility cost} \\ \text{per year} \end{bmatrix} \qquad (5.4\text{--}6)$$

This implies that we know how much energy we will need to make our product. As a result, we must have finished at least initial estimates of all four steps of our design hierarchy. Again, to proceed we want this potential to be positive. If it is, we are ready to estimate how much the equipment will cost, and how much working capital we will need.

ropes (i.e. 4500 m additional rope in total) to implement our solution. Because 200 m of 32 mm nylon rope costs £550 we will have a cost of extra rope of £24 750. The ropes will need to be raised every two weeks for a couple of days during the tube-worm breeding season (May–August). This takes about 32 man days of labor, or around £3200 per year. (Casual labor is cheap in Scotland.)

A pilot study shows that raising the ropes achieves a 10% increase in mussel yield per year. This gives an increase in income of $150 \times 0.1 \times £810$ or £12,150 per year. Finally, we expect that the ropes will need replacing after five years.

We are now in a position to calculate the net present value of the project over its five-year life cycle. We assume that the cost of money is 8% per year. To begin, we see that

$$net\ income\ gain\ per\ year = £12\,150 - 3200 = £8950.$$

Though capital is spent at the start, income is gained in each of the successive five years, before the ropes must be renewed. Thus

$$NPV = -24\,750 + \frac{8950}{1.08} + \frac{8950}{1.08^2} + \frac{8950}{1.08^3} + \frac{8950}{1.08^4} + \frac{8950}{1.08^5}$$

$$= £10,985$$

$$ROI = \frac{NPV}{investment \times 5\,years} \times 100\%$$

$$= \frac{10,985}{24,750 \times 5} \times 100\% \approx 9\%\ per\ year$$

$$IRR = 24\%$$

because

$$-24,750 + \frac{8950}{1.24} + \frac{8950}{1.24^2} + \frac{8950}{1.24^3} + \frac{8950}{1.24^4} + \frac{8950}{1.24^5} \approx 0$$

We expect the investment to pay off, with a modest return on investment, of around 9% per year. Note that because we have taken the time value of money into account in calculating the net present value, this is the real rate of return, i.e. the rate of return above the assumed interest rate of 8%. We would need to achieve an interest rate in excess of 20% to gain the same profit as is predicted. The return is not huge, but the risk is also low, so it is a project worth investing in. The actual profit will be affected by factors such as the increased yield of mussels and the labor required to raise the ropes.

EXAMPLE 5.4–2   A SAMPLE EXECUTIVE SUMMARY

Imagine that we wish to build a business around the control of zebra mussel bio-fouling, described in Example 5.1–3. What would the executive summary for the business plan for this effort look like?

SOLUTION

A possible executive summary follows:

Silver Bullets for Zebra Mussels

Silver Bullets is a company founded to provide an environmentally safe and effective control method, initially for zebra mussels but with subsequent diversification into a wider range of products for targeted solutions in aquatic environments.

Our business plan is framed around a method for controlling zebra mussels, an invading species which blocks industrial raw-water systems, notably power-station cooling-water supplies. The key to our product is encapsulating a toxin in microscopic particles of edible material. The mussel's natural filtering then concentrates the toxin from the water into their own bodies. Compared to simply dumping the toxin into the water supply (the current method) this allows a much lower quantity of toxin to be used; our estimates suggest at least 1000 times less. This represents an enormous saving both economically and in terms of environmental damage. This last point is particularly important in freshwater environments, such as the Great Lakes, which is where the zebra mussel problem is at its worst.

We have strong patent protection for our product. The patent may be defended on the use of food-encapsulated particles in aquatic environments, and on the particle size range appropriate for mussel filtration. This gives us broad coverage, not only of our first product but also a large range of other potential products for use on filter feeding freshwater or marine species. We plan to diversify our product range under this patent umbrella. We have already identified another invading species, the Asian clam, which we believe our product will efficiently control. We plan to position the business in the broad area of specific, environmentally friendly pest control. Another area we are planning to begin testing in is enhancement of edible bivalve production and quality. One product we plan to develop is a feed containing minerals or other nutrients which limit growth of farmed oysters, clams, or mussels; this is covered by our existing patent.

We have tested our silver bullet for zebra mussels successfully at the laboratory scale. Research and development work is still required to identify a manufacturing route, but this type of encapsulated particle is common in pharmaceutical and agrochemical applications, so we anticipate this being achieved within a year. Manufacturing will be on the scale of a few tons per year and is thus best subcontracted out. The key hurdle we need to cross is market penetration. We anticipate this being possible only with the aid of an established sales and marketing organization and so a key plank of our business plan is the establishment of an industrial partnership with a power, water, or environmental-control company.

Our market is large and growing fast geographically. Its core is the Great Lakes region, where we will concentrate our efforts initially. According to one source the total cost of zebra mussel infestations is five billion dollars annually. While this estimate is necessarily crude, it does give an indication of the scale of cloth would you wish to protect by patenting? Design a series of experiments to optimize the design of the cloth, to verify its effectiveness, and to determine if the project is likely to be commercially viable.

7. *Bacteriophages.* There is an increasing problem in the world of resistance of bacteria to penicillins. This is largely caused by over-prescription in the West. A possible solution is the use of bacteriophages. Discovered independently by a Canadian biologist d'Herelle and an English microbiologist Twort, bacteriophages are a class of virus which "eat" bacteria.

There are many types of bacteriophage, each of which attacks a specific bacterium. Bacteriophages found limited use in treating dysentery in the First World War but were never used widely in Western medicine, and completely disappeared following the discovery of penicillin. However their development in the Soviet Union continued, particularly in Georgia. In Tbilisi, bacteriophages have been used on a large scale and to treat a wide variety of diseases very successfully over decades. However the Soviet documentation does not meet FDA standards and the scientific resources in the Tbilisi labs are primitive by Western standards. Recently, there have been a few high-profile cases of successful use in the West in cases where the alternative was certain death. Your company proposes to develop bacteriophages in collaboration with Tbilisi scientists for use on penicillin-resistant bacteria. Consider the risks involved in this project and the steps required for commercialization.

8. *Pizza vending machine.* In the US pizza bought from a vending machine is microwave heated from frozen. This is quick and convenient. However in the European market, and particularly in Italy, it is believed to produce a low-quality product. In particular, it is seen as crucial to a high-quality pizza that the base be freshly mixed and kneaded. Your company is developing a pizza vending machine which will make a fresh pizza starting from raw ingredients, in under three minutes. Set final specification for this device and critically assess its viability.

9. *Plastic, disposable, sterile fermenters.* Many useful pharmaceutical products, such as insulin, are grown by genetically modified bacteria such as *E. Coli* under carefully controlled conditions. The primary unit operation that is used is batch fermentation with "traditional" fermenters is expensive, heavy, and requires stringent sterilization between batches to avoid contamination.

Your company would like to produce an inexpensive, disposable fermenter for use in developing markets. The product is to be made entirely of polyethylene, one of the cheapest and most widely available thermoplastic resins. Previous work by your colleagues has drawn up a quantitative list of needs (see Table 5.9A) along with identifying a novel technology that appears to be ideal for making the plastic heat transfer elements. As yet, no experimental work that is specific to your project has been carried out

---

Table 5.9A *Needs for a disposable fermenter.*

Essential
- Main fluid container volume is 750 ml.
- Must be able to heat 750 ml of water from 20 °C to 30 °C in 5 minutes.
- Must be able to remove heat at a rate of at least 20 W during the bacterial growth phase to maintain water at 37 °C.
- Must be inexpensive and sterile.

Desirable
- Heating/cooling element to consist of a strip of plastic ribbon containing microfluidic channels, immersed in the fluid.
- Plastic ribbon to contain 50 capillaries.
- Heat transfer fluid flowing through capillaries to be water, supplied at 30 °C.
- Capillaries within plastic ribbon to be elliptical, major axis 400 μm, minor axis 100 μm.
- Pressure drop through ribbon must be no more than 1 bar.
- Only 1 ribbon per fermenter to be used.

Useful
- Fluid container must be transparent
- Minimum approach temperature in the heat exchanger to be 4 °C.

---

and all information has been sourced from available literature. You have been asked to join the development team, consisting mostly of control and electrical engineers, to provide design expertise for the heat exchange components and also to give your opinion on some patents that could inhibit product development.

You have been asked to investigate the following matters:

(1) The business team is concerned about patents. For supply chain control, the plastic ribbon is to be manufactured in-house using technology under license from Cambridge University, which would give rights to patent applications US2009011182 and WO2008044122. A recent patent survey has uncovered what appears to be prior art in the form of US4440195, US5427316, and US3372920. Examine the claims from each of the above patents and patent applications and comment on whether or not you think the prior art invalidates the patent applications.

(2) The manufacturing team needs to know the exact geometric specifications of the plastic heat exchanger for the final design drawings. The electrical design team also needs to know whether they can use a cheap pump for the heating/cooling fluid flow that provides 1 bar positive pressure, or whether they have to use a more expensive unit that can provide 2.5 bar pressure. Stating clearly any assumptions you make, perform calculations to provide the requested information.

(3) Outline for the manufacturing team where the missing information is in this project and what further experimentation or investigation is required to verify the design prior to committing capital to the project.

Financial year	2000/1	2001/2	2002/3	2003/4	2004/5
Income	0	0	200,000	800,000	2,000,000
Costs	54,500	228,850	291,800	314,800	350,800
Profit before tax	−54,500	−228,850	−91,800	485,200	1,649,200

Calculate the IRR over this five-year period.

What type of assumptions are likely to have been made in producing these estimates of income and costs? What are the key risks you would like to evaluate further before putting your money into this project.

14. *CO₂ capture and sequestration.* One of the threats to our current civilization is believed to be global warming caused by carbon dioxide emissions. About half of the human-produced emissions are due to domestic heating and cooking, which is why the US government is urging better home insulation. The other half of emissions comes from cars, manufacturing, and electric power generation. Of these sources, electric power generation may be the easiest to change because it includes a small number of large generating plants burning coal. This is why governments will be considering a carbon tax on emissions from these plants.

Because you work as a technical resource person for an environmental group, you want to write a short report for a non-technically trained audience explaining how the CO_2 produced in making electric power can be captured and stored. This could be used as the basis of an elevator pitch or as an executive summary for representatives in your organization wishing to lobby for the use of carbon capture and sequestration. You also want to report the increased cost of electricity if this is done. This idea is currently a major emphasis in the US Department of Energy, with a budget of $1bn. Because of the enormous amount of literature on this subject, you plan to focus on just two sources: the September 25, 2009 issue of *Science*, and the Bhown Report (AIChE National Meeting, November 10, 2009) for the Electric Power Research Institute (EPRI).

Write a technical summary and your best estimate of costs. While you may express your personal opinions, please carefully identify these after describing the best technology.

15. *Kimchi fridge.* Kimchi is a Korean delicacy, made from fermented cabbage. Traditionally it is stored by burial in earthenware pots. It can be kept for years and improves with age. Most Koreans now live in flats and so do not have access to burial space for their kimchi, but its pungent aroma makes it undesirable to keep it in the fridge. Moreover, optimal storage conditions differ somewhat from those of normal refrigeration. Home manufacture and storage of this prized delicacy remain a central part of Korean culture and for this reason the dedicated kimchi fridge is popular. It is projected that within five years 60% of South Korean households will own a kimchi fridge,

a market approaching one billion dollars annually. A major manufacturing company wishes to enter this lucrative market and in order to do so is looking for a design with a unique selling proposition. One suggestion is the environmentally friendly kimchi fridge, incorporating a solar panel to be externally mounted (perhaps on a balcony or air-conditioning unit) to power the fridge. Your consultancy has been hired to explore possibilities. Investigate existing competitor products and suggest possible unique selling propositions for a product.

16. *NanoProducts plc.* NanoProducts plc has developed a hand-held genetic testing device for home use or application in the doctor's surgery. The test takes less than half an hour to administer, requires a drop of blood and is at least 95% reliable. The device uses disposable lab-on-a-chip technology and a re-usable reader. The unit cost of the tests is envisaged to be small compared to the value of the test to the customer. Write an elevator pitch for the company. The presentation should take less than 10 minutes and the aim is to persuade potential investors to read your business plan.

17. *Fresh spot.* Your company has developed an irreversible indicator of meat freshness. It is a small tab, 1 cm in diameter which can be pinned to the surface of the meat and which changes color in response to a combination of thermal history and reaction to the enzymes present in the early stages of perishing. It is not a fail-safe device, but the color change is a pretty good indicator that meat should not be consumed; a lack of color change cannot be taken as a guarantee of freshness, however. Different indicators are required for different meats: your company has already developed indicators for chicken, beef, and lamb, and is working on ones for pork and turkey. Fish is proving problematic: an indicator for salmon has been developed but a different indicator will be needed for each species of fish. Write an executive summary aimed at potential investors.

18. *Altitude Shoes.* Altitude Shoes is a company recently formed by two bright engineering graduates, who have found it irksome frequently to have to change shoes as they move between different working environments. In particular, they have frequently found that they need to change from flat to heeled shoes, or vice versa, during the working day. They have set up Altitude Shoes to address this problem and are developing a shoe which will have an adjustable heel height. Write a business plan for Altitude Shoes.

REFERENCES AND FURTHER READING

Azapagic, **A.**, **Perdan**, **S.**, and **Clift**, **R.** (2004) *Sustainable Development in Practice*, Wiley-Blackwell, London.
Boon-Long, **S. Laguerie**, **C.**, and **Couderc**, **J. P.** (1978) Mass transfer from suspended solids to a liquid in agitated vessels. *Chemical Engineering Science* 33, 813–819.
Edwards, **M. F.** (2006) Product engineering: some challenges for chemical engineers. *Chemical Engineering Research and Design* 84, 1–16.

Peters, M. S. and **Timmerhaus, K. D.** (1991) *Plant Design and Economics for Chemical Engineers*. McGraw-Hill, New York, NY.

Pratt, J. J. and **Banholzer, W. F.** (2009) Improving energy efficiency in the chemical industry. *The Bridge* **39**, 15–21.

Pressman, D. and **Elias S.** (2008) *Patent It Yourself*, 13th Edition. Nolo Press, New York, NY.

Sandler, S. I. (2006) *Chemical and Engineering Thermodynamics*, 4th edition. Wiley, New York, NY.

Sartori, G., Ho, W. S., Savage, D. W., Chludzinski, G. R., and **Wiechert, S.** (1987) Sterically-hindered amines for acid-gas absorption. *Separation and Purification Methods* **16**, 171–200.

Sefton, M., Hunkeler, D. J., Prokop, A., Cherrington, A. D., and **Rajotte, R.** (1999) *Bioartificial Organs*. New York Academy of Sciences, New York, NY.

Sonesson, U., Mattsson, B., Nybrant, T., and **Ohlsson, T.** (2005) Industrial processing versus home cooking – an environmental comparison between three ways to prepare a meal. *Ambio* **34**, 411–418.

Swift, J. (1726) *Gulliver's Travels*. B. Motte, London.

Turton, R., Bailie, R. C., Whiting, W. B., and **Shaeiwitz, J. A.** (2008) *Analysis, Synthesis, and Design of Chemical Processes*, 3rd Edition. Prentice-Hall, Upper Saddle River, NJ.

West, G. B., Brown, J. B., and **Enquist, B. J.** (1997) A general model for the origin of allometric scaling laws in biology. *Science* **276**, 122–126.

6

Commodity Products

Up to now, we have presented a general strategy for the design of chemical products. This strategy centers on a four-step template of needs, ideas, selection, and manufacture. In the needs step, the project team converts marketing information into specifications. These specifications, which should be as quantitative as possible, are the first major application of science and engineering in the product design process. In the second step, the project team generates and assembles ideas which can potentially meet this need. Professionals suggest that many ideas, perhaps about 100, are needed for successful development.

The third step, selecting the best idea, is often the most difficult in applying the template. Normally, the team will be choosing among three or four product ideas, all of which look good. Usually, the science underlying these good ideas will be incomplete, and so testing the ideas will include risk mitigation. Often, choosing among the ideas will involve factors like environmental legislation and public response, which may reflect incomplete science. The last step in the template, manufacturing, is less disquieting not because it is easier, but because it often has more technology and less public relations. Throughout this process, the organization's management will review the process. These reviews, or gates, should be critical, because many studies show that most product failures can and should be identified earlier than they are.

This product development template works well, but it is applied differently to different types of chemical products. This is because these different products have different key steps and because they rest on different aspects of science and technology. In the remainder of this book, we will explore in detail how these different types of products are best developed. We will emphasize which steps are the most difficult because this is when the product's core team will have trouble.

We have idealized chemical products as four types: commodities, devices, molecular products, and microstructures. These are characterized by different physical scales and have different difficult steps in their design. Commodity products, made in large chemical plants which usually make only one product, are treated as continua, without any molecular structure. Devices, which are basically small chemical plants, are designed macroscopically, with key

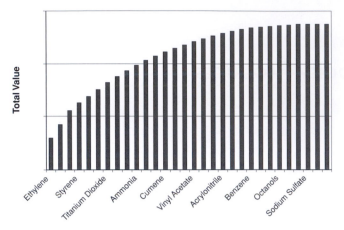

Figure 6.0–1 *Cumulative sales of chemicals* The number of commodity chemicals made continuously in dedicated equipment is small relative to the number of chemicals known.

dimensions from a centimeter to a meter. The key step in their design is selection between good alternatives, all of which will work well. Molecular products depend for their effectiveness on molecular structure, so their key scale is nanometers. The key step in their design is their discovery, which is beyond the scope of this book; however, once the target molecule is identified, the key to their design also becomes selection. Not surprisingly, microstructured products depend for their value on structures at the micrometer scale. The key step in their design is needs, and especially in rephrasing these needs as scientifically based specifications.

In this chapter, we summarize commodity product design. We will discuss other types of products in subsequent chapters. In our discussions of commodities, we should remember that we are discussing a small number of chemicals, fewer than 50 of the more than 30,000,000 compounds which are known. To illustrate this, we plot in Figure 6.0–1 the cumulative sales of chemicals like ethylene, styrene, and sulfuric acid. We note that the incremental value of these chemicals is under 10^{10} per year after the largest thirty or so. We do not assert that other chemicals are not useful or important: far from it. We do want to remind ourselves that the number of different commodities is far smaller than the prominence of commodity processing would suggest.

This chapter gives only the briefest summary of commodity process design, because this subject is detailed in dedicated books restricted to this subject. The chapter's structure is the same as that used in Chapters 7–9 for other types of chemical products. In Section 6.1, we describe the key characteristics of these commodities. In Section 6.2, we discuss how to begin commodity designs to make these products. This effort is based on material flow, the core of chemical process design. Sections 6.3 and 6.4 review, respectively, the toolboxes of reaction engineering and of unit operations, which are most useful for commodity design. Finally, Section 6.5 gives brief examples illustrating these designs. The entire

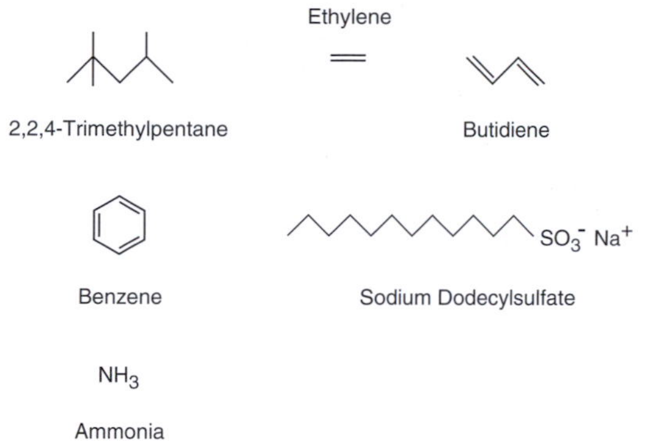

Ethylene

2,2,4-Trimethylpentane

Butidiene

Benzene

Sodium Dodecylsulfate

NH₃

Ammonia

Figure 6.1–1 *Typical commodity species* These examples, derived from fossil fuels, have low molecular weights and simple chemical structures.

chapter is short, because engineers will need only a review and chemists will be focused on other types of products.

6.1   Characteristics of Chemical Commodities

Commodity chemicals, the core of the chemical industry, are simple molecules produced in large quantities at the lowest possible cost. Their chemical structures are known, and are exemplified by the examples shown in Figure 6.1–1. The molecular weights of these products are typically less than 100 daltons. The king of these products is liquid fuel: gasoline, "petrol," and diesel. These are represented in the figure by 2,2,4-trimethylpentane which is the standard for "100 octane." Liquid fuel has value because it contains so much energy in such a small volume. We can illustrate this by imagining that while carrying four liters of gasoline, we push our car over 50 km of flat road. We then put the gasoline in the car and drive home. It takes the same energy for both the outbound and the return journeys.

The other commodity chemicals shown in Figure 6.1–1 are largely based on carbon. Ethylene and butadiene are olefins, used as building blocks for many other products, including polyethylene and butyl rubber. Benzene is an aromatic, a key to products like polystyrene and nylon. Sodium dodecylsulfate represents synthetic detergents; and ammonia is central to the manufacture of synthetic fertilizers and hence to the "Green Revolution" in agriculture. While some may feel uncomfortable with such fertilizers, we should remember that about two billion people depend on these for food. Without these chemicals, it is doubtful there would be enough food for the world's population.

The purity required for these commodity products varies widely. A typical specification for a technical-grade commodity is around 95%. However, the propylene purity for the manufacture of polypropylene may be 99.9%, while the fraction of hexane in petroleum ether can be less than 50%. Not surprisingly, higher purity costs more.

Sulfur-containing compounds, hydrogen sulfide (H_2S) and sulfur dioxide (SO_2), are less important in themselves than for the problems that they cause with manufacture. Hydrogen sulfide poisons catalysts. As a result, it is often separated

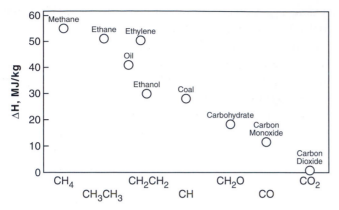

Figure 6.1–2 *Enthalpy of feedstocks* The higher enthalpy of natural gas and petroleum liquids means these feedstocks are a more energy-intense resource than biologically derived feedstocks.

using amines as described below. The separated H_2S is split into two streams. One is burned:

$$H_2S + \tfrac{3}{2}O_2 \rightarrow SO_2 + H_2O \qquad (6.1\text{–}1)$$

This stream is then reacted with more H_2S to make elemental sulfur:

$$SO_2 + 2H_2S \rightarrow 3S + 2H_2O \qquad (6.1\text{–}2)$$

These two reactions, called the Claus process, are a fixture of petroleum refining.

Commodity chemicals like those in Figure 6.1–1 are currently made from fossil fuels, and are likely to be made from such sources for the foreseeable future. The best feedstock is natural gas – methane – because it is abundant and has extra hydrogens to supply additional energy. The best alternative, petroleum, contains primarily hydrocarbons, with around five to thirty carbons. These hydrocarbon mixtures are often described as fractions with a rough range of number of carbon atoms. Naphthas have 5–9 carbons; kerosene has 9–16; and gas oil, including diesel, has 16–25. The fractions with fewer carbons are called "lighter"; those with more are called "heavier." This description refers less to their density or even to their molecular weight than to their viscosity.

Crude oil has more heavy molecules than light ones, but light molecules have a greater market. Thus, petroleum is reacted to reduce its molecular weight, a process called "cracking." Because this cracking usually reduces the molecular weight excessively, at least for the production of liquid fuels, the small molecules made by cracking are recombined, a process called "reforming." Cracking and reforming are part of the core of the commodity chemicals business.

Commodity chemicals can also be made from coal or from renewable sources. Indeed, in 1950, the Du Pont company asserted that their chemicals were "pure because they were made only from air, water and coal." In the future, limited petroleum reserves and abundant coal reserves will generate market pressure to use more coal. However, as Figure 6.1–2 shows, the enthalpy contained in coal is much less than that in petroleum or natural gas, so that making commodity chemicals from coal will be more expensive.

In the same way, making commodity chemicals from renewable feedstocks will cost still more. To show why, we remember that methane is CH_4, petroleum can be considered $(CH_2)_n$, coal is roughly $(CH)_n$, and biomass is approximately $(CH_2O)_n$. The enthalpies drop as the H/C ratio drops and as oxygen is included. It is certainly possible to make ethylene and benzene from biomass. However, for the next 50 years, the energy required for such transformations is probably going to be excessive. To phrase this in other terms, Figure 6.1–2 implies that, after the coming 50 years, methane and petroleum should be seen as too valuable to burn.

We now want to review the key chemical and physical properties of the commodity products typified by the compounds in Figure 6.1–1. We can organize this review around three ideas: the products' key scale, their chemical reactivities, and their volatilities. The key scale for the products would at first seem to be nanometers, for this is the scale of the molecules themselves. However, we routinely ignore this scale, and just treat the products as continua. We will only rarely invoke the molecular scale in designing the processes for the manufacture of commodities.

The second idea for review is that the products' chemical reactivities vary widely. Simple alkanes like ethane and octane are not very reactive, and require highly energetic reactants for even routine substitutions. On the other hand, butadiene is easily polymerized, and ammonia–air mixtures are highly explosive. We will normally avoid chemical details in describing the reactions of these commodities, treating these phenomenologically:

$$A + B \rightarrow C \tag{6.1–3}$$

Many joke that this implies that argon reacts with boron to make carbon. However, this willful neglect of chemistry in commodity chemical manufacture does not seem to cause major problems, as evidenced by the fact that many European chemical engineers study almost no chemistry after high school.

Third, commodity chemicals usually are highly volatile, a consequence of their low molecular weights. Examples in Figure 6.1–1 include ethylene and benzene, whose boiling points at atmospheric pressure are $-104\,°C$ and $+80\,°C$. The solubilities of commodities in water or heavy oils also vary widely, especially when they contain acid or amine groups. This means that these commodities will most frequently be purified by distillation or absorption, as reviewed in Section 6.3 below. Before this review, we turn to the issue of how we begin commodity product design.

6.2 Getting Started

Commodity chemical products are made in enormous quantities, well over 10,000 tons per year. Several synthetic routes for making each commodity are known, but one will normally be preferred because it has the lowest cost. For these products, cost is the undisputed king. Changes in the chemical process which reduce costs by a few percent can be worth millions and can completely alter the industry. We expect that most of our effort will be on manufacture. Still, we can use the same design template of needs, ideas, selection, and manufacture to improve the manufacturing itself.

Needs. Typically, our need is for a particular chemical at a purity above 95%, at an amount over 10,000 tons/year and at a price already defined by the market. We normally assume that our entering the market will not affect the price, which may not be true; however, subject to this assumption, we will normally continue, assuming a production cost perhaps 20% less than the price. Cost is our religion.

Ideas. The ideas for making this product may derive from the four-step outline suggested by Douglas (1988). Douglas recommends first deciding whether we want to use a batch process or a continuous one. Experience suggests that for commodities a continuous process will almost always be cheaper. Second, Douglas urges drawing a flow diagram, a chart illustrating the flow of different chemical streams in the reactor. For example, we may want to make ammonia from nitrogen and hydrogen:

$$N_2 + 3H_2 \rightleftharpoons 2NH_3 \qquad\qquad (6.2\text{--}1)$$

In our first guess of a flow diagram, as given in Figure 6.2–1(a), we show the nitrogen and hydrogen streams entering a process and leaving as ammonia. Note that this flow diagram includes both reaction and separation; note also that it is a flow diagram in space, echoing our intention to use a continuous, steady-state process.

Third, Douglas suggests recognizing that this process will not involve complete reactions. For example, for the conventional Haber process, we only get perhaps 20% conversion, even with a stoichiometric feed; thus, the process must have a reactor followed by some sort of separator, as shown in Figure 6.2–1(b). We are beginning to recognize the need for recycles.

Because ammonia synthesis is a classic process dating from 1910, we can find many details for this particular example given in the open literature. The temperature used is 400 °C; the pressure is about 150 bar. Many other details are concerned with the source of nitrogen and hydrogen, which are normally produced by burning natural gas in air. This means that the process must include separations to remove the carbon oxides produced. In addition, the use of air means that we must also somehow purge our process of argon, originally present in the air from which the nitrogen is made. This argon does not cause trouble in the reactor, but it must be somehow flushed from the process.

Fourth, Douglas suggests more detailed identification of the separations required. This leads to the more complex flow sheet shown in Figure 6.2–1(c). The gases from the reactor are chilled to 10 °C to condense liquid ammonia, and the non-condensables are recycled. Part of this recycle is purged to get rid of the argon. Unfortunately, significant amounts of nitrogen and hydrogen dissolve in the liquid ammonia. Once the pressure is released (to about

(a) The Overall Process

(b) Splitting Reaction and Separation

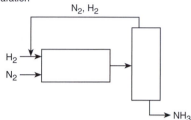

(c) A Simplified Flow Sheet of an Existing Process

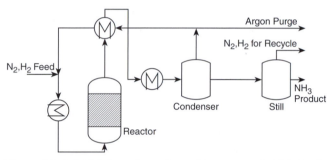

Figure 6.2–1 *Initial flow charts for making ammonia* This fertil-izer, basic to the Green Revolution, is currently made from natu-ral gas.

20 bar), the dissolved gases can be removed by simple distillation for recycle and recompression.

We seek ideas to improve this process. In the ammonia case, three targets seem obvious. First, we can seek better catalysts. Second, we can somehow get rid of the argon and so not waste the hydrogen and nitrogen necessarily discarded with the purge. Third, we can invent a better, more selective separation.

Selection. We must now select among our best ideas to choose the best one for detailed design. Because the Haber process is so important and a century old, each of these three areas has already been the target of enormous investment. The catalyst, the first area for study, has received the greatest effort. The original iron oxides used by Haber were found under reactor conditions to in fact include surface clusters of metallic Fe. The activity of these catalysts can be enhanced by trace quantities of many metals, especially ruthenium. Removing the argon, which is the second area of interest, would certainly be possible by liquefying air and distilling off the nitrogen, another highly developed, century-old process. While the separation of argon from oxygen is difficult because the difference in boiling points is so small, the separation of argon and oxygen from nitrogen is relatively easy. Because this has been well known for at least 50 years, we can infer that this distillation is too expensive, costing much more than purging the argon and wasting some nitrogen and hydrogen. The separation, the third area for effort, was originally achieved by cooling the reactor's product gases and washing them with water. Today, when reactor pressures are somewhat higher, the ammonia is condensed directly.

We may decide to focus on the third area, a better separation of ammonia from the other gases; past extensive efforts on ammonia catalysts make it unlikely that we will find anything new and we have no good ideas to eliminate argon. In this case, we may select three good ideas for better ammonia separations:

(1) a membrane 100 times more permeable to ammonia than to nitrogen and hydrogen at room temperature;
(2) a similar membrane selective and stable at reactor temperatures (c. 400 °C); and
(3) an adsorbent selective for ammonia at reactor conditions.

If any one of these ideas is successful, it could make the production of ammonia cheaper and more energy efficient.

Manufacture. The selection will normally lead to a single process improvement for making the target commodity. Such an improvement will most frequently be a change in catalyst, but could also be an improved separation, as in the ammonia example given above. The next step is a detailed process design incorporating this improvement, a design usually supported by commercially available process simulators like ASPEN and HYSIS. When this design is complete, it will be compared with a well-defined benchmark, which is normally the existing process. This comparison will most often depend almost completely on one criterion: will the new process cost less?

To illustrate this, imagine that we use matrix scaling and risk analysis to decide that an ammonia-selective membrane offers the greatest chance of improving the existing ammonia synthesis process, shown in simplified form in Figure 6.2–2. The membrane should separate the reactor output. Ideally, it will operate near the reactor temperature of 400 °C, and so will probably not be polymeric but ceramic. Because most ceramic membranes separate on the basis of size, we will actually need two membranes. The first will be permeable to the smallest species, hydrogen, which must then be recompressed. The second membrane will be permeable to the second largest species, the product ammonia, but retain the largest species, nitrogen, which can be returned directly to the reactor. We must compare the membrane selectivity needed and estimate the cost of recompression. Alternatively, if we are willing to cool the reactor effluent, we can use known membranes highly selective for ammonia over hydrogen and nitrogen. This replaces the cost of hydrogen recompression with the requirements of cooling the reactor output and reheating the recovered nitrogen and hydrogen. Again, we need to estimate the flux and selectivity needed to make this process cheaper than the existing one. Either membrane solution must also be as safe as the current process, a formidable challenge.

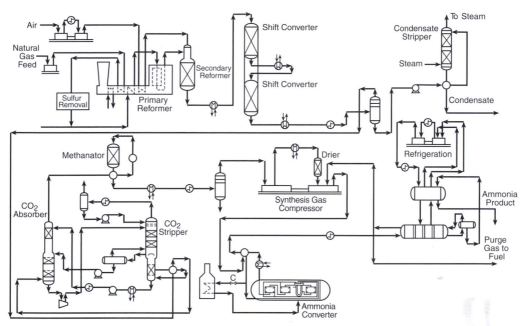

Figure 6.2–2 *A current process for making ammonia* The complexity of this process reflects the need to produce large amounts of ammonia at the lowest possible cost.

6.3   The Commodity Toolbox: Reactors

At this point, we need to review the tools which we will use to select the best ideas for commodity chemicals. Not surprisingly, these tools are the core of the discipline of chemical engineering. They are well covered in texts on chemical process design, and so we give only the briefest synopsis here. Basically, there are three important sets of tools: thermodynamics, reaction engineering, and unit operations. Other topics like fluid mechanics and process control are involved, but they are usually less critical to the design process.

Thermodynamics is in this context the science of what can happen, founded on the first and second laws. The first law is just an energy balance: the energy into a steady-state process equals the energy out. The second law is an entropy balance: the entropy of the universe must increase. In more popular terms, the first law is sometimes paraphrased as "you get what you pay for;" and the second as "there is no free lunch." For commodity chemical products based on hydrocarbons, thermodynamics is especially useful because extensive experimental measurements allow detailed, machine-assisted calculations. For example, we can calculate the equilibrium conversion of ethane into ethylene and hydrogen or the vapor–liquid equilibrium of propylene and propane. These values are helpful in selection among ideas because they can tell us what is possible and what is its minimum possible cost.

The next key group of tools comes from the engineering of continuous reactors. The three basic idealizations are batch reactors, plug flow reactors, and

A Batch Reactor

Volume V

A Plug Flow Reactor

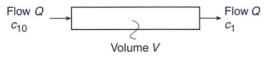

Flow Q
c_{10}

Flow Q
c_1

Volume V

Figure 6.3–1 *Basic chemical reactors used for commodities* The plug flow reactor, often with solid catalyst, is the most common.

A Continuous Stirred Tank Reactor (CSTR)

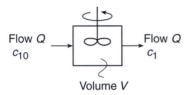

Flow Q
c_{10}

Flow Q
c_1

Volume V

continuous stirred tank reactors (CSTRs or CSRs). These three types are shown schematically in Figure 6.3–1. Normally, plug flow reactors are preferred because they give greater conversion per reactor volume and so are usually cheaper. Stirred tanks are sometimes useful for non-isothermal reactions. Think of a wood-burning stove: the logs that are already burning help to get the newly added logs to burn. No one would think of burning a log as a plug flow, lighting one end of the log and hoping that it will burn to the other end.

To put these ideas on a more quantitative basis, we begin by considering a first order batch reactor. We make a mass balance on the species "1," which is being consumed:

$$V\frac{dc_1}{dt} = 0 - rV = -k_R c_1 V \qquad (6.3\text{--}1)$$

where c_1 is the reagent's concentration; V is the batch reactor's constant volume; r is the reaction rate, in moles per volume per time; and k_R is the reaction rate constant, in units of reciprocal time. This mass balance is subject to the initial condition

$$t = 0 \quad c_1 = c_{10} \qquad (6.3\text{--}2)$$

where c_{10} is the initial reagent concentration. Integrating, we find

$$\frac{c_1}{c_{10}} = e^{-k_R t} \qquad (6.3\text{--}3)$$

Thus the reagent concentration drops exponentially with time. Almost everyone remembers this base case.

Alternatively, we can make a mass balance on a small volume of a plug flow reaction operating at steady state, which for the same reaction gives

$$\text{accumulation} = \text{flow(in} - \text{out)} - \text{consumption}$$

$$0 = -Q dc_1 - r dV = -Q dc_1 - k_R c_1 dV \qquad (6.3\text{--}4)$$

where Q is the constant volumetric flow through the reactor. This mass balance is not subject to an initial condition, because it is operating in steady state. Instead, it is subject to a boundary condition at the reactor inlet:

$$V = 0 \quad c_1 = c_{10} \qquad (6.3\text{--}5)$$

where c_{10} is now the reagent concentration entering the reactor. Integrating, we obtain

$$\frac{c_1}{c_{10}} = e^{-k_R \left(\frac{V}{Q}\right)} \qquad (6.3\text{--}6)$$

where (V/Q) is a "residence time" or "space time." Since Equations 6.3–3 and 6.3–6 have the same mathematical form, the conversion in a batch reactor and in a steady-state plug flow reactor will show the same exponential decay.

The stirred tank reactor behaves differently. A mass balance on this type of reactor yields for steady state:

$$\text{accumulation} = \text{flow(in} - \text{out)} - \text{consumption}$$

$$0 = Q(c_{10} - c_1) - rV = Q(c_{10} - c_1) - k_R c_1 V \qquad (6.3\text{--}7)$$

where c_{10} is again the reagent concentration at the reactor inlet. Thus

$$\frac{c_1}{c_{10}} = \frac{1}{1 + \dfrac{k_R V}{Q}} \qquad (6.3\text{--}8)$$

Now the reciprocal of the exiting reagent concentration c_1 is proportional to the space time (V/Q).

This simple picture is often useful not so much because reactions are frequently first order but because all but one reagents are frequently present in excess. The one limiting reagent not in excess may be expensive. It may be much harder to separate so that processing is easier if it is all consumed. In either case, no matter what the exact kinetics are, the reaction will behave as if it is first order. For example, imagine the actual reaction rate r is

$$r = k'_R c_2 c_1 \qquad (6.3\text{--}9)$$

where c_2 is a second reagent. If c_2 is 100 times bigger than c_1, its concentration won't change much while c_1 drops from its initial value to near zero. Thus we can write

$$r = (k'_R c_2) c_1 \qquad (6.3\text{--}10)$$

where the quantity in parenthesis is an effective first-order reaction rate constant. Thus non-first-order reactions will often behave as if they are first order.

Table 6.3–1 *Variations of limiting reagent concentration*
These reactions are assumed to be irreversible.

Kinetics	Batch	Plug flow	Stirred tank
First-order	$\ln c_1 \propto t$	$\ln c_1 \propto \dfrac{V}{Q}$	$\dfrac{1}{c_1} \propto \dfrac{V}{Q}$
Second-order	$\dfrac{1}{c_1} \propto t$	$\dfrac{1}{c_1} \propto \dfrac{V}{Q}$	$\dfrac{1}{c_1} \propto \left[\dfrac{V}{Q}\right]^{\frac{1}{2}}$ (a)
Zero-order	$c_1 \propto t$	$c_1 \propto \dfrac{V}{Q}$	$c_1 \propto \dfrac{V}{Q}$

(a) High conversion only.

In a few cases, the reactions will not be first order, and cannot be approximated as such. The two most common cases are second- and zero-order reactions. For second-order reactions,

$$r = k_R c_1^2 \tag{6.3–11}$$

where k_R is a second-order reaction rate constant, whose units are often volume per mole per time. For zero-order reactions,

$$r = k_R \tag{6.3–12}$$

where this rate constant k_R now has dimensions of moles per volume per time. To the novice, zero-order reactions may seem highly unlikely: after all, the reaction rate surely depends on the concentration of reagent. However, for many catalytic reactions, especially the gas–solid catalyst reactions so common for making commodity chemicals, the reaction rate may often be described by a rate like

$$r = \frac{k_R K c_1}{1 + K c_1} \tag{6.3–13}$$

where k_R and K result from the detailed chemistry. When $K c_1 \gg 1$, which is often the case, the rate will appear to be zero order. The behavior of first-, second-, and zero- order reactions in batch, plug flow, and stirred tank reactors is summarized in Table 6.3–1, and will be useful in some of the examples treated below.

The analysis above is restricted to dilute, irreversible, isothermal reactions. These restrictions are frequently not serious. To see when they might be, we imagine an isomerization reactor fed with one pure gaseous isomer "1" which is converted to a second gaseous isomer "2" at constant pressure and temperature. The rate of the reversible reaction is

$$r = k_R \left(c_1 - \frac{c_2}{K}\right) \tag{6.3–14}$$

where k_R is the rate constant for the forward reaction and K is the equilibrium constant for the reaction, i.e.

$$c_2^* = K c_1^* \tag{6.3–15}$$

where c_1^* and c_2^* are the equilibrium concentrations of 1 and 2, respectively. But at constant pressure and temperature, the total concentration c will be the sum of the concentrations of the two isomers:

$$c = c_1 + c_2 \tag{6.3–16}$$

Thus the reagent concentration c_1 varies with the conversion both because it is diluted by the product c_2 and because the reaction is reversible. This complicates the simple analysis which led to Equations 6.3–3, 6.3–6, and 6.3–8.

A second serious complication occurs because concentrated reactions are usually not isothermal but adiabatic. The resulting temperature changes may alter the reagent and product concentrations. Much more seriously, they change the rate constants, which are typically strong functions of temperature

$$k_R = k_R^0 e^{-E/RT} \tag{6.3–17}$$

where E is the activation energy. This energy is often roughly constant, independent of temperature. For many gas–solid catalytic reactions, however, the activation energy will be larger at lower temperatures and smaller at higher temperatures. This is because the reaction rate is governed by chemical kinetics at lower temperatures, but becomes influenced by diffusion of the gaseous reagents into the porous solid catalysts at high temperature. This complication and the others mentioned above are important for the cost of commodity chemicals, but are less important for other types of chemical products. We now turn to simple examples.

EXAMPLE 6.3–1 ESTER HYDROLYSIS IN DIFFERENT REACTORS

Batch laboratory results show that we can hydrolyze half of an ester in one minute. We want to react 95% of this ester in a reactor fed at 25 L/sec. If the reaction is first order and isothermal, how large a plug flow reactor is needed? How large a stirred tank is needed? Why are these results different?

SOLUTION

The rate constant can be found using Equation 6.3–3:

$$\frac{c_1}{c_{10}} = e^{-k_R t}$$

$$0.5 = e^{-k_R (60\,\text{sec})}$$

$$k_R = 1.16 \times 10^{-2}\,\text{sec}^{-1}$$

For the plug flow reactor, Equation 6.3–6 gives

$$\frac{c_1}{c_{10}} = e^{-k_R \frac{V}{Q}}$$

$$0.05 = e^{-\frac{1.16 \times 10^{-2}}{\text{sec}} \frac{V}{25 \times 10^{-3}\,\text{m}^3/\text{sec}}}$$

$$V = 6.5\,\text{m}^3$$

For the stirred tank, Equation 6.3–8 predicts

$$\frac{c_1}{c_{10}} = \frac{1}{1 + \frac{k_R V}{Q}}$$

$$0.05 = \frac{1}{1 + \dfrac{1.16 \times 10^{-2} \ V/\text{sec}}{25 \times 10^{-3} \ \text{m}^3/\text{sec}}}$$

$$V = 41 \ \text{m}^3$$

The stirred tank must be much larger than the plug flow reactor because the tank immediately dilutes the incoming feed and thus slows the reaction. This is why plug flow reactors are common for commodity chemical manufacture.

EXAMPLE 6.3–2 METHANOL SYNTHESIS FROM BIOMASS

Methanol is often suggested as a renewable fuel which can be made from biomass. After the biomass is dried, it is pyrolyzed to make carbon dioxide and hydrogen, which are then reacted over a Cu/ZnO catalyst at 50 bar

$$CO + 2H_2 \rightleftharpoons CH_3OH$$

The equilibrium constant for this reaction is 1.75×10^4 at 25 °C, but is 4.75×10^{-4} at the expected reaction temperature of 350 °C. What is the equilibrium conversion under each of these conditions?

SOLUTION

This problem is about equilibrium conversion, not about chemical kinetics. It is about how much we can possibly make, not how fast we can make it. From the stoichiometry, the equilibrium constant is defined as

$$K = \frac{p_{CH_3OH}}{p_{CO} p_{H_2}^2}$$

where each partial pressure, p_i, has units of bar.

We now define a conversion X. If we feed one mole of CO and two of H_2, we will have $(1 - X)$ of CO, $(2 - 2X)$ of H_2, and X of CH_3OH, for a total of $(3 - 2X)$ moles. Thus

$$K = \frac{\dfrac{3pX}{(3 - 2X)}}{p^3 \dfrac{3(1 - X)}{(3 - 2X)} \left[\dfrac{3(2 - 2X)}{(3 - 2X)} \right]^2} = \frac{X(3 - 2X)^2}{36p^2 (1 - X)^3}$$

We know K at a particular temperature; we know p is 50 bar; so we can find X. At 25 °C, X is greater than 0.98. At 350 °C, it is 0.65. But if conversion is so much better at 25 °C than at 350 °C, why do we ever choose to operate at the higher temperature? The reason is kinetics. At low temperature, the reaction takes far too long. Thus we will choose to run hot, even though the incomplete conversion implies complex separations and recycles. We turn to these separations next.

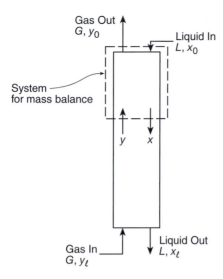

Figure 6.4–1 *Mass balance in gas absorption* The gas and liquid concentrations y and x are of the species being absorbed.

6.4   The Commodity Toolbox: Separations

Commodities are chemically well-defined compounds of modest molecular weight and high volatility. Thus commodity separations center on distillation and gas absorption. While distillation is much more important, gas absorption is simpler, and so is described first. As in our summary of reactors, our description begins with dilute solutions, because that limit is much simpler.

6.4.1 ABSORPTION

Absorption is most often used to separate gases, especially removing carbon dioxide and sulfur oxides. The mixed gas is pumped into the bottom of a packed tower, which is basically a piece of pipe stood on its end. The tower is filled with inert packing, which often looks like ceramic macaroni. A liquid – often an aqueous solution of reactive amines – trickles down the column, countercurrent to the gas flowing upwards. The target species, for example sulfur dioxide, dissolves in the amine, so the gas coming out of the top of the tower is purified. This is the technique used, for example, to prevent acid rain.

Designing an absorption tower is like finding a partner on an Internet dating service. We must answer two questions: how tall and how fat. To find how tall a tower is needed, we make a mass balance, an energy balance, and a rate equation for the tower. The mass balance, also called the "operating line," is made on gas and liquid flowing in and out of the dashed box in Figure 6.4–1:

$$yG + x_0 L = y_0 G + x L \tag{6.4–1}$$

where G and L are the gas and liquid fluxes, normally in moles per tower cross section per time; and y and x are the concentrations at the positions shown, normally expressed as mole fractions.

The energy balance, called the "equilibrium line," is just

$$y^* = mx + b \qquad (6.4\text{–}2)$$

where y^* is the concentration in the gas which would be in equilibrium with the liquid concentration x. The proportionality constant m is a solubility, closely related to a Henry's law constant; b is another constant. This simple equation is a linear approximation to the vapor–liquid equilibrium, which is not at all linear over large variations in concentrations. For this reason, this analysis is only valid if the concentration of the gas being absorbed does not vary very much over the length of the tower. The most important case where this is valid is for absorption of dilute gases, in which case $b = 0$ and m is Henry's constant; CO_2 and SO_2 stripping, the most common examples, usually both involve dilute gases.

The rate equation is the difficult part of the analysis of tower height. Like the operating line, this is also a mass balance, but written on only one phase. For example, for the gas phase,

$$[\text{accumulation}] = \text{flow(in} - \text{out)} - [\text{absorbed into liquid from gas}]$$

$$0 = -G\frac{dy}{dz} - K_y a\,(y - y^*) \qquad (6.4\text{–}3)$$

where G is still the convective gas flux flowing in the direction z, y is the mole fraction of the species being absorbed, z is the tower length, K_y is an overall mass transfer coefficient, a is the interfacial area between gas and liquid per tower volume, and y^* is as defined for Equation 6.4–2. The three quantities K_y, a, and y^* are tricky, and merit careful thought.

The interfacial area a, which is a function of the tower packing, has the greatest affect on the speed of the process and hence on the height of the tower. For a short tower, we want a large a and hence a small tower packing; but such packing may require a very fat tower, which is hard to operate. The overall mass transfer coefficient K_y includes the resistance to diffusion both in the gas and in the liquid. These resistances are functions of the mass transfer coefficients in the gas and the liquid phases, weighted by a Henry's law coefficient. Because the units of these quantities vary widely, the weighting needs care, as detailed in texts on this subject. For many cases of gas absorption, the dominant resistance is in the liquid. However, when the liquid reacts with the gas being absorbed, the mass transfer in the liquid is accelerated, and the two resistances may be more similar.

The remaining variable in Equation 6.4–3 is y^*, which is hardest to understand. This is the concentration that would exist in the gas if the gas were in equilibrium with the liquid, which it isn't. To understand this in more physical terms, imagine that the gas and liquid fluxes in the tower were everywhere in equilibrium. Then the actual concentration y would equal y^*. However y^* would not equal x: the concentration in the gas would certainly not equal that in the liquid. As an example, the mole fraction of oxygen in air is 0.21, but that in water in equilibrium with air is about 30 times smaller.

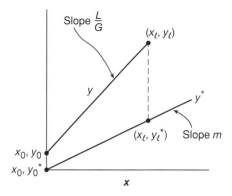

Figure 6.4–2 *Graphical analysis of gas absorption* This analysis uses a mass balance, called an "operating line," and an energy balance, called an "equilibrium line," to give a useful picture of the process.

Equations 6.4–1 to 6.4–3 can be combined and integrated between the boundary conditions shown in Figure 6.4–1 to give the tower height l for dilute gas absorption:

$$l = [HTU] \times [NTU] = \left[\frac{G}{K_y a}\right] \times \left[\frac{1}{1 - \frac{mG}{L}} \left(\ln \frac{y_l - mx_l}{y_0 - mx_0}\right)\right] \qquad (6.4\text{–}4)$$

The second quantity in square brackets, called the *NTU* or the "number of transfer units," is a dimensionless measure of the difficulty of the absorption process. The first quantity in square brackets, the *HTU* or "height of a transfer unit," measures the efficiency of the equipment; *HTUs* range from less than 0.1 m to 1.0 m, clustering around 0.3 m.

The equations given above are often supplemented by a graphical representation like that shown in Figure 6.4–2. The upper "operating line," of slope (L/G), graphs the actual gas concentration y vs. the actual liquid concentration x. The lower "equilibrium line," of slope m, plots y^* vs. x. The vertical distance between the lines $(y - y^*)$ is the driving force causing absorption from the gas into the liquid. If the operating line were below the equilibrium line, the process would involve transport from the liquid into the gas, which is called "stripping." Drawings like that in Figure 6.4–2 are key to extending this analysis to studies of concentrated gas absorption, as detailed in many chemical engineering texts.

This analysis allows estimation of the tower height l. To find out how fat the tower should be, we must consider the three-phase fluid mechanics of a gas moving upwards and a liquid moving downwards through the tower's solid packing. Such flows will be smooth only if the pressure drop along the packing is modest. Estimates of these effects, detailed in more specialized books, suggest changing the tower's diameter to alter its capacity while keeping the fluxes constant. Our estimates of tower height and diameter establish the absorber's design.

This analysis of gas absorption is about 80 years old, a foundation of commodity chemical processing. We might conclude that absorption had been optimized and that further improvements were unlikely. This conclusion would be

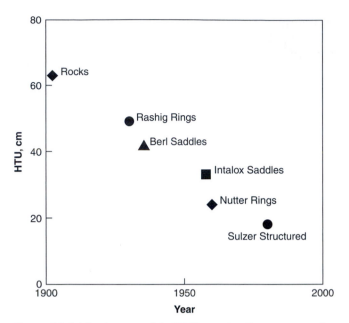

Figure 6.4–3 *The decrease of the HTU over time* Because we seek a small HTU, this graph illustrates the increasing efficiency of commodity chemical separations.

incorrect: absorption equipment continues to evolve, to allow more efficient separations. As evidence of this, we consider the evolution of estimated values of the *HTU* over the last century, shown in Figure 6.4–3. The figure shows that the *HTU* continues to decrease, so that the height of the tower required has got smaller. There is still considerable merit in trying to improve equipment for commodity chemical processing.

EXAMPLE 6.4–1   FINDING THE RATE CONSTANT $K_y a$

As part of a study of CO_2 sequestration, we are absorbing carbon dioxide with a new highly reactive amine. The gas flux of 0.23 mol/m^2 sec contains 8.6% CO_2; the liquid flux is large and can hold a lot of CO_2. The liquid enters free of CO_2. If 93% of the CO_2 is removed in a 2.6 m column, what is $K_y a$?

SOLUTION

From the data given, x_0 is zero, L is large, and m is small. Thus Equation 6.4–4 becomes

$$l = \left[\frac{G}{K_y a}\right] \ln \frac{y_l}{y_0}$$

$$2.6\,\text{m} = \left[\frac{0.23\,\text{mol/m}^2\,\text{sec}}{K_y a}\right] \ln \frac{0.086}{0.07(0.086)}$$

$$K_y a = 0.24\,\text{mol/m}^3\,\text{sec}$$

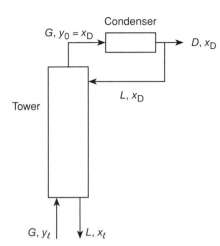

Figure 6.4–4 *Distillation producing an extremely pure product* This example is mathematically a close parallel with gas absorption.

We cannot separate the values of K_y and a without knowing details of the packing.

6.4.2 DISTILLATION

Distillation, which is the most important commodity separation process, boils a liquid mixture and sends the vapor into the bottom of a tower. At the top of this tower, the vapor is condensed, and much of it is returned to flow back down the tower. The result is that the condensate at the top becomes enriched in the more volatile species. Such a process is familiar to the general population because it is how liquor is made.

The contents of a distillation tower are frequently the same as those of an absorption tower, that is, ceramic macaroni or farfalle. They can also have periodic structure, like stacks of ceramic lasagna noodles. Then the analysis of this "differential" distillation is much like that of gas absorption. Alternatively, the distillation tower's contents may consist of a series of trays or stages, which encourage liquid and vapor to approach local equilibrium. While such a "staged column" is the norm in the analysis taught in chemical engineering, a "packed column," especially one with "structured" lasagna-like packing, is becoming the norm in practice.

As with absorption, we will focus our discussion of distillation on dilute solutions, because this is the easiest case. Initially, we consider distillation to make an extremely pure product, as shown in Figure 6.4–4. The column itself is the same as that for absorption: it has steady vapor fluxes G and liquid fluxes L. However, as described above, the exiting vapor is condensed completely. Part of this condensate is the distillate product D, equal to $(G - L)$. Part is returned to the top of the column. The composition of the vapor y_0, of the distillate x_D, and of the returned liquid x_0 are all the same.

We must now consider the contents inside the distillation column. If the contents are packing, then the analysis closely parallels that for absorption, and results in an estimate for tower height l:

$$l = \left[\frac{G}{K_y a}\right] \left[\frac{1}{1 - \dfrac{mG}{L}} \left(\ln \frac{y_l - y_l^*}{y_0 - y_0^*}\right)\right]$$

$$= \left[\frac{G}{K_y a}\right] \left[\frac{1}{1 - \dfrac{m[R_D + 1]}{R_D}} \left(\ln \frac{y_l - (b + mx_l)}{x_D - (b + mx_D)}\right)\right] \qquad (6.4\text{--}5)$$

where again the first quantity in square brackets is the *HTU* and the second is the *NTU*. The reflux ratio R_D, which is the liquid refluxed per distillate collected (L/D), is our chief means of controlling the column. The tower height l will be minimized when the amount of distillate collected is near zero, i.e. when R_D is infinite. We have again used an equilibrium line of the form of Equation 6.4–2; again this is only valid if the concentrations do not vary very much over the length of the distillation column. Hence, our restriction to the case of dilute distillation. Now the most common case of this is when the product is concentrated and we are removing dilute impurities. Then we are at the other concentration limit as that in the gas absorption case: the product mole fraction is close to one throughout the column. We are no longer in the Henry's law regime, b is not zero and m is not the Henry's law constant.

Alternatively, the contents of the column may be stages, small mixed volumes typically 0.6 m high, where liquid and vapor can approach equilibrium. In our analysis, we will assume that liquid and vapor actually do reach equilibrium. In fact, this is a terrible assumption, and often underestimates the number of stages by a factor of two or more.

Nevertheless, the normal analysis assumes equilibrium stages. By convention, the first stage is at the top of the column, and the stages are numbered going down the column. The concentrations are identified by the stage that they are leaving. Thus y_2 is the vapor concentration leaving stage "2" and entering stage "1"; x_0 is the concentration of liquid entering stage "1" at the top of the column. The analysis estimates the number of stages N from the Kremser equation:

$$N = \frac{\ln \left(\dfrac{y_{N+1} - y_N^*}{y_1^* - y_0^*}\right)}{\ln \left(\dfrac{y_{N+1} - y_1^*}{y_N^* - y_0^*}\right)} \qquad (6.4\text{--}6)$$

where y_{N+1} is the vapor concentration coming into the column, i.e. y_l in Figure 6.4–2; y_N^* is in equilibrium with the liquid leaving, i.e. it is $(mx_l + b)$; y_1^* is x_D; and y_0^* is the vapor concentration in equilibrium with x_D, i.e. $(mx_D + b)$. Like the analysis of absorption, staged distillation is also represented with graphs like those in Figure 6.4–2, even through the column actually does not contain continuously varying concentrations, but rather discrete values for the individual stages.

Unlike absorption, distillation usually involves concentrated solutions. The analysis of concentrated distillation is well established but complicated by different feeds, different feed locations, and highly non-linear vapor–liquid equilibrium. Because it is clearly and completely described elsewhere, we cover only the special case of constant relative volatility α, defined as

$$\alpha = \frac{y^*}{x}\left(\frac{1-x}{1-y^*}\right) \tag{6.4-7}$$

where the mole fractions are those of the more volatile component. When such a system is distilled at total reflux, i.e. at $R_D = \infty$, we can estimate the number of ideal stages as

$$N = \ln\left(\frac{x_D}{1-x_D}\frac{1-x_l}{x_l}\right)\Big/\ln\alpha \tag{6.4-8}$$

This "Fenske equation" is a mainstay of early approximate estimates of distillation. The similarities between differential and staged distillation are illustrated in the following example.

EXAMPLE 6.4–2   PURE BENZENE PRODUCTION

We want to separate a vapor containing 95% benzene and 5% toluene to yield a product which is 99.9% benzene. The equilibrium line under these conditions is

$$y^* = 0.58 + 0.42x$$

What is the minimum number of transfer units needed? What is the minimum number of stages, estimated from the Kremser equation?
      What is it estimated from the Fenske equation?

SOLUTION

The minimum number of transfer units or stages implies a very large reflux i.e. R_D is infinity. In other words, the vapor flux G equals the liquid flux L. Thus

$$y_1^* = x_D = x_0 = 0.999$$

Note that y_0^* $(= 0.58 + 0.42x_0)$ equals 0.99958. From a mass balance on the entire column,

$$y_l = x_l = 0.95$$

Again, from the equilibrium line, y_l^* $(= 0.58 + 0.42x_l)$ is 0.979. With this basis, the number of transfer units is

$$\begin{aligned}
NTU &= \frac{1}{1 - \frac{mG}{L}}\left(\ln\frac{y_l - y_l^*}{y_0 - y_0^*}\right) \\
&= \frac{1}{1 - 0.42}\left(\ln\frac{0.95 - 0.979}{0.999 - 0.99958}\right) \\
&= 6.7
\end{aligned}$$

The number of ideal stages found from the Kremser Equation 6.4–6 is

$$N = \frac{\ln\left(\dfrac{y_{N+1} - y_N^*}{y_1^* - y_0^*}\right)}{\ln\left(\dfrac{y_{N+1} - y_1^*}{y_N^* - y_0^*}\right)}$$

$$= \frac{\ln\left(\dfrac{0.95 - 0.979}{0.999 - 0.99958}\right)}{\ln\left(\dfrac{0.95 - 0.999}{0.979 - 0.99958}\right)} = 4.7$$

While the number of ideal stages is smaller than the *NTU*, the number of real stages will be larger because they won't reach equilibrium. Finally, to use the Fenske equation, we first use Equation 6.4–7 to find the relative volatility when x equals 0.97, an average value in the column:

$$\alpha = \frac{y^*}{x}\left(\frac{1-x}{1-y^*}\right) = \frac{0.9874}{0.97}\left(\frac{0.03}{0.0126}\right) = 2.42$$

Then the number of stages is estimated from Equation 6.4–8 as

$$N = \ln\left(\frac{x_D}{1-x_D}\frac{1-x_l}{x_l}\right)\Big/\ln\alpha$$

$$= \ln\left(\frac{0.999}{1-0.999}\frac{1-0.95}{0.95}\right)\Big/\ln 2.42 = 4.5$$

This agrees with the estimate of the Kremser equation, although the Fenske equation is not restricted to dilute solutions.

6.4.3 DISTILLATION EFFICIENCY

Distillation as reviewed above is an important process consuming one million barrels of oil per day – 4% of the total – in North America alone. Its efficiency of energy utilization is normally estimated as about 12%. This would seem to offer an enormous opportunity to increase the efficiency of commodity chemical manufacture and to reduce the cost of commodity products. This is not the case. In the next few paragraphs, we examine this issue, which requires more intellectual depth than the earlier parts of this section. Still, we feel that because this important point is infrequently discussed, we should do so here.

We define the efficiency of distillation η as

$$\eta = \frac{\text{energy of unmixing}}{\text{energy added to boil liquids}} \tag{6.4–9}$$

We now consider a column at 1 atm fed at a rate F and composition x_F to produce a distillate D of composition x_D and a bottoms B of composition x_B. An overall mass balance gives

$$F = D + B \tag{6.4–10}$$

A balance on the more volatile species yields

$$x_F F = x_D D + x_B B \tag{6.4-11}$$

Thus

$$\frac{D}{F} = \frac{x_F - x_B}{x_D - x_B} \tag{6.4-12}$$

The minimum free energy of unmixing to make nearly pure distillate ($x_D \approx 1$) and nearly pure bottoms ($x_B \approx 0$), is

$$\text{energy of unmixing} = -FRT\left[x_F \ln x_F + (1 - x_F)\ln(1 - x_F)\right] \tag{6.4-13}$$

The reboiler heat is $G\Delta \tilde{H}_{vap}$, where G is the vapor flux up the column. From the Clausius Clapyron equation,

$$\ln \frac{1\,\text{atm}}{p_2^{SAT}} = \frac{\Delta \tilde{H}_{vap}}{R}\left(\frac{1}{T} - \frac{1}{T_B}\right) \tag{6.4-14}$$

$$\ln \frac{1\,\text{atm}}{p_1^{SAT}} = \frac{\Delta \tilde{H}_{vap}}{R}\left(\frac{1}{T} - \frac{1}{T_D}\right) \tag{6.4-15}$$

where p_1^{SAT} and p_2^{SAT} are the pure component saturated vapor pressures at temperature T of the more volatile and the less volatile species, respectively; and T_D and T_B are their boiling points at 1 atmosphere pressure. Note that we have assumed that the enthalpy of evaporation $\Delta \tilde{H}_{vap}$ of the two components is the same. Combining these relations, we may show

$$\Delta \tilde{H}_{vap} \approx \frac{RT^2}{T_D - T_B}\ln \alpha \tag{6.4-16}$$

where T is an average temperature in the column, and the relative volatility α is defined at the feed concentration.

We now consider the case in the column where the reflux is a minimum, which is the opposite of that used in the previous subsections where the reflux was large. This case of minimum reflux will require the minimum energy, though it will also require a very large investment in equipment. Now the ratio of the liquid L and vapor G in the column is

$$\frac{L}{G} = \frac{L}{L + D} = \frac{R_D}{1 + R_D} = \frac{x_0 - y_F}{x_D - y_F} \tag{6.4-17}$$

This assumes a saturated liquid feed. The relative volatility, taken as constant, is

$$\alpha = \left(\frac{y_F}{1 - y_F}\right)\left(\frac{1 - x_F}{x_F}\right) \tag{6.4-18}$$

Combining Equations 6.4–9, 6.4–12, 6.4–13, 6.4–16, 6.4–17, and 6.4–18, we find after rearrangement,

$$\eta = -\left[\frac{T_B - T_D}{T}\right]\left[\frac{\alpha - 1}{(1 - x_F + \alpha x_F)\ln \alpha}\right]\left[x_F \ln x_F + (1 - x_F)\ln(1 - x_F)\right] \tag{6.4-19}$$

This is the result we seek.

Thus, the efficiency of this column is the product of the three terms in square brackets. The first is closely related to the Carnot efficiency: it is the temperature difference from reboiler to condenser, relative to the average temperature of the column. The second term in square brackets, which is an effect of enthalpy, is almost constant, of order one. The third term in square brackets is a new source of inefficiency, a theoretical upper bound in addition to the Carnot limit. The new term is due to the entropy of mixing, an efficiency beyond the thermal restraints covered by Carnot. To illustrate this, we consider the distillation of benzene and toluene, whose relative volatility is 2.42 and whose boiling points are 80 °C and 110 °C, respectively. For a feed of 50% benzene, the efficiency is

$$\eta = -\left[\frac{110 - 80}{95 + 273}\right]\left[\frac{2.42 - 1}{(1 - 0.5 + (2.42 \times 0.5))\ln 2.42}\right][0.5\ln 0.5 + 0.5\ln 0.5]$$
$$= -[0.082][0.90][-0.69] = 0.051 \qquad (6.4\text{–}20)$$

Because of the small temperature difference and the free energy of unmixing the process is inherently inefficient. Thus the chances of dramatically improving the efficiency of distillation are limited.

6.5   Using the Commodity Toolbox

We are now in a strong position to design a process to make our desired commodity product. We have chosen a product, a purity, and an amount which we want to make. We have assumed that we know the product's price. We have quickly generated ideas for making these products, summarizing these ideas as flow sheets. We have quickly selected the best flow sheet, which will often parallel that of our chief competition. We will often have an advantage over that competition, for example with a new catalyst.

We must then define the details of the product's manufacture. Most often we will design dedicated equipment with non-stoichiometric feeds, with the limiting reagent being the most expensive one. Occasionally, we may choose the limiting reagent to facilitate the separation. The reactors will usually be plug flow. The separations will be dominated by dedicated distillation towers. In these designs, we will depend heavily on widely available software developed specifically to optimize these designs. This software is especially dependable for estimating the physical properties of the low-molecular-weight hydrocarbons which are the core of the commodity chemical business.

Processes with these characteristics are carefully developed and lovingly detailed in texts on chemical process design. Note that the designs almost always imply equipment dedicated to a single product. Such dedication is key for commodities, as the following examples show. It is not key for other types of products described in later chapters.

EXAMPLE 6.5–1   HYDRODEALKYLATION OF TOLUENE

Benzene is the chief precursor of styrene and cyclohexane, the building blocks of polystyrene and nylon. Its synthesis from toluene is one of the most common

(a) The Overall Process

(b) Adding Reactions and Recycle

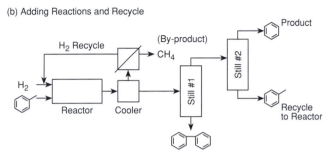

Figure 6.5–1 *Initial flow charts for hydrodealkylation of toluene*
The benzene produced is a feedstock for making polystyrene and
nylon.

examples studied in chemical process design. It is a magnificent example, dating
from a design problem set by the American Institute of Chemical Engineers in
1957.

The basic chemistry uses toluene and hydrogen at 700 °C and 34 bar:

Conversion is about 50%. The chief side reaction is:

To reduce this second reaction, we expect to need a ratio of hydrogen to toluene
over 5:1. We want to produce 1400 kg mol/hr, with a purity of over 99.9%. Design
this process.

SOLUTION

The needs are clearly given in the problem statement: 1400 kg mol/hr, with a
purity of three nines. The ideas will center around the preliminary sketch in
Figure 6.5–1(a); hydrogen and toluene go into the process, and benzene and
methane come out.

We next split the reaction and add the necessary recycles and separations.
The ideas for doing so will vary, and one possibility is given by the sketch in
Figure 6.5–1(b). We have cooled the reaction products to remove the gases
methane and hydrogen. We hope to somehow separate these gases – perhaps
with a membrane – so that the unreacted hydrogen can be recycled. Because the

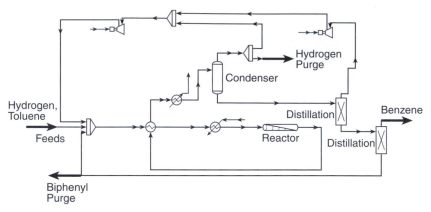

Figure 6.5–2 *A possible process for hydrodealkylation of toluene* This chart identifies the major pieces of equipment but does not size them.

boiling points of the benzene, toluene, and biphenyl are so different, we plan to distill these liquids in two stills. The first still removes the biphenyl in a bottoms stream; the second still produces the benzene as a distillate, a good idea for any final product. The bottoms stream of this second still, which is largely toluene, is recycled to the reactor.

Before we start to add important details, including heat integration, we can see how our ideas will develop. For example, how will we separate hydrogen and methane? While the figure implies using a membrane, membranes selective for methane over hydrogen are not known. Membranes selective for hydrogen over methane are known, but they are expensive and have a low flux. Pressure swing adsorption is a possibility but is of unknown cost. As a second example, the first column boils everything just to remove a little impurity. Is this column necessary? At the same time, methane and hydrogen dissolve in benzene–toluene mixtures. Will we need a third column to get these species out of our benzene product?

Solving the problems will take a month or more; however, doing so uses established calculations, correlations, and heuristics well known within chemical engineering. One possible answer, shown in Figure 6.5–2, has no gas separator, but just purges part of the hydrogen–methane mixture. This solution has no still #1, but does have a still #3 to remove dissolved gases from the liquid product. The biphenyl is allowed to build up, and purged from the toluene recycle.

The moral behind this example is that generating ideas about established process alternatives is easy. Deciding which alternatives are best and turning them into a viable manufacturing process is hard work, but uses well-established tools widely available to those engaged in process design. Because the key for commodity chemical producers is cost, the key to commodity process design may be influenced less by design details than by the cost of feed.

EXAMPLE 6.5–2 ACETIC ACID CONCENTRATION WITHOUT DISTILLATION

As a second example, we consider the recovery of acetic acid from dilute aqueous streams. Acetic acid is a major by-product of commodity chemical manufacture

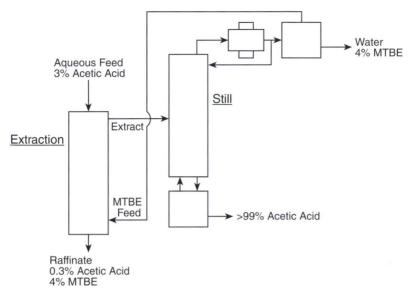

Figure 6.5–3 *An existing process for recovering acetic acid* The process first extracts the aqueous feed with an ether, then distills the extract.

produced by, for example, the manufacture of terepthalic acid and of cellulose acetate. This stream is often said to be one of the by-products with the highest unrealized value in the entire chemical industry. Because acetic acid boils at 118 °C and water boils at 100 °C, we would normally expect distillation to work well; however, the relative volatility is so small that such distillation requires a very high reflux with many stages. Because the acetic acid by-product is commonly a 3% stream in water, its separation is too expensive to justify conventional distillation.

Suggest an alternative separation which dramatically reduces the energy required.

SOLUTION

To make this need more quantitative, we seek to reduce the energy required by a factor of ten compared with a current benchmark. As a benchmark, we choose not a simple distillation but an existing alternative process, shown schematically in Figure 6.5–3. In this alterative, a feed containing 3% acetic acid is treated in a countercurrent extractor fed with methyl-*t*-butyl ether (MTBE). The acid, largely extracted into the ether, is then fed to a still. The bottoms stream out of the still is nearly pure acetic acid. The distilled vapors are condensed to form two phases: an MTBE-rich phase, which is recycled to the extractor, and a water-rich phase, which contains too much MTBE to be discarded. As a result, the water phase is stripped with steam to recover as much MTBE as practical. While this process is much cheaper than simple distillation, the increasing price of energy makes it less attractive economically. In particular, from mass and energy balances on this process, we expect that we need about 1000 kJ of energy to recover a mole of acetic acid.

Our idea generation leads to three alternative technologies: freezing, pervaporation, and adsorption. We need to evaluate which, if any, of these merits further effort. This evaluation is summarized next.

(1) *Freezing*. The first of the possible separation methods is freezing: we simply chill acetic acid solutions to form solid crystals. With a feed containing 3% acetic acid, the first crystals formed will be water ice. Fifty years ago, many investigated the possibility of purifying seawater in this way, removing pure water crystals from the resulting brine as a source of pure water. Interestingly, the best way to separate the crystals turned out not to be filtration but flotation. However, freezing was quickly eclipsed by reverse osmosis as a means of separating seawater.

Here, too, freezing looks like a weak option. After the water is removed, the remaining eutectic mixture contains about 20% acetic acid, which would then need to be separated by distillation. Such a eutectic feed means that over 80% of the water has been removed so that the energy required in distillation has been reduced. Still, the reduction is less than that achieved by azeotropic rectification in Figure 6.5–3. One Dow Chemical patent (US Patent 3,677,023, July 18, 1972) does teach this technology, but only for feeds containing over 80% acetic acid. For our case, freezing is a weak idea.

(2) *Pervaporation*. The second possible separation is pervaporation across a membrane. In pervaporation, we heat the liquid feed so that it evaporates through a membrane, producing a concentrated vapor. The separation can be dramatic, unaffected by azeotropes, a result of both differences in solubility and differences in diffusion coefficient. In our dreams, we can thus imagine heating a feed of 3% acetic acid and collecting a vapor highly enriched in acetic acid. We would not need to evaporate much water and so could drastically reduce the energy required for the separation.

This charming dream has spurred an enormous effort. Much of this effort has used polyvinylalcohol (PVA) membranes, because these are successful in the dehydration of alcohol solutions. Membranes made of grafted PVA, polycarbonate, polydimethylsiloxane, and supported organic liquids have been tried, as have composites with added zeolites and carbon particles. The scope of this effort is a testament to the real charm of this dream.

However, pervaporation and other membrane separations have two huge disadvantages. First, each membrane separation is one stage, and efforts at staged membrane separations have been uneconomic. Thus, the selectivity to separate the acetic acid completely must be high. Normally, it will not be. If the membrane's selectively were 20, a high but not impossible value, then the permeating vapor from a 3% acetic acid feed would still be only 40% acetic acid. On the one hand, this means that most of the water has been removed; on the other, it means that this permeate will still need to be distilled or separated in some other way.

The second disadvantage of pervaporation for separating acetic acid is more serious. We believe that all of the membranes studied to date are more permeable to water than to acetic acid. This is exactly the opposite of what we seek. Even if we have a highly selective membrane, we are evaporating water, the major component. We therefore will not have the significant energy savings which we seek. Unless we identify a pervaporation membrane which is more permeable to acetic acid than to water, membrane separations seem a poor prospect.

(3)   *Adsorption.* The third attractive separation method, which has the most promise, is adsorption on nanoporous solids. Unlike membranes, adsorption is easily staged; unlike extraction, adsorption does not produce an effluent which must itself be treated. Although adsorption is complicated because it is an unsteady process, its analysis is well understood. However, in investigating this alternative, we are implicitly accepting the production of an acetic acid concentrate, perhaps with 30% acid in water. We will need to separate this concentrate further, almost certainly by distillation.

Our challenge is to identify a suitable adsorbent. As before, we want an adsorbent for acetic acid, not for water: we want to remove the dilute component, not the more concentrated one. While ion-exchange resins are an obvious high-capacity choice, these must be regenerated by adding another acid, which is expensive. Though activated carbon could work, it is of limited capacity. The best choice is probably a solid, nano-porous amine. Liquid amines, especially in aqueous solution, are the standard method for treating acid gases like H_2S and CO_2 and are feasible for acetic acid. By adjusting the chemical structure of the amine, we can hope to make it easy to regenerate by changes in temperature. (Pressure-sensitive amines are also available, which could offer still lower operating costs.) However, such solid amines are not widely available.

We suggest investigating solid amines as a means of concentrating aqueous acetic acid from 3% to at least 20%. To do so, we must identify an amine adsorbent with a capacity of 10^{-3} mmol/cm^3. While this is a small capacity compared with an ion exchange resin, it suggests a surface area of over 100 m^2/cm^3. If the adsorbent were to adsorb acetic acid at 25 °C and release it at 50 °C, then the energy required would be 2 kJ/mol acid separated, which meets our need. However, we would still need to distill the concentrate produced by adsorption. We are closer to fulfilling our need, but we are not there.

6.6   Conclusions for Commodity Products

Commodity chemicals are low-molecular-weight species made in dedicated equipment in quantities of 10,000 tons or more per year. While the molecular structure of these products is exactly known, it is not carefully considered; instead, the commodities are treated as continua. The design of processes for making these compounds uses the same design template given earlier in the book: needs, ideas, selection, and manufacture. However, because the commodity processes are approximately known and cost of manufacture is so important, the first two steps take little time. The third step, selection, is harder, but will normally quickly lead to three or so different process steps which we feel are the most promising. Usually, we will be able to use matrix scaling and risk analysis to select one process change with the greatest potential.

Then our real work begins. We must design the details of the process needed to make the chemical product. The results will be compared in detail with a benchmark, which will normally be an existing competitor process. This comparison will use well-developed process simulators and physical property packages. We will include other criteria, like safety, in this evaluation. However, the most important criterion for a commodity chemical will almost always be cost.

Designs for chemical commodities depend on process details, and are well developed in texts dealing with chemical process design. Problems illustrating these designs are detailed and complicated. Because including these designs in this book would unbalance our treatment, we have omitted problems on this topic here. We refer those interested to the more conventional process-oriented texts.

The result of these efforts, efficient manufacture of large amounts in dedicated equipment, is a contrast with the design of other types of chemical products detailed in the remaining chapters of this book. There, the products are not treated as continua, and they are rarely made in dedicated equipment. Most importantly, the key step in the process design template is not manufacture. How these other designs proceed is described next.

REFERENCES AND FURTHER READING

Agrawal, R. and **Herron, D. M.** (1997) Optimal thermodynamic feed conditions for distillation of ideal binary mixtures. *AIChE Journal* **43**, 2984–2996.

Cussler, E. L. (2009) *Diffusion*, 3rd Edition. Cambridge University Press, Cambridge, ch. 10, 12, 13.

Douglas, J. M. (1988) *Conceptual Design of Chemical Processes*. McGraw-Hill, New York, NY.

Peters, M., Timmerhaus, K., and **West, R.** (2002) *Plant Design and Economics for Chemical Engineers*, 5th Edition. McGraw-Hill, New York, NY.

Pratt, J. J. and **Banholzer, W. F.** (2009) Improving energy efficiency in the chemical industry. *The Bridge* **39**, 15–21.

Seader, J. D., Henley, E. J., and **Roper, D. K.** (2010) *Separation Process Principles*, 3nd Edition. Wiley, New York, NY, ch. 7.

Seider, W. D., J. D. Seader, J. D., Lewin, D. R., and **Widagdo, S.** (2008) *Product and Process Design Principles: Synthesis, Analylsis, and Design*, 3rd Edition. Wiley, New York, NY.

Schmidt, L. D. (2005) *Engineering of Chemical Reactions*, 2nd Edition. Oxford University Press, Oxford, ch. 2, 3.

Turton, R., Bailie, R. C., Whiting, W. B., and **Shaeiwitz, J. A.** (2003) *Analysis, Synthesis, and Design of Chemical Processes*, 3rd Edition. Prentice-Hall, Upper Saddle River, NJ.

7

Devices

This chapter focuses on chemical devices which measure, make, or purify chemicals on a much smaller scale than the commodity chemicals described in the previous chapter. For example, we may wish to make oxygen-enriched air for a single emphysema patient, not for making steel. We may seek to remove water from the lubricating oil in one truck, not from a feed in an oil refinery. We may seek a ten-minute analysis of blood cholesterol from one drop of a patient's blood so the doctor has the analytical result before the patient's appointment is over.

In each of these cases, the device itself will be our desired product. The best device will not necessarily be the cheapest. This is a major change from commodity process design, where cost is king. The device which we seek to design will normally be much smaller than a conventional process: our device will usually be smaller than a meter and may be as small as a few hundred micrometers. However, the tools which we use will still assume that the chemicals involved form continua. We will still base our thinking on thermodynamics, chemical kinetics, and unit operations, just as we did in designing a process for chemical commodities.

As in the earlier chapters, we will base our designs on a four-step design template of needs, ideas, selection, and manufacture. As before, we will rewrite our needs as specifications, which will be quantitative. We will normally have no trouble in generating a lot of good ideas to meet these needs. Indeed, we are always impressed with how easily both professionals and students invent good possible designs for chemical devices. Thus, the first two design steps for chemical devices are often quick and easy, just as they are for chemical commodities.

The third "selection" step is the difficult one for chemical devices. As we will show, we will often invent three or more good processes meeting our specifications. All of these will work. When we choose among them, we will choose the "most convenient" or the "easiest to use" or the "hardest to misuse." We often won't choose the cheapest. Such choices involve more than economics, and so are uncomfortable for many technical professionals. While these professionals may be happier designing processes for commodities, the growth of the chemical enterprise is not expected to be in the commodity sector.

> **Table 7.1–1** *Examples of chemical devices*
> These are often designed to meet specifications beyond minimum cost.
>
> Small processes
> Membrane bioreactors (for sewage treatment)
> Used motor-oil purification
> Alcohol production from waste paper
> Artificial kidneys
> Home storage of solar energy
> Nerve-gas destruction
> Coffee makers
> Analysis (lab-on-a-chip)
> Pregnancy tests
> Specific ion electrodes
> Swimming-pool test kits
> Fast cholesterol analysis in a doctor's office
> Wearable calorie counters
> Controlled-atmosphere packaging
> Osmotic pump
> Breathable bottle seal
> Hot-flashes bra

We begin our discussion of devices with a brief description of their general characteristics in Section 7.1. In Section 7.2, we suggest starting device design with a synthesis of material flow that is parallel to the synthesis used for commodities. This may lead to a flow sheet in time, rather than one in space. The lines on the flow sheet may be sequential steps, like the steps of a recipe for making a cake. The lines on the flow sheet may not represent pipes, continuously moving material from one operation to the next.

Next, we will describe the toolbox used to select between good alternative designs. Essentially the same toolbox that is used for commodities, this is based on reaction engineering (Section 7.3) and unit operations (Section 7.4). The difference between devices and commodities is in emphasis. Stirred batch reactors are more important than plug flow reactors; homogeneous catalysis is more common than gas–solid reactions; distillation takes a backseat to adsorption. In other words, devices can be a "world turned upside down," a revolution in the sense of a reversing of the common choices. We end in Section 7.5 with more complex examples illustrating the combination of these various ideas to select the best design from a small group of good choices.

7.1   Properties of Devices

Chemical devices are often like small chemical plants, taking a chemical feedstock and producing a product. Some examples, given in Table 7.1–1, illustrate the wide range of processes involved. Many have the same mission as larger processes: water purification, solar-energy storage, or waste utilization. Often, however, they will produce pure water or fuel or reusable motor oil from a dispersed, local feedstock for a local market. Biomass utilization for renewable energy is one good example. China plans to collect biomass locally on a grid of unit area of 100 km^2. Biomass from each grid element will be processed locally.

A second group of chemical devices aims at fast, small-scale chemical analysis, examples of which are also given in Table 7.1–1. While some analyses like swimming-pool test kits and pregnancy test kits are aimed at individual consumers, the greatest market for these analyses is probably in medical testing. Current testing, often based on large samples of patient blood, uses large volumes of expensive reagents, some of which are highly toxic. Moreover, because these medical analyses are slow, performed on expensive equipment in centralized locations, the results are not available when the patient is talking to his doctor. If the analysis could be made fast on a drop of blood in the doctor's office, the cost would drop, waste solvents would be reduced, and communication between the patient and the doctor could be improved. This dream has fuelled the many research programs pursued under the umbrella of "lab-on-a-chip."

A third type of chemical device, also listed in Table 7.1–1, provides the consumer with a particular benefit. One example is "controlled-atmosphere packaging" or "modified-atmosphere packaging." Such packaging wraps fruits or vegetables in films which are unequally permeable to different gases. It exploits the metabolism of these foods to enhance their freshness and extend their shelf lives. In the same sense, osmotic pumps release drugs into a living system at a low, constant rate, chosen to maximize the drug's efficacy. They thus reduce the amount and the cost of drug needed and suppress undesirable side effects.

These examples illustrate that chemical devices operate at much smaller scales than commodity chemical processes. Instead of making 10,000 tons per year, they may process only millilitres per hour. Their design does use similar tools to commodity processes. However, it uses a different rationale based on different criteria, as explained in the next section.

7.2 Getting Started

The design of chemical devices begins just as the design of commodity processes does, with a synthesis of material flow. One such synthesis, based again on the ideas of Douglas (1988), is shown in Table 7.2–1. This synthesis contains four steps, shown in the first column of the table. In the first step, we decide if the process will be batch or continuous. In the second, we draw a flow diagram. Next, we choose the key reactor and separator. Finally, we decide on separations, supporting operations, and recycles. This four-step synthesis parallels the "synthesis of material flow," an excellent way to initiate the design of processes which make commodity chemicals.

For chemical devices, the synthesis in Table 7.2–1 makes choices which are broader and less predictable than those for commodities. Normally chemical

Table 7.2–1   *Device design vs. commodity design.*

Design sequence	Devices	Commodities
Type of process	Batch	Continuous
Process diagram	In time	In space
Essential step	Reactor or separation	Reactors with recycles
Supporting steps	Large variety	Distillation, gas absorption
Energy use	Rarely important	Integrated, optimized

devices will be batch. Normally we will be processing only a small amount, so we will operate the process only periodically. For example, we will reprocess motor oil only when we have collected enough oil to make it worth our while. We will analyze well water for coliform bacteria only when we have samples of well water. Sometimes, we will operate our device in steady state, as we will for a kidney dialysis patient whose blood we must purify. Even then, however, we keep an average patient on dialysis only for four hours per week, so the equipment must frequently be started, stopped, and cleaned.

In the second step of device synthesis, we will sketch a process diagram, little boxes representing specific operations and connected by arrows showing a sequence. For example, to show the treatment of an oily waste, we might show adding acid to convert soaps to droplets of insoluble fatty acids, followed by a loose bed filter which increases the size of oil droplets. The sequence tends to describe steps in time, like a recipe for making cookies. The cookie recipe might have a box saying "Cream (mix) the butter and sugar." It might next show an arrow to a second box saying "Add four eggs." The boxes in the recipe and in our device design represent a sequence in time, not flows in space.

The third step in the sequence in Table 7.2–1 is the choice of the key reactor or separator. Normally, there is only one because the devices strive for simplicity. If the key is a reactor, we will try to operate with a stoichiometric feed and a high conversion. We will accept a long reaction time to avoid additional separation and recycle.

The final steps of device synthesis are the identification of other useful operations. The choice of these operations involves a much greater spectrum of options than for commodity chemicals. In particular, distillation is much less important, and adsorption is a good choice more frequently. Heat integration is rarely central because our objective is convenience, more than cost.

The contrast of chemical devices with commodity processes is made more explicit by the comparison in Table 7.2–1. Commodity chemicals are normally made continuously, and for as long as possible without interruption, to use the equipment as efficiently as possible and to minimize the capital cost per unit of product produced. Commodity chemical processes are easily represented by a diagram showing flows from one operation to another. Thus, the lines on this diagram represent pipes, and the flows are in space, not in time. Reactors may be stirred tanks, plug flow, or fluidized beds, but all are fed continuously, producing steady streams of products. Separations are dominated by distillation; gas absorption and stripping are common; and other separations – even extraction – may be important but are always regarded as just a little abnormal. Heat integration, a real opportunity when energy prices increased in the 1970s, is now an essential optimization for any internationally competitive commodity chemical plant.

We illustrate these generalizations in the examples which follow. We then review the tools needed for these designs in the next section.

EXAMPLE 7.2–1   LOCAL AMMONIA PRODUCTION

Ammonia is made by combining nitrogen and hydrogen under a pressure of 120 atm and at a temperature of 400 °C, as outlined in Section 6.1. In fact, the common process is complex for two main reasons. First, natural gas is burnt in air to produce a mixture of hydrogen, nitrogen, and carbon oxides. Because the carbon oxides will foul the catalyst in the ammonia synthesis reactor, they must be carefully removed, reflected in a complicated process diagram. The second reason for the complex process is that the synthesis reaction typically only goes to 20% completion, so the ammonia produced must be separated and the unreacted nitrogen and hydrogen recycled. If the ammonia were adsorbed, this recycle would not be necessary.

Our company is interested in making an ammonia plant suitable for use at one large farm or at an agricultural cooperative of perhaps forty farms. The ammonia will be made from hydrogen from the electrolysis of water, using electricity generated on the farm. The nitrogen will be made from air, using a commercially available hollow-fiber membrane module.

What will our new process look like?

SOLUTION

To make the process as simple as possible, we choose a batch reactor containing catalyst and solid magnesium chloride to capture the ammonia made. We will feed hydrogen and nitrogen to the reactor at high pressure and temperature until the ammonia saturates the solid chloride. We will then stop the feed and release the pressure to recover the ammonia. Such operation can give 90% conversion, over four times greater than that of the conventional process. This means that our new process can be much simpler than the existing process, which was shown in Figure 6.2–2.

This example illustrates the simplifications possible for chemical devices. Here, the simplifications occur because our feed is not natural gas, and our batch reactor gets higher conversion. We should note that while this conversion is observed in the lab, it has not been demonstrated over long operation. Still, chemical devices may sometimes offer these major simplifications.

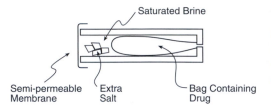

Saturated Brine

Semi-permeable   Extra   Bag Containing
Membrane   Salt   Drug

Figure 7.2–1 *Schematic drawing of an osmotic pump* This device delivers a steady dose of drug, independent of the properties of the drug.

EXAMPLE 7.2–2 DESIGNING AN OSMOTIC PUMP

Drugs are often given orally, as pills taken every few hours. In many cases, this means that drug concentrations in the blood can fluctuate widely. Just after the pill is taken, the drug concentration jumps, sometimes briefly beyond the toxic limit. (In France, this is called "le burst effect.") After an hour or two, the drug concentration wanes, often dropping below the concentration where it is effective. Thus, for many drugs, the blood concentration is occasionally too high and often too low, only periodically passing through the desired range.

These concentration variations have sparked many inventions aimed to provide a steady drug release. One such invention, shown in Figure 7.2–1, is called an "osmotic pump." The pump consists of a rigid housing capped with a semi-permeable membrane. The housing is partially filled with a balloon which in turn is filled with a solution of the drug whose delivery is to be controlled. The rest of the housing is filled with saturated brine in which sodium chloride crystals are suspended.

When this device is surgically implanted in the human body, the difference in osmotic pressure between the blood and the brine causes a water flow across the semi-permeable membrane into the device. The flow is proportional to the concentration difference across the membrane. The concentration difference of brine across the membrane stays constant because the brine inside the pump contains suspended salt crystals and these crystals dissolve as water flows in. Because the concentration difference stays constant, the flow is constant; because the flow is constant, the balloon is squeezed to constantly release the drug solution. The beauty of this device is that the constant release of drug solution does not depend on the drug's properties, but only on those of brine and the membrane.

To design such a device, we plan to use a semi-permeable membrane which has a reported permeability (defined as volume flux per unit pressure difference across the membrane) of 10^{-10} cm/sec kPa. What membrane area do we need to supply a release of 0.8 μL of drug per hour?

SOLUTION

This osmotic pump depends for its operation on water transport through a semi-permeable membrane. The amount of this transport, J_1 is given by

$$J_1 = L_P A \Delta\pi = L_P A R T c_2$$

where L_P is the permeability, A is the membrane area, $\Delta\pi$ is the osmotic pressure, R is the gas constant, T is the temperature, and c_2 is the salt concentration,

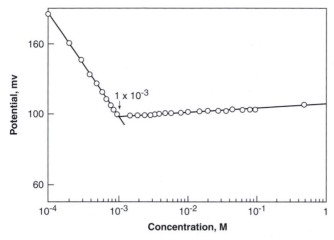

Figure 7.2–2 *Potential of dodecylsulfate in 0.1M sodium chloride* The potential drops by the amount expected until the critical micelle concentration. Above that, the activity of the dodecylsulfate ion is nearly constant (Kale *et al.*, 1980).

which at the body temperature of 37 °C equals 5.4 mol/dm^3. Thus the area is found from

$$\frac{0.8 \times 10^{-9}\,\text{m}^3}{3600\,\text{sec}} = \left(\frac{10^{-10}\,\text{cm}}{\text{sec}\ 10^3\,\text{kg/msec}^2}\right) A \left(\frac{8.31\,\text{kgm}^2}{\text{sec}^2\,\text{K mol}}\right) 310\,\text{K} \times 5.4$$
$$\times\, 10^3\,\text{mol/m}^3$$
$$A = 1.6 \times 10^{-7}\,\text{m}^2$$

This corresponds to a circular disc of membrane about 0.05 cm in diameter.

EXAMPLE 7.2–3   AN ELECTRODE FOR MEASURING DODECYLSULFATE

The increasing concern with pollution has generated a market for measurements of detergents in ground water. Our core team has decided that one opportunity in this area is for specific ion electrodes which measure the presence of one species like dodecylsulfate, a component of common dry laundry detergents.

What would such a device be? How would it work? What would its readings, illustrated by the data in Figure 7.2–2, actually mean?

SOLUTION

A specific ion electrode for this use would be a batch reactor consisting of three basic parts: an inner filling solution, a liquid membrane, and the solution being tested. The inner solution is 0.01 M NaCl around an Ag/AgCl electrode. The liquid membrane consists of a chlorinated aromatic solution containing tributyl-hexadecylammonium chloride stabilized in a microporous support. The measurement also uses a second reference electrode which need not concern us here. The

electrode is operated by applying a potential which stops any current: the reactor
is run so there is no reaction. If a reaction did occur, it would involve transport
of the dodecylsulfate ion across the liquid membrane, because this is the only ion
present that is hydrophobic enough to get through the liquid membrane.

The data in Figure 7.2–2 give the potential required vs. the concentration of
dodecylsulfate. To see what these data mean, we turn to the Nernst–Plank equations for the flux of an ion j_1:

$$-j_1 = \frac{Dc_1}{k_B T} \nabla (\mu_1 + z_1 \varphi)$$

where D is the ion's diffusion coefficient, c_1 is its concentrations, μ_1 is its chemical potential, z_1 is its charge, φ is the applied potential, and $k_B T$ is Boltzmann's
constant times the temperature. Because the transport is zero, the dilute solution
is nearly ideal, and the dodecylsulfate ion has a charge of (–1), we have

$$\mu_1 = \mu_1^0 + k_B T \ln c_1 = \varphi$$

where μ_1 is the chemical potential, a function of a reference value in the standard
state and the concentration c_1. For a concentration change from c_1 to $10\,c_1$, we
thus have the potential change

$$\varphi (10c_1) - \varphi (c_1) = k_B T \ln \frac{10c_1}{c_1} = 2.303 k_B T$$
$$= 0.059 \,\text{volts}$$

Thus each decade change in concentration of dodecylsulfate will yield a change
of 59 mV in the potential which we need to stop the reaction. This slope is called
the Nernst limit.

At low concentrations, the data are exactly what we expect, showing a slope
close to the Nernst limit. At a concentration close to 10^{-3} M, however, the data
change abruptly, and the potential difference seems to become almost independent of concentration. These data are consistent with micelle formation. At low
concentrations, the dodecylsulfate ions are unassociated and behave like other
normal anions. Above a "critical micelle concentration" (cmc), in this case about
10^{-3} M, these anions associate into micelles, aggregates of ions. When more
dodecylsulfate anion is added, it does not go into solution but into this aggre-
gated phase. Thus the dilute data show that this detergent electrode is working,
and the concentrated data give a detailed picture of micelle formation.

7.3   The Device Toolbox: Chemical Reactors

The intellectual tools used for chemical device design are much the same as those
used for the process design of commodity chemicals. Now, the tools are used
more easily, more approximately. Now, we are interested in about how long a
chemical reaction will take, or about how hard a separation will be. Now, we seek
convenience and simplicity. In contrast, for chemical commodities, we focused

on exact rates of reaction and separation efficiencies because these needed to be optimized to minimize cost.

Thus, the analysis in this section is mathematically simpler than that used for commodities. However, while it is tempting to relax mentally, remember that the use of engineering judgment will be more frequent and more subtle. That is why this subject is exciting.

When we carry out a batch chemical reaction, the amount of reactants drops with time. We will use this time to estimate the reactor volume. Normally, we will be given the amount we want to produce M. We must choose the reactant concentration in the reactor and the fraction of this reactant which we want to convert. In the absence of side reactions, we will normally choose the initial concentration c_{10} to be as high as is convenient. The minimum volume of the reactor needed V is thus

$$V = \frac{M}{\nu c_{10}} \tag{7.3–1}$$

where ν is a stoichiometric constant, the amount of product made per reagent consumed. Because we want to avoid recycles, we will often choose the fractional conversion to be high.

Estimating the time that we must run the reactor depends on the detailed chemistry, for which there are three important cases: first-order kinetics, second-order kinetics, and zero-order kinetics. For first-order kinetics, the rate of reaction per reactor volume r is proportional to the reactant concentration c_1:

$$r = kc_1 \tag{7.3–2}$$

where k is the reaction rate constant, with dimensions of $(\text{time})^{-1}$. For a batch reactor of volume V, this leads to the mass balance

$$V\frac{dc_1}{dt} = -kc_1 V \tag{7.3–3}$$

The initial reagent concentration is often known

$$t = 0 \quad c_1 = c_{10} \tag{7.3–4}$$

Integrating and combining these equations gives

$$\frac{c_1}{c_{10}} = e^{-kt} \tag{7.3–5}$$

The characteristic time which we seek is just the reciprocal of the first-order rate constant k.

First-order chemical kinetics occur commonly in chemical devices, but the time $(1/k)$ may not always involve just the chemistry of the system. For example, consider a chemical reaction which is so fast that it is controlled by diffusion of the reagent to the point of reaction. This diffusion could access an adjacent gas phase or a suspension of catalyst pellets. The amount of reagent diffusing j_1, given as mass per time per surface area, is

$$j_1 = k_D c_1 \tag{7.3–6}$$

where k_D is a mass transfer coefficient, roughly equal to a diffusion coefficient divided by a short characteristic distance (this film thickness). A mass balance on the reactor now gives

$$V\frac{dc_1}{dt} = j_1 A \tag{7.3–7}$$

where A is the total interfacial area of the gas phase or the catalyst. Because this mass balance is also subject to the initial condition in Equation 7.3–4, we can integrate Equations 7.3–6 and 7.3–7 to obtain

$$\frac{c_1}{c_{10}} = e^{-\left[\frac{k_D A}{V}\right]t} \tag{7.3–8}$$

where $[V/k_D A]$ is the characteristic time for this case. This new time is a function of the physics in the device, of variables like stirring, viscosity, and particle size. In contrast, the characteristic time $[1/k]$, in the same mathematical form in Equation 7.3–5, is a function of the system's chemistry and doesn't depend on variables like stirring.

Other reaction orders occur less frequently. The one most commonly mentioned in textbooks is the second-order reaction, for which the reaction rate per volume r is

$$r = kc_1 c_2 \tag{7.3–9}$$

where c_1 and c_2 are the concentrations of two different reactants. Two cases of the second-order reaction are common. First, when a batch reactor is fed stoichiometrically, then c_1 and c_2 are equal and the mass balance gives

$$V\frac{dc_1}{dt} = -kc_1^2 V \tag{7.3–10}$$

Because the initial concentrations c_{10} and c_{20} are also equal, this can be integrated to give

$$\frac{1}{c_1} = \frac{1}{c_{10}} + kt \tag{7.3–11}$$

Thus the characteristic time in this case is $[1/kc_{10}]$. The research director of one international chemical company used knowing this time as the sole criterion for hiring new chemists.

The second important case for a second-order reaction occurs when one reagent is present in excess, e.g. c_{10} is always much less than c_{20}. In this case, c_{20} will be nearly constant, and the reaction rate r is

$$r = (kc_{20})c_1 \tag{7.3–12}$$

where the quantity in parentheses is now a new pseudo-first-order rate constant, the product of the second-order constant and the concentration of the excess reagent. The characteristic time is thus $(1/kc_{20})$.

In the final mechanism, zero-order chemical kinetics, the reaction rate per volume r is independent is the concentration of reagent c_1

$$r = k \tag{7.3–13}$$

where k is the reaction rate constant. If we again use this reaction rate in a mass balance, we get

$$V\frac{dc_1}{dt} = -kV \qquad (7.3\text{--}14)$$

Integrating from an initial concentration c_{10}, we find

$$c_1 = c_{10} - kt \qquad (7.3\text{--}15)$$

Now the characteristic time is (c_{10}/k). While this type of reaction kinetics may seem at first blush to be unlikely, it occurs commonly in catalytic reactions, when the rate is limited not by the amount of reagent but by the number of catalytically active sites. The determination of these times and their use in device design is illustrated in the examples which follow.

EXAMPLE 7.3–I   ETCHING A PHOTORESIST

Imagine that we are etching a silicon wafer coated with a new optically sensitive photoresist with a dilute solution of aqueous sodium hydroxide. The reaction rate shows an activation energy around 30 kJ/mol, suggesting the reaction may be strongly influenced by chemical kinetics. However, the reaction rate in 0.16 M NaOH also depends on the spinning speed of the wafer, as shown by the isothermal data below.

Wafer rotation, rpm	Rate constant, sec^{-1}
6	0.53
9	0.61, 0.61
15	0.70, 0.66
30	0.81, 0.79, 0.88
70	1.16

What is the reaction mechanism?

SOLUTION

In some cases, including this one, the rate is affected both by chemical kinetics and by mass transfer. In cases like these, we may show the overall reaction rate constant, k, is given by

$$\frac{1}{k} = \frac{1}{k_s} + \frac{1}{k_D a}$$

whre k_s is the rate constant for surface reaction, k_D is the rate constant for diffusion (i.e., the mass transfer coefficient), and a is the surface area per volume. Because these two reactions occur sequentially, this result is often said to correspond to two chemical resistances in series and is compared to Ohm's law of electrical resistance in series.

At constant temperature, we expect k_s to be a constant, but k_D to vary with stirring. In most cases, k_D varies with stirring speed to the power of 0.3 to 0.8.

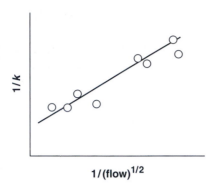

Figure 7.3–1 *Mass transfer analyzed with a Wilson plot* The intercept at infinite flow is the reciprocal of the surface reaction rate constant.

The most common power is 0.5, which is that for the dissolution above. Thus we can expect

$$\frac{1}{k} = \frac{1}{k_s} + \frac{B}{\omega^{1/2}}$$

where ω is the speed of rotation and B is a constant. Thus a graph of $(1/k)$ vs. $(1/\omega^{1/2})$ should be linear, with an intercept proportional to the reciprocal of the rate constant k_s, and a slope related to the diffusion constant and area per volume. Such a graph, sometimes called a Wilson plot, does work in this case, as shown in Figure 7.3–1.

EXAMPLE 7.3–2   DISPOSAL OF TETRAPHENYLBORATES

The Savannah River Site (SRS) is one of two principal sites for manufacturing atomic weapons in the USA. As such, it has produced large amounts of dangerous radioactive waste, including cesium-137 and strontium-90. These salts are unusually dangerous because they are water-soluble. To reduce the risk of accidental release, these species have been mixed with sodium tetraphenylborate to form precipitates. While this made the cesium and strontium isotopes insoluble, the precipitates unexpectedly decayed to produce flammable benzene. This method of treating the Cs-137 and the Sr-90 has been abandoned.

However, 22,000 kg of these precipitates remain in one storage tank. The buried tank, about 85 ft in diameter and 33 ft tall, was designed to hold 1.3 million gallons of waste. SRS would like to destroy the precipitates so that it can regain use of the tank to process new radioactive wastes produced by the nuclear power industry. These wastes will be highly toxic, but water insoluble.

SRS has proposed four methods for disposing the precipitates and recovering use of the tank. The first of these methods, "aggregation," is the cheapest and the simplest. It simply requires mixing small amounts of the waste with large amounts of concrete and adding the resulting solid to other, low-level radioactive waste stored on the site itself. This method is relatively safe, it is inexpensive, and it is legal under existing federal law. It was the planned method until the State of South Carolina announced that it would oppose this much additional radioactivity being left within the state. Accordingly, it has been abandoned, at least for now.

The remaining three methods of precipitate disposal chemically destroy the precipitates. These three methods are steam reforming, wet-air oxidation, and reaction with peroxides (Fenton's reagent.) After reaction, the radioactive residue will be added to other waste and "vitrified," i.e. made into glass. The glass will be stored in a repository like that in Yucca Mountain, Nevada. Details of these three methods are below.

Steam reforming. This fluidized bed reaction method, which is the best-developed technology, burns the precipitates at about 700 °C and 1 atm to produce carbon dioxide. It has already been used to process radioactive waste for the Hanford Site, the second major weapons laboratory. It is a strong, well-developed method.

Steam reforming does have major disadvantages. The first of these is that the precipitates do not supply enough heat of combustion by themselves. As a result, more carbon must be added to the burner, possibly as coke, as sugar, or as polyethylene chip. In some cases, all of this carbon is not burned but forms small black carbon particles in the ash removed from the reformer. Carbon particles in the ash are benign in themselves but may cause problems in the vitrification. This point requires careful attention.

The other disadvantages of steam reforming concern the exhaust gases. First, any traces of mercury in the radioactive waste will be evaporated and leave the reactor as vapor. Ways to capture mercury vapor are well known and may be needed. Second, the exhaust gases may entrain radioactive dust. Again, similar dust is removed by filters in other applications of this technology. Neither disadvantage is crippling.

Wet-air oxidation. This method pumps bubbles of oxygen at about 100 atm through an aqueous slurry of radioactive waste at perhaps 300 °C. The tetraphenylborate salts are decomposed to produce an unknown palette of intermediates, probably including benzene and phenols. These intermediates may require separation and further reaction. Thus, if steam reforming can be described as a "burner," wet-air oxidation is a "cooker." While the chemical reactions are perhaps 100 times slower than those for steam reforming, the lower temperature and the lack of exhaust gases are significant advantages. These characteristics almost certainly are why wet-air oxidation is a frequent choice for getting rid of chemical munitions, including nerve gas.

Wet-air oxidation also has major disadvantages. The chief of these are the high pressure required and the less-complete process development. The high pressure on the gas–slurry mixture means that the reactor could suddenly rupture and spray its contents on the surroundings. The incomplete development means uncertainty, especially in the carbon compounds produced by the process. This uncertainty is serious because the Savannah River Site wants a disposal method as soon as practical.

Fenton's reagent. The third reaction method, Fenton's reagent, uses a combination of hydrogen peroxide and a ferrous iron catalyst at room temperature and

Table 7.3–1 *Three methods for tetraphenylborate destruction* Steam reforming was the method recommended.

	METHOD		
Variable	Steam reforming	Wet-air oxidation	Fenton's reagent
Reactor	Fluid bed	Bubble column	Stirred tank (batch)
Temperature	700 °C	300 °C	Ambient
Feed	>10% slurry	>10% slurry	>10% slurry
Carbon fate	CO_2, C, Na_2CO_3	Unknown (includes benzene)	Unknown (includes benzene)
Time reaction	~10 seconds	~1 hour	~12-hour cycle
Extra energy	Sugar or polyethylene	High pressure	Hydrogen peroxide
Problems	Residual carbon in solid product; fines in exhaust gas	Missing kinetics; missing composition of product	Explosive reagent; large reactor size, large waste volume, missing kinetics, composition of product

pressure. This classical chemistry, dating from the 1890s, is used to treat organic compounds in waste water. It doesn't require increasing the temperature, like the alternatives; it doesn't require increasing the pressure, like wet-air oxidation; it doesn't produce fines, like steam reforming. Instead, it supplies chemical energy to destroy the tetraphenylborates.

Fenton's reagent also has serious, almost crippling, disadvantages. Its kinetics are slow, at least four times slower than wet-air oxidation. This implies a larger reactor. The compounds produced are unknown. Most seriously, peroxides are notoriously unstable and explosive.

Compare these three technologies and recommend one as the preferred route.

SOLUTION

A comparison of the important characteristics of the three possible technologies is given in Table 7.3–1. Right away we see good reasons to reject use of Fenton's reagent. It is slow, dangerous, and not well understood. Politically, the idea of combining radioactive waste and an explosive reagent seems folly. We will not consider Fenton's reagent further.

The other data in Table 7.3–1 suggest that steam reforming is the best method currently available. Wet-air oxidation is a good backup, successfully used for nerve gas but missing significant technical information.

We should test this conclusion by assessing and planning management of the risk associated with both technologies. Preliminary assessments of this risk are shown in Table 7.3–2. Steam reforming wins high marks as an established method exhausting the carbon in the precipitates as CO_2. Its fast kinetics are a real asset. It could be compromised by residual carbon, mercury, and radioactive fines. The chief risk with steam reforming is the elemental carbon in the solid product,

Table 7.3–2 *Risk assessment of two methods of tetraphenylborate destruction* The risks center on uncertain chemistry.

Risk	Probability	Consequence	Level
Steam reforming			
Carbon in product	0.5	0.5	0.25
Mercury vapor	0.5	0.1	0.05
Radioactive dust	0.3	0.1	0.03
Wet-air oxidation			
Slow kinetics	0.6	0.7	0.42
Unknown products	0.3	0.3	0.09
High pressure	0.1	0.7	0.07

because this might compromise vitrification. We should promptly plan experiments to explore this. The other problems with this technology are significant but are within current practice.

The biggest problem with wet-air oxidation is our greater ignorance of the chemistry. The kinetics are believed to be slow, and the products made are unknown. Some quick experiments with non-radioactive potassium tetraphenylborate would go a long way to resolving this ignorance. At present, we have not set a standard which the kinetics must meet; nor have we identified possible separation methods for the products. These standards should be set and modified as more experimental results become available. While the use of high pressure is a significant risk, it seems within current practice. It would be useful to see what the reaction of the State of South Carolina and of environmental advocates would be to this possibility.

7.4   The Device Toolbox: Separations

The choice of possible separations useful in chemical devices depends on the same unit operations as are used to make commodity chemicals. However, because the objective is convenience and not just cost, the more important operations are different. Gas absorption and distillation, the core of commodity operations, are much less frequently used. Extraction is used most often in crossflow mode, not as a countercurrent process. Adsorption from both liquid solutions and gas mixtures is often chosen. Membranes, filtration, and decantation can be reasonable alternatives.

In reviewing these alternatives, our goal is rarely a detailed design suitable for a manufacturing process. Instead, we will most often just be trying to see if a particular separation makes any sense at all. We are not concerned with the separation's cost, at least not yet. We are trying to estimate whether a particular technique is reasonable.

In the synopses below, we review the common separation techniques. These most common operations are listed in Table 7.4–1. The first column identifies the process, the second gives characteristic separations, and the third identifies the key variables. The fourth column in this table gives the basic equipment. The fifth column estimates the flexibility of operation, an important criterion for a device not often intended to operate in steady state, but as needed, with varying amounts and concentrations of feed.

The different processes are listed in rows not in order of their frequency of use, but rather in the order in which they are often learned and hence cataloged in the mind of the designer. The first is gas absorption, or its converse, gas stripping. In this process, a gas mixture is fed up a packed tower filled with a solid designed to give a large surface area. The solid is plastic or ceramic, often shaped like pasta and called after its inventors. Penne-shaped packaging is called "Rashig rings," farfalle shapes are "Berl saddles," and ceramic lasagna is "structured packing." Liquid poured into the top of the column absorbs some component of the gas as it flows down the tower. For example, ammonia is absorbed into water, or high concentrations of sulfur dioxide are absorbed into oil.

Three additional characteristics of gas absorption are common. First, the amount absorbed is usually greater when the gas flows upward and the liquid flows downward, i.e. when the gas and liquid flow countercurrently. Second, the liquid often reacts chemically with the species being absorbed. Third, the absorbing liquid must somehow be regenerated, usually by heating to release the absorbed component. This "stripped" liquid can then be re-used.

Gas absorption is simple in concept and broadly applicable in large-scale practice, but it is often too complex for a chemical device. One reason is the need to regenerate the absorbing liquid, which increases the equipment required for the process. A second reason is that for effective operation, the fluxes of gas and liquid, that is, the flows per cross-sectional area, must be within a narrow range. The range is bounded at low flow by "loading," i.e. by a minimum flow to give a large interfacial area between gas and liquid. The range is bounded at high flow by "flooding," where no flow will occur countercurrently. These constraints are much easier to meet for a continuous process run in steady state than for an occasionally operated device. Absorption is frequently not a good choice for a chemical device.

Distillation is not a frequent choice for a device either. Distillation, which requires two liquids of different volatility, separates these by heating to release the more volatile as vapor. Ethanol and water are the classic example. The amount of separation is greatly enhanced by letting the vapors flow up a packed column, condensing the vapors at the top, and sending most of them back down the column. Even moonshiners know how to do this; if they begin with a beer containing 7% ethanol, they can only make a distillate of 15% if they collect the vapors directly, but they can make 70% ethanol if they send some condensate back down the column. Large columns may contain "stages," separate compartments where vapor and liquid equilibrate; such columns are a standard academic example. However, small columns

Table 7.4–1 Common separation processes used in chemical devices These contrast with those used for commodities, where absorption and distillation are the norm.

Process	System separated	Key variables	Flexibility of operation	Basic equipment	Remarks
Absorption, stripping	Gas mixtures, volatiles dissolved in liquids	Solubility	Poor	Two packed columns, two heat exchangers, and two pumps	Simple; useful for environmental control
Distillation	Volatile liquid mixtures	Relative volatility	Fair	Packed column, reboiler, condenser	Can give very pure products
Extraction, leaching	Dilute solutions of valuable solute	Relative solubility	Good	Stirred tank	Good for valuable, non-volatile products
Adsorption	Solutes from liquid or gas solutions	Isotherm, breakthrough curve	Excellent	Packed column, pump	Hard to understand; underused
Membranes	Solutes from liquid or gas solutions	Permeance	Fair	Membrane module, pump	Good, but oversold
Filtration/decantation	Suspensions, emulsions	Particle size, density differences	Good	Tank	Simple but slow

almost never contain stages, but rather packing similar to that used for gas absorption.

Distillation is used at small scales, not only by moonshiners but also by the makers of flavors and fragrances. However, it is not a common choice for devices because it requires two solutes which have different volatility. This will not work for separating coffee flavor from coffee beans. Distillation columns are slow to start and stop, and are subject to similar constraints about the countercurrent vapor and condensate flows that restrict gas absorption. Distillation just isn't that useful for devices.

Extraction and leaching are good choices for devices. The obvious example is a coffeemaker, where hot water flows through ground coffee to extract caffeine and flavors. Similarly, to make salad dressing, oil, vinegar, and herbs are shaken together to extract flavors from the herbs. In each case, the process could be significantly improved with a more complex device. For example, in the coffeemaker, the bed of coffee grounds could be more uniform to prevent channelling, and the temperature of the hot water could be more closely controlled to prevent extraction of bitter chlorogenic acids. However, at least at present, these improvements imply a more complex device which will not justify an increased device price.

Adsorption, the capture of a specific solute on a bed of solid particles, is a separation especially well suited to chemical devices. It is particularly attractive for water purification, both in the common Brita filters sold in supermarkets and the arsenic removal units under development for Bangladesh. It is also useful for air purification, like odor removal for cheese plants or oxygen-enriched air for emphysema patients. The process is easily started and stopped and can handle a reasonable range of feeds.

Adsorption is not carefully explained in many books on separations. This is not because it is unimportant but because it is mathematically complex and hence hard to describe with equations. For those unfamiliar with adsorption design, we suggest starting by calculating the amount of adsorbent needed, which is just the amount of solute to be adsorbed divided by the adsorbent's maximum experimentally measured capacity. If this amount is reasonable, then the concentration exiting from the adsorbent bed should be measured as a function of time. This is enough for a preliminary design.

The other separation techniques which are listed for devices in Table 7.4–1 are membranes, filtration, and decantation. These methods normally involve a membrane or filter medium, some kind of solid sheet across which a solution is forced. For membranes, the sheet is often non-porous and solutes must dissolve in and diffuse across the membrane. This means that membranes do not necessarily separate by size: in one oft-quoted example, large phenol molecules pass across reverse-osmosis membranes more easily than smaller water molecules or sodium and chloride ions. Such membranes are also used for non-cryogenic gas separations, where large carbon dioxide molecules pass through membranes more easily than smaller methane molecules.

When the membranes have pores, the process is usually called filtration. When these pores are smaller than a micrometer, the process usually is operated in "crossflow," where the flow tangential to the membrane's surface is much larger than the flow through the membrane. When the membrane's pores are larger than perhaps 30 μm, the chief flow is not across but through the membrane. Such flow generates an accumulated bed of filtered particles, called a "cake." The membrane's pores can be larger than the particles being captured, which is the case for furnace filters and for the human nose. In these cases, the capture mechanism is not only size but also electrostatics. In still other cases, the membrane just serves as a framework for coagulating liquid droplets and hence in facilitating phase separations by decantation. All of these membrane methods can work well in devices, as the examples below suggest.

Before looking at the examples, we should review our objectives. We seek devices which effect a single chemical separation conveniently and flexibly. We understand that this separation may not be accomplished in the same way as if we were seeking the separation for a large-scale chemical process, where we would look for a minimum cost. Here we seek convenience.

EXAMPLE 7.4–1   A BETTER COFFEEMAKER

We want to explore the possibility of making a better home coffeemaker. To do so, we seek the simplest possible model for separating coffee flavors from ground coffee. What model should we use?

SOLUTION

If we watch the coffee being made in an existing machine, the coffee's color is darkest initially, and then gets lighter and lighter as the water flow continues. If we sample the coffee as it is made, the first drops are strongest and the last have the least flavor. This suggests that color and flavor are correlated.

To make this more quantitative, we assume that the concentration of flavor varies with time t as

$$\frac{c_1}{c_{10}} = e^{-t/\tau}$$

where c_{10} and c_1 are the flavor concentrations initially and at time t, and τ is a time constant for making coffee. We must now estimate the value of τ. Two cases seem limiting. First, if all the flavor quickly dissolves as soon as the coffee is wet, then

$$\tau = \frac{V}{Q}$$

where V is the volume of liquid trapped in the coffee grounds, and Q is the flow rate of water through these grounds. In this mode, making coffee is instantaneous, largely involving washing the wet grounds. No major improvements in design are possible.

The second model assumes that the time constant τ varies with the rate of flavor removal from each particle of ground coffee

$$\tau = \frac{1}{ka}$$

where k is a mass transfer coefficient out of the coffee, and a is the particle area. In this case, k probably represents flavor diffusion within the particles, which is often estimated as

$$k = \frac{4D}{R}$$

where D is the diffusion coefficient in the particles, and R is their average radius. Since the area a is given by

$$a = \left[\frac{(1 - \varepsilon)}{\frac{4}{3}\pi R^3} \right] (4\pi R^2) = \frac{3(1 - \varepsilon)}{R}$$

where ε is the void fraction of grounds, we now have

$$\tau = \frac{R^2}{12D(1 - \varepsilon)}$$

This model is beginning to have reassuring features, including the prediction that flavor extraction depends on the size of coffee particles.

To go further, we can follow two different threads. First, we can think about the shape of the filter basket holding the coffee grounds. Two common shapes are conical and cup-shaped. Which is better? Second, we can think about the chemistry, because particular flavors are best extracted at specific temperatures. We need to think about the best way to control the temperature of the water, while always thinking about a simple device.

EXAMPLE 7.4–2   RECOVERING COOKING OIL FROM RESTAURANT WASTE

One major challenge for sewage works is oil and grease in restaurant waste water. Typically, a restaurant produces about 10,000 L of waste water per day, containing about 0.1 vol% oil and grease. A typical government's waste disposal limit is around 0.01 vol% oil.

Methods of removing the oil include gravity settling, filtration, hydrocyclones, and centrifuges. These technologies are often too expensive for a typical restaurant. For example, ultrafiltration, suitable for small particles or droplets, works well, typically giving a flux of 1000–2000 L/m^2/day. Cost figures are hard to come by: $100 per m^2 is one guess. This is about ten times more expensive than the mechanical methods.

We seek a technology for cleaning the water for one restaurant. To provide a basis for our answer, we should calculate the times for gravity separation of the various sizes of oil droplets in water. Then we must recommend a specific treatment technology to meet the government's standard.

SOLUTION

To meet the government standard, we must capture 9 L/day of oil and grease, which is 90% of the amount fed. How we do so depends on the state of the mixed oil. A fraction is actually dissolved in the water, but this fraction is small and so is ignored here. The largest fraction, about three quarters of the total, is in droplets larger than 200 μm. The remaining quarter is present as much smaller droplets (<0.5 μm), often hard to separate because the droplets are stabilized by detergents. These two types of oil require different separation technologies.

The fraction containing the larger drops can be separated by gravity settling. To estimate the time t required for this settling, we remember from Stokes' law that

$$\text{velocity} = \frac{l}{t} = \frac{\text{force}}{6\pi\mu R}$$

where l is the settling distance, μ is the viscosity of water, and R is the droplet radius. The force, due to the oil's buoyancy, is proportional to the density difference $\Delta\rho$ between the oil and the water, about 0.1 g/cm^3. This leads to the equation

$$t = \left(\frac{9\mu l}{2g\Delta\rho}\right)\frac{1}{R^2}$$

where g is the acceleration due to gravity. If the distance l is 4 cm and the drops are 0.2 cm in diameter, they separate in a few seconds; if they are 20 μm across, they are separated in a few hours. If they are submicron, they settle too slowly and mix too easily (i.e. by Brownian motion) to be separated by gravitational settling.

Nonetheless, the larger droplets – three quarters of the total – can be separated in this way. Commercially available decanters can do this job effectively. A decanter, typically costing around $3000, can handle 5 L/sec, well beyond our requirement. Some of the units skim the liquid air interface so that the distance that the droplets need to travel can be less than the 4 cm assumed above.

However, even if gravity settling is completely effective, the restaurant effluent will still contain 250 ppm oil, beyond the legal limit of 100 ppm. To remove more oil, we must use a different method. One solution is simply to add another 15,000 L of fresh water per day to our effluent, expensively diluting our problem away. Alternatively, to remove more oil we can use a hydrocyclone or ultrafilter. The latter seems to be much more frequently used, giving fluxes of around 1500 L/m^2/day. This suggests that we need an ultrafilter with about 7 m^2. While the membranes in such a filter will cost less than $1000, the entire unit will cost perhaps ten times that much. We expect that most restaurants will be unwilling to bear this cost to recover two more liters of oil per day.

We note that the problem can be significantly reduced by altered housekeeping. In particular, we suggest using detergents only when absolutely necessary. By doing so, we will increase the number of larger oil droplets in the effluent and reduce the amount of oil chemically stabilized as small micelles. Thus we urge a combination of gravity settling and restricted detergent use to remediate the oil in restaurant waste.

EXAMPLE 7.4–3   WATER REMOVAL FROM MOTOR OIL

Our core team has been asked to seek equipment to remove water from lubrication systems for cars and heavy trucks. The equipment must meet the following specifications:

(1) keep the total concentration of water in the petroleum-based oil below 1%;
(2) handle flows of 7 L/min from –30 to 100 °C;
(3) have a volume of less than 3 L; and
(4) run without maintenance for at least 300 hr of continuous operation, or for 30 days of intermittent operation, whichever is longer.

The units should cost less than $25 each to manufacture.

To evaluate possible technologies, our core team plans to focus on a heavy truck having a volume of lubricant of 20 L. Operating the engine adds about 200 g of water to the lubricant every hour. The lubricant contains water in three forms: dissolved, emulsified, and suspended. The dissolved water is less than perhaps 0.5%; the suspended droplets can be as much as 5%, typically as 100 μm droplets. However, most of the water, and the hardest to remove, is emulsified in micelles, perhaps 0.01 μm in diameter. These micelles are formed with detergents added to motor oils to reduce corrosion.

Our core team has focused on four technologies. The first is some form of loose bed filter, like that for a furnace or a fish tank. This filter does not attempt to capture the droplets but only to increase their size to perhaps 0.1 cm so they will fall to the sump of the oil reservoir. The second technology is adsorption in packed beds of silica gel, which can absorb up to 35% water by weight. The third method is air stripping, which is just blowing air through the oil to evaporate and remove the water. The fourth idea is a water-permeable membrane. When oil-containing water is pumped past such a membrane, the water diffuses across the membrane and hence is separated as vapor.

We have a management review soon, at which we are to recommend the best technology of the four. However, almost as soon as we start this selection, we discover a patent which has many of the features we seek (US Patent 6,517,725 issued February 11, 2003). What should we do?

SOLUTION

Our core team must select a system for removing 200 g water per hour from 20 L lubricating oil. The team has considered four technologies which have potential for this separation: loose-bed filtration, silica-gel adsorption, low-pressure stripping, and membrane dehydration. The team believes that each of the first three technologies is seriously flawed, as detailed below. It believes that the membrane dehydrator will work well, but that effective, established competition already exists. As a result, the core team advises against developing a new product for this task.

The shortcomings of the first three technologies are as follows. The loose-bed filtration assumes that 100 μm oil droplets will coalesce to give droplets that are at least ten times larger. This is unproven and the literature is contradictory. Even if this is feasible, the time for the larger droplets to settle out is comparable with the residence times for the oil in the separator, so that the fluid mechanics involved will be difficult. This looks risky.

The second technology, use of silica gel or another desiccant, should work easily. However, meeting the requirement of 300 hr operation without maintenance requires 3000 L of desiccant, far beyond the 3 L specified. While we can imagine some form of pressure swing adsorption, this will be complex and expensive. This technology is also risky.

The third idea, air stripping, is feasible and has been occasionally used in the past. However, the bubble area per volume is normally only around 3 cm^2/cm^3, more than ten times smaller than that observed in the membrane unit. This means that we will struggle to meet the specified limit of a 3 L separator volume. Nevertheless, because we expect to operate at close to steady state, we should be able to avoid problems of loading, flooding, and turn-down ratio, which often occur in stripping. (These problems do not occur for the membrane separator, either.) However, because lubricating oils contain large amounts of detergent, we risk significant foaming problems. This technology seems marginally feasible but inferior to the membrane separator described next.

The fourth idea, the membrane separator, has considerable merit. The total flow J_1 of water is related to the concentration difference by

$$J_1 = kA\Delta c_1$$

where k is the mass transfer coefficient, about 10^{-3} cm/sec; A is the total membrane area; and Δc_1 is the concentration difference across the membrane. If we assume that the water concentration difference is about 50 percent of the maximum possible,

$$\frac{200\,\text{g/hr}}{3600\,\text{sec/hr}} = (10^{-3}\,\text{cm/sec})\,A\,(0.5\,(0.01\,\text{g/cm}^3))$$
$$A = 1.1\,\text{m}^2$$

Typical membrane areas per volume are about 1000 m^2/m^3, so this suggests a membrane unit of perhaps 1 L, less than our specification of 3 L. We expect the assembled cost to be under $20/m^2$. We believe that with an oil flow of 7 L/min, we can easily keep the concentration below 1% water.

However, current patented membrane technology for this separation already exists. Thus we must have a significant advantage over this technology to be successful. We can find no such advantage. More specifically, we have considered the following:

(1) *A more water-permeable membrane.* Because mass transfer is not membrane limited, this offers no advantage.
(2) *A cheaper membrane.* Because assembly of the separator costs more than the membrane, this offers no advantage.
(3) *Better membrane module design.* While this is possible, the current vendors and others have significant intellectual property in this area. A breakthrough seems unlikely.

We can hope for advantages in membrane module manufacturing if this is a strength of our current business. However, our recommendation is not to pursue the new product.

7.5   Using the Devices Toolbox

In this chapter we have outlined the design of chemical devices, which focuses on convenience. This is different to designs of processes for making commodity chemicals, which are chosen to minimize cost. The device designs still use the same design template of needs, ideas, selection, and manufacture. They still center on rewriting the needs as quantitative specifications. They still develop ideas using a design sequence that chooses the type of process and synthesizes a process diagram.

The differences between device design and commodity process design are greatest in selection and manufacture steps. For devices, selection between promising alternatives usually involves both quantitative and qualitative features. For example, for oxygen supplied in the home to an emphysema patient, it involves both oxygen cost and oxygen accessibility by a patient who is probably elderly and possibly physically impaired. Thus, selecting the best device design routinely involves "comparing apples and oranges." Selection for devices is harder and riskier than manufacture of devices.

The relative difficulty of selection and manufacture for commodity products is different. The selection of a commodity process depends on minimizing cost, a definite and quantitative criterion. Once the selection is made, the manufacturing details become central. These details remain difficult even in the face of effective computer-aided simulations, a major accomplishment of chemical engineering.

While selection is key for devices and manufacture is central for commodities, the tools remain the same: reaction engineering and unit operations. Chemical devices normally use batch reactors, and commodity products are commonly made in continuous reactors, but the same ideas of reaction kinetics are used. The tools to describe pressure swing adsorption are the same for a device supplying oxygen in the home and for a process separating xylene isomers for the manufacture of polyamides. The tools are the same.

However, the ways in which the tools are used are different, as the examples given below demonstrate. Device design is simply a broader subject, involving questions that cover a greater scope. This makes device design more difficult, more challenging, and potentially more rewarding.

EXAMPLE 7.5–1 POLYPEPTIDE SYNTHESIS

On May 25, 1959, Bruce Merrifield began development of a new method of polypeptide and protein synthesis. In simple terms, the method involved attaching one particular amino acid A to the surface of solid 100 μm beads in a packed bed. After the bed was washed, a second amino acid B was added in excess and reacted, making AB. Again the bed was washed; again another acid was added, making ABC. The procedure could be repeated again and again. Finally, the molecule was removed from the beads. This let Merrifield make entire polypeptides. For this work, he won the Nobel Prize for chemistry in 1984.

Many drug companies are now interested in making pharmacologically active peptides by a similar route. Commercial machines to make these peptides are available but produce only 5 g of a decapeptide in a day. We want to make 500 g per day of such a material.

The main reason that the current method is less productive is that the reactive surface area is small, so that the effective concentration of molecules being synthesized is 2×10^{-6} M. We believe that if we can increase the effective concentration, we can make a more productive machine. To do so, our core team suggests the following:

(1) Use 1 μm beads to increase capacity.
(2) Carry out the reaction on the walls of 1 μm pores in 100 μm particles.
(3) Attach the active acid A to a water-soluble polymer, and put it inside a 100 μm microcapsule. Other acids – B, C, etc. – can diffuse into the capsule, but the polymer can't diffuse out.

We want to select the best of these three ideas.

SOLUTION

We first calculate the characteristics of the current system as a benchmark. While we must make significant assumptions in this calculation, the results allow us to evaluate the merits of the new ideas more clearly.

The current technology produces 5 g/day of a decapeptide. If this peptide has a molecular weight of 1000 daltons, this is 5×10^{-3} moles. Such a molecule requires ten steps, each of which includes a reaction followed by a washing. We estimate these steps take 20 minutes each, or 200 minutes per molecule. If setting up and shutting down the apparatus takes the same amount of time, we can make

$$\left[\frac{1\,\text{molecule}}{\text{site}\,400\,\text{min}} \right]$$

Because we make 5×10^{-3} moles per day, we need

$$\frac{5 \times 10^{-3}\,\text{mol}}{24\,(60\,\text{min})} \left(\frac{6 \times 10^{23}\,\text{molecules}}{\text{mol}} \right) \bigg/ \left(\frac{1\,\text{molecule}}{\text{site}\,400\,\text{min}} \right)$$
$$= 8 \times 10^{20}\,\text{sites}$$

This is our current capacity.

Our current process uses 100 μm beads in a packed bed of 0.4 void fraction. Such a bed has an area per volume a equal to

$$\frac{4\pi R^2 (1 - \varepsilon)}{\frac{4}{3}\pi R^3} = \frac{6(1 - \varepsilon)}{d} = \frac{6 \times 0.6}{10^{-2}\,\text{cm}}$$

$$= \frac{360\,\text{cm}^2}{\text{cm}^3}$$

where R and d are the particle radius and diameter. We expect to have a site on every 1 nm × 1 nm square. Thus we have

$$\frac{360\,\text{cm}^2}{\text{cm}^3} \left(\frac{\text{site}}{\left(1 \times 10^{-7}\,\text{cm}\right)^2} \right) = \frac{4 \times 10^{16}\,\text{sites}}{\text{cm}^3}$$

Our reactor volume will be

$$\frac{8 \times 10^{20}}{4 \times 10^{16}} = 2 \times 10^5\,\text{cm}^3$$

$$= 20\,\text{L}$$

If such a reactor were 2 m long, it would have a diameter of about 12 cm. The pressure drop Δp in such a reactor is related to the velocity v by

$$v = \frac{l}{t} = \frac{\Delta p d^2}{150 \mu l} \frac{\varepsilon^3}{(1 - \varepsilon)^2}$$

where d is the bead diameter, l is the reactor length, and μ is the viscosity, taken to be close to that of water. The time t should be the time to flow through the bed, taken as 10% of the reaction time of 20 minutes, or 120 sec. Thus the pressure drop is found from

$$\frac{200\,\text{cm}}{120\,\text{sec}} = \frac{\Delta p\,(0.01\,\text{cm})^2}{150\left(\dfrac{0.01\,\text{g}}{\text{cm sec}}\right)200\,\text{cm}} \left(\frac{0.4^3}{0.6^2}\right)$$

$$\Delta p = 3 \times 10^7 \frac{\text{g}}{\text{cm/sec}} \equiv 3000\,\text{kPa}$$

The pressure drop is about 30 atm, typical of that in a chromatographic bed.

With this benchmark as a basis, we now turn to the three ideas suggested by the core team and summarized in Table 7.5–1. The first idea is to use 100 times smaller beads, which have 100 times more surface area per bead volume, and hence 100 times more sites per reactor volume. Thus, the reactor volume will be unchanged, even though the reactor capacity is one hundred times greater. This part of the small-particle idea is successful. However, if the length and bed diameter of the reactor are unchanged, the pressure drop will increase 10,000 times. While such a pressure drop could be realized with thick-walled equipment, it would almost certainly deform the particles, change the void fraction, and reduce the flow through the bed.

Table 7.5–1   *Three alternative technologies for making decapeptides* We are trying to increase reactor capacity one hundred times.

Method	Reactor volume	Pressure drop	Risk	Mitigation
Benchmark 100 μm beds	20 L Reactor	30 atm	None; existing equipment	
100 × Smaller beads	Unchanged	Up 10,000 times	Beads deform	Use pancake bed; use pores; or use larger beads
Porous beads with 100 × larger capacity	Unchanged	Unchanged	Bead synthesis; slower rates	Make beads; measure rates
Microcapsules with 100 × larger capacity	Unchanged	Unchanged	Uncertain microcapsule permeability; uncertain rates; bed destroyed after each synthesis	Check microcapsule literature

We can try to mitigate this crippling problem by changing the reactor dimensions. For example, to keep the pressure drop the same, we could increase the bed diameter by a factor of 30. Such pancake-shaped beds have been used for the chromatographic separation of human insulin. Alternatively, we could reduce the required pressure drop about ten times by replacing the bed of spherical beads with a monolith having cylindrical pores. This would still require a short, fat bed. In our judgment, neither a pancake bed nor a monolith will allow the rapid washing and reaction demanded by this synthesis. We conclude that using 1 μm particles is a bad idea.

The second idea, 100 μm particles with 1 μm pores, is more promising. The presence of 1 μm pores in a bead with 0.40 void fraction gives a surface area per volume of

$$\frac{6\,(1-\varepsilon)}{d} = \frac{6\,(1-0.4)}{10^{-4}\,\text{cm}} = \frac{36\,000\,\text{cm}^2}{\text{cm}^3}$$

one hundred times larger than the benchmark. Such a bed has 100 times more capacity for the same reactor volume but an unchanged pressure drop, as Table 7.5–1 reports. This looks good.

However, the idea of using porous particles has significant problems. First, we are unsure how to make these, and we will need to explore the chromatographic literature to see how this might be realized. Second, we now have an additional resistance to rapid processing: we must wait for each amino acid to diffuse from the particle external surface into the sites deep within its core. If the amino acid's diffusion coefficient D in the pore is 10^{-6} cm²/sec, the time for this wait is about

$$\frac{d^2}{4D} = \frac{(100 \times 10^{-4}\,\text{cm})^2}{4\,(10^{-6}\,\text{cm}^2/\text{sec})} = 25\,\text{sec}$$

This is less than the 20 minutes we plan for each step. Still, we will want experimental verification of this estimate before we pursue this idea much further.

The third idea replaces the porous particle with a 100 μm microcapsule. The walls of the microcapsule must be permeable to the amino acids but impermeable to a polymer scaffold trapped within the microcapsule. To test this idea, we assume a 1% solution of the scaffold polymer of molecular weight 10,000 with one site per molecule. Then the capacity of sites per volume of solution is

$$\left(\frac{0.01 \text{ g}}{\text{cm}^3 \text{ microcapsule}} \right) \frac{\text{mol}}{10^4 \text{ g}} \left(\frac{6 \times 10^{23} \text{ sites}}{\text{mol}} \right) = \frac{6 \times 10^{17} \text{ sites}}{\text{cm}^3}$$

This is 15 times the capacity of the benchmark, so the new reactor volume can even be smaller, although the pressure drop is about the same. This idea has potential.

However, it too has problems. We do not know how to make the microcapsules. While there is a large literature on microcapsule manufacture, ours must have walls permeable to amino acids but impermeable to the 10,000 molecular weight scaffolding polymer. We do not know if this is possible. Most seriously, after we make the decapeptide, we need to recover it from inside the microcapsules, probably by destroying the microcapsules. We are uncertain how to do this.

We conclude that the idea of solid 100 μm particles with 1 μm pores has the most merit and deserves further development. We suggest three months of this development, focusing on the fabrication of these particles. At that point, we recommend another critical review.

EXAMPLE 7.5–2   HOME OXYGEN SUPPLY

Fifty- to seventy-year-old smokers can require hospitalization because they develop emphysema. In this condition, their lungs collapse during exhalation, which impedes flow and prevents the air in their lungs being refreshed. To overcome this, they can breath 2 L/min of air with 30% oxygen so that the driving force for respiration is increased. Thus they can survive with a smaller lung area.

These patients are often supplied with supplemental sources of oxygen so they can be sent home from hospital. While the patients' requirements at home vary, they are typically on oxygen-enriched air for about 15 hours per day. Various sources are available, as outlined in Table 7.5–2.

Most obviously, oxygen can be delivered in cylinders by gas companies. If it is delivered as gas, it is easy to use. However, a typical cylinder won't last a full day so the elderly patient may have to wrestle 40 kg cylinders to sustain her supply. Liquid oxygen cylinders are cheaper, lighter, and supply 35 L of liquid, which is the equivalent of 30,000 L gas, enough for over nine days. However, these cylinders require refrigeration. Liquid oxygen is explosive and can cause serious burns due to frostbite. Directly supplying oxygen has significant disadvantages.

Table 7.5–2 *Possible methods of home oxygen supply* Oxygen cylinders are cheapest, but can be hard for elderly patients to handle. Pressure swing adsorption units are most successful.

Method	Capital cost*	Operating cost*	Purity	Mobility	Remarks
Oxygen cylinders (Gas)	$0	$2000 per year	100%	No	A 40 kg cylinder of 1360 L provides 11 hours at 2 L/min.
Oxygen cylinders (Liquid)	$0	$1200 per year	100%	Yes	A 25 kg cylinder of 35 L liquid provides 9 days at 2 L /min. Frostbite hazard; explosive.
Membranes	$800 initial investment	$50 per year	30–40%	Yes	Must be used without dilution.
Pressure swing adsorption	$1500 initial investment	$100 per year	90%	Yes	Noisy; high initial cost.

*Costs are based on a person requiring 2 L/min treatment for 15 hours a day. If this were liquid, it would take up less than 200 cm^3 but require refrigeration.

Alternatively, we can make oxygen using membranes or pressure swing adsorption (PSA). For a membrane separation, we just compress the air on one side of the membrane, producing a permeate with 30% oxygen on the other side. We send this permeate directly to the patient. For PSA, we force air into a bed which adsorbs nitrogen but not oxygen. Periodically, the pressure is lowered to release the trapped nitrogen. The patient can again breath the oxygen-enriched gas passing through the bed. The advantages and disadvantages of these technologies are given in Table 7.5–2.

The current market opportunity is to modify one of these systems so that it can be portable. Such a portable system should supply at least three hours of oxygen, enough for the patient to go out to dinner, go to a movie, or just go shopping. This would be a significant improvement in the patient's quality of life.

Which technology offers the best product?

SOLUTION

Of the four technologies listed in Table 7.5–2, oxygen gas cylinders are the least attractive. Such a cylinder is clumsy, weighing about 15 kg. This is heavy for an elderly patient; indeed, the use of oxygen gas for this purpose is obsolete. The use of liquid oxygen is more promising, and small, insulated oxygen bottles have been tested for this purpose. However, liquid oxygen is explosive, so the idea of bringing it into a movie theatre seems risky.

Thus the possibilities are PSA and membrane-based separations. The membrane-based separations are quieter, cheaper, and potentially more compact. However, the current membrane is more permeable to oxygen than

Table 7.5–3 *Risk mitigation for portable oxygen.*

Process	Probability	Consequences	Level	Mitigation
Membrane				
Battery too big	0.7	0.7	0.49	Check current power densities for lithium batteries.
Membrane selectivity inadequate	0.5	0.7	0.35	Review existing literature for membranes which are more permeable to nitrogen.
PSA with oxygen liquid				
Refrigeration too costly	0.5	0.5	0.25	Investigate small cryogenic refrigerators.
Safety issues	0.5	0.9	0.45	Examine fire codes.
PSA with oxygen gas				
Gas cylinder weight is excessive	0.3	0.3	0.09	Explore different cylinder materials and higher pressures

nitrogen, so that the oxygen-enriched permeate may need to be recompressed before use. Alternatively, if we can invent a membrane which is more permeable to nitrogen than to oxygen, we could give the compressed retentate gas to the patient directly.

The current PSA system is too bulky and too noisy to be easily portable or used in social situations. On the other hand, it does produce over 90% pure oxygen. To supply 2 L/min of 30% oxygen for a three-hour movie, we thus need the PSA unit to supply

$$\frac{2\,\text{L}}{\text{min}} \left(\frac{0.30}{0.90}\right) 180\,\text{min} = 120\,\text{L}\ O_2 \text{ at STP}$$

These alternative processes are compared in terms of risk in Table 7.5–3. The clear winner is PSA used to store oxygen gas. This probably requires the least work to reach a marketable product. The other two technologies seem much more speculative. The use of liquid oxygen by itself is already being driven out of the marketplace by PSA alone. The use of membranes, which was the original non-cryogenic method, seems even less attractive. We should focus our efforts on PSA plus storage of oxygen gas.

EXAMPLE 7.5–3   A BREATHABLE BOTTLE CAP

One major consumer products company produces liquid shampoo in high-density polyethylene (HDPE) bottles. The bottles, filled 90% full at 40 °C, are then capped. When they are cooled to 15 °C, the bottles frequently collapse.

The company seeks an improved bottle cap which prevents collapse. The improved container should not spill if inverted. It should remain effective for up to six months and should stand "thermal cycling." The improvements should cost less than $0.05 per bottle.

We want to design such a bottle cap. In our design, we will generate five or so ideas which have real potential. We will choose the best idea for further development, supporting our choice with engineering calculations. We must explore the risks involved in this choice, and explain how we plan to mitigate these.

To give our calculations a basis, we assume as a benchmark an HDPE bottle 7 cm in diameter and 20 cm tall. The bottle is cooled from 40 °C to 15 °C in 30 minutes. To estimate the shelf life, we assume the shampoo contains 0.1% limonene and that we cannot lose over 10 percent of this in six months.

SOLUTION

Our goal is to design a bottle cap which prevents collapse of a 750 cm^3 bottle filled with 600 cm^3 of hot liquid as it cools from 40 °C to 15 °C. Our strategy is to design the cap to reduce the pressure difference between the cooling liquid and the surroundings.

Needs. Our design should meet three essential needs and two desirable needs. The essential needs are:

(1)   The bottle doesn't collapse in 30 minutes of cooling.
(2)   The bottle can be inverted without losing liquid.
(3)   The new cap costs less than $0.05.

The two desirable needs are:

(4)   The capped bottle loses less than 10% of its aroma in six months
(5)   It can stand thermal cycling caused by temperature swings in a warehouse or a truck.

How we will meet these needs is described below.

To make these needs still more definite, we suggest that the new cap must meet two specifications. First, when the temperature drops from 40 °C to 15 °C, the vapor pressure of water drops from 55.3 mm Hg to 12.8 mm Hg. To make up for water condensation, we must add

$$150 \, \text{cm}^3 \left[\frac{55.3 - 12.8}{760} \right] = 8 \, \text{cm}^3 \, \text{air}$$

Because of gas cooling, we must add

$$150 \, \text{cm}^3 \left(1 - \left[\frac{273 + 15}{273 + 40} \right] \right) = 12 \, \text{cm}^3 \, \text{air}$$

Thus our cap must allow transport of $(8 + 12) = 20 \, \text{cm}^3$ of air in the 30 minutes of cooling. At the same time, we must stop evaporation of scents like limonene over the six months of storage. If we assume that the product contains

---

Table 7.5–4 *Three promising ideas for a breathable bottle cap.*

Idea	Structure	Advantages	Disadvantages
Membrane	30 μm PDMS film	Cheap; easy technology; won't leak liquid	Won't pass 20 cm^3; may leak limonene excessively
Nanoporous plug	30 μm polypyropylene with 10 nm pores	Cheap; easy; meets flow specification	Leaks limonene excessively; may leak liquid
Microporous gel plug	1 mm porous plug with tunable pores	Cheap; meets flow specification; meets limonene specification	Technology requires development; may leak liquid

---

0.1% limonene and that we can afford only a 10% loss, then we must have less than

$$0.001(600 \text{ g product})(0.1) = 0.06 \text{ g loss}$$

We now turn to product ideas.

Ideas. After considerable effort, our core team decides that five ideas have special merit. These are as follows:

(1) *A permeable membrane.* We suggest a non-porous, highly permeable polymer membrane which will allow diffusion but not convection.

(2) *A nanoporous plug.* We are interested in a highly hydrophobic, nanoporous plug, which will allow vapor flow but prevent liquid flow.

(3) *A microporous gel plug.* We are intrigued by a plug with gel-coated pore walls. At low temperature, the gel swells, plugging the pores. At high temperature, the gel collapses and the pores open.

(4) *A one-way valve.* We can imagine a one-way valve, perhaps nothing more than a crimped piece of tube. High pressure outside the bottle causes the crimp to open and flow to occur. High pressure inside the bottle squeezes the crimp shut.

(5) *A shape-memory polymer.* This cap would have a valve permanently open until it is heated and cooled. The valve then closes permanently.

After considerable discussion, our core team decides to reject the last two ideas. The one-way valve is judged too expensive, especially when considering the cost of assembly. The shape-memory polymer seems vague, without specifics.

Selection. The three ideas selected for further development are summarized in Table 7.5–4. The membrane cannot leak 20 cm^3 air. It will have trouble not leaking limonene, but it will never leak liquid. The nanoporous plug will leak enough air but will leak too much limonene. The microporous gel plug can leak air but not limonene. However, this technology is relatively underdeveloped and so will require more investment before it is ready for implementation.

Membrane. We imagine a 30 μm thick film of polydimethylsiloxane (PDMS) supported by a porous, non-woven backing layer and covering the cap area of 2 cm². We choose PDMS as the most permeable, readily available polymer film. We choose 2 cm² area as typical for a cap of the 750 cm³ bottle chosen as our benchmark. The total amount of transported across this membrane is

$$J_1 t = \left(\frac{DH}{l}\right) A \Delta c_1 t$$

where J_1 is the total flux; t is the cooling time; DH is the permeability, the product of the diffusion coefficient D and the solubility H; l is the membrane thickness; A is the cap area; and Δc_1 is the concentration difference, expressed as a partial pressure. For the PDMS film,

$$J_1 t = \left(\frac{7 \times 10^{-7}\,\text{cm}^2/\text{sec}}{30 \times 10^{-4}\,\text{cm}}\right) 2\,\text{cm}^2 \times \left(\frac{4.2\,\text{cm Hg}}{76\,\text{cm Hg}}\right) (1800\,\text{sec})$$

The thickness of 30 μm is that for a typical plastic bag. The driving force of 4.2 cm Hg is that for the total pressure outside (76 cm Hg) minus the total pressure inside (71.8 cm Hg) after the water is chilled to the final temperature; the actual average driving force will be less. Thus

$$J_1 t = 0.05\,\text{cm}^3$$

We believe that it will be difficult to meet the target for air flow. The obvious ways are to increase the permeability by using a different polymer, to reduce the membrane's thickness, or to increase the permeable area. At the same time, the driving force will be smaller than that assumed here. This idea will probably not work.

Nanoporous plug. We next turn to a nanoporous plug. This could be made of a 30 μm layer of stretched polypropylene film containing 10 nm pores. Because gas transport through the film is by Knudsen diffusion, the total amount of transported $J_1 t$ is given by

$$J_1 t = \left(\frac{\varepsilon D}{\tau l}\right) A \Delta c_1 t$$

where ε and τ are the void fraction and tortuosity of the film, about 0.3 and 3, respectively; and the other variables are those defined above. The diffusion coefficient D in this case is

$$D = \frac{2d}{3}\sqrt{\frac{k_B T}{m}}$$

where d is the pore diameter, k_B is Boltzmann's constant, m is the molecular weight, and T is the absolute temperature. For the 10 nm pores used here, D is 0.06 cm²/sec for air, and 0.01 cm²/sec for limonene. Thus, for air

$$J_1 t = \left(\frac{0.3\,(0.06\,\text{cm}^2/\text{sec})}{3\,(30 \times 10^{-4}\,\text{cm})}\right) 2\,\text{cm}^2 \times \left(\frac{4.2\,\text{cm Hg}}{76\,\text{cm Hg}}\right) (1800\,\text{sec})$$

$$= 400\,\text{cm}^3$$

This exceeds our criterion for 20 cm^3 flow. However, because limonene has a vapor pressure of about 2 mm Hg, the amount of limonene which would be lost is

$$J_2 t = \left(\frac{0.3 \, (0.01 \, \text{cm}^2/\text{sec})}{3 \, (30 \times 10^{-4} \, \text{cm})} \right) 2 \, \text{cm}^2$$

$$\times \left(\frac{0.2 \, \text{cm Hg}}{76 \, \text{cm Hg}} \right) \frac{136 \, \text{g}}{22.4 \times 10^3 \, \text{cm}^3} \left(\frac{86400 \, \text{sec}}{\text{day}} \right) 180 \, \text{days}$$

$$= 170 \, \text{g lost}$$

The bottle only contains about $(0.001 \times 600) = 0.6$ g so this plug does not meet the criterion for limonene loss.

Microporous gel plug. The third idea, the most promising and the most risky, consists of a 0.1 cm thick plug mounted in the cap. The plug, made of a tough polymer, contains 30% 100 μm pores whose walls are coated with a partially hydrolyzed hydrogel like poly-*N*-isopropylacrylamide. This gel, which absorbs water, expands below 30 °C to block the pores and prevent convection. Above this temperature, the gel collapses so that the pores are open, to perhaps 50 μm in diameter. (This critical temperature can be adjusted by changing the degree of gel hydrolysis.) The gas flow is now given approximately by the Hagen–Poiseuille law:

$$J_1 t = \left(\frac{\varepsilon d^2 \Delta p}{32 \mu l} \right) A t$$

where Δp is now the pressure drop across the gel plug and μ is the viscosity of air. Thus

$$J_1 t = \left(\frac{0.3 \, (50 \times 10^{-4})^2 \left(\frac{4.2}{76} \right) \left(\frac{10^5 \, \text{kg}}{\text{m sec}^2} \right)}{32 \, (2 \times 10^{-5} \, \text{kg/m cm}) \, 0.1 \, \text{cm}} \right) 2 \, \text{cm}^2 \times (1800 \, \text{sec})$$

$$= 2 \times 10^6 \, \text{cm}^3$$

We are way over the 20 cm^3 needed. At the same time, the limonene flux will be by diffusion across the water-swollen hydrogel:

$$J_2 t = \left(\frac{\varepsilon D H}{l} \right) A \Delta c_2 t$$

$$J_2 t = \left(\frac{0.3 \times 10^{-5} \, \text{cm}^2/\text{sec} \, (0.01)}{0.1 \, \text{cm}} \right) 2 \, \text{cm}^2$$

$$\times \left(\frac{0.2}{76} \right) \frac{136 \, \text{g}}{22.4 \times 10^3 \, \text{cm}^3} \left(\frac{86400 \, \text{sec}}{\text{day}} \right) 180 \, \text{days}$$

$$= 1.5 \times 10^{-4} \, \text{g}$$

This easily meets the criterion for limonene loss. The gel plug looks good, but the manufacturing details are unknown. The gel plug is the most promising, most risky idea.

We should focus future efforts on the gel plug, with the nanoporous plug as a backup. We suggest reducing the risk of the gel-coated plug by experiment. We will first choose a commercially available microporous plug. We will coat this plug with temperature-responsive gels, beginning with poly-*N*-isopropylacrylamide. At the same time, we suggest re-examining the limonene specification and the estimation for the nanoporous plug because this is an unusually simple solution.

EXAMPLE 7.5–4   HOT-FLASHES RELIEF

Women undergoing menopause are frequently bothered by "hot flashes" (or "hot flushes"), brief periods where their chest, neck, and face become uncomfortably flushed. These women are also bothered by "night sweats," a similar phenomenon. Many women are resentful that the mostly male medical establishment has not taken their complaints more seriously.

The causes of hot flashes are known, but cures are much less certain. The causes are changes in hormones, especially estrogen, progesterone, and testosterone; the results include small changes in skin temperature, for example, from 34.9 to 35.3 °C. This temperature increase is caused not by increased metabolism but by vasoconstriction, that is, by decreased blood flow. In other words, the hot flashes are not due to the body making more heat, but are due to changes in the system for removing heat.

The relief for menopause centers on hormone replacement therapy. This treatment, which uses hormone mixtures administered orally (i.e. as pills), does relieve symptoms but is implicated in increased breast cancer, heart disease, and stroke. This has led many women to seek alternative remedies like black cohosh and red clover, which have uncertain results. Other suggested cures are as simple as baby wipes, liquid-soaked tissues which wet the skin, which is then cooled by evaporation. Such a solution is effective but not necessarily practical. Imagine a woman at a dinner party, told to strip to the waist, wipe herself with a wet rag, and just relax.

The team plans to design a cooler that a woman can wear under her clothing and which will keep her comfortable. The team has decided to investigate two alternatives:

(1) *Peltier coolers.* These well-developed, solid-state devices get cool when a current is passed through them. While they operate without any moving parts, using the Peltier effect, they will require the woman to carry a battery.

(2) *Heat storage devices.* These systems, developed for solar-energy storage, would encapsulate compounds like waxes or Glauber's salt in polymer sponges. The materials melt over a specified temperature range. While these systems also have no moving parts, they must be tuned to a particular temperature.

Other ideas, like reversible desiccants, are also possible.

Table 7.5–5   A comparison of two wearable coolers.

	Specification	Pelter cooler	Controlled melter
Cost	<$100	$100[a]	$5
Weight	< 500 g	200 g	100 g
Energy/hot flash	1.7 kJ	2.4 kJ	1.7 kJ
Power	30 W	180 W	30 W
Chief advantage		Wearer can control cooling	Automatically regenerates
Chief disadvantage		Higher cost and weight	No separate control

[a] This price assumes a thermostat costing under $20.

To begin this design, we must decide on the cost, the weight, the energy, and the power required. As initial guesses, we assume a manufacturing cost under $100. We assume that the device should weigh less than 500 g and be wearable. To estimate the energy, we assume ten hot flashes per day, each lasting one minute and each raising by 1 °C a tissue volume of 20 cm × 20 cm × 1 cm. Thus the energy required is

$$(20 \times 20 \times 1)\,\text{cm}^3 \frac{4.18\,\text{J}\,(1\,°\text{C})}{\text{cm}^3\,°\text{C}} \frac{10\,\text{flashes}}{\text{day}} = \frac{17\,\text{kJ}}{\text{day}}$$

The power required is

$$400\,\text{cm}^3 \frac{4.18\,\text{J}}{\text{cm}^3\,°\text{C}} \frac{1\,°\text{C}}{60\,\text{sec}} = 28\,\text{W}$$

We must now select the best design.

SOLUTION

Both the Peltier cooler and the melter meet the desired specifications, as shown in Table 7.5–5. However, they do so in very different ways, as explained in the following paragraphs.

Peltier cooler. This device uses ten solid-state semiconductors which get cool when an electrical current flows through them. The devices, based on bismuth telluride, have no moving parts, achieving their cooling from the Peltier effect. They are inefficient, less than 10% of the Carnot limit, and are mostly used in cooling computers. They are also used as refrigerators for expensive camping equipment. Many of the available devices are smaller than 1 cm.

We imagine a short tee shirt containing these devices and worn under conventional clothing. The devices could also be incorporated into a bra. Ideally, we would like ten or more devices so that the cooling is well distributed. We can buy six devices each providing 30 W of cooling for about $10 each. A rechargeable lithium ion polymer battery supplying the required 17 kJ will cost about $20 and weigh around 40 g. Moreover, this device can either be made to have an automatic thermostat or can cool when the wearer demands cooling.

If we can purchase such a thermostat for under $20, we can meet our price target of $100. The devices weigh very little, and the battery will weigh around 40 g. Thus, if we can develop a thermostat weighing less than 100 g, we should be well under our target weight of 500 g.

Controlled melter. This device is a polymer sponge filled with a compound which melts at body temperature. When a woman has a hot flash, her skin temperature rises, melting the compound and keeping her skin cool. When the flash recedes, the body temperature drops and the compound refreezes. Thus the energy needed is only that required for one flash because the system is constantly self-regenerated. Again, the system has no moving parts. While it is not completely solid, any liquid solutions are contained within the polymer.

The key design decision for this system is the choice of the compound. The obvious choices are inorganic hydrates like those suggested for solar-energy storage. Examples are Glaubers salt ($NaSO_4 \cdot 10H_2O$), $MgSO_4 \cdot 6H_2O$, and $(Na_2S)_2 \cdot H_2O$. While these have large energies of fusion, they are pure compounds and, hence, melt at a single temperature. For example, Glauber's salt melts at 33 °C. However, women's skin temperatures vary between individuals, and hot flashes occur over a range of conditions. Thus these pure compounds may not be appropriate.

As an alternative, we choose petroleum waxes, somewhat lighter (i.e. lower melting) than those used for candles. These mixed materials can melt from 33–37 °C, the range of interest here. Their heat of fusion is typically more than 30 kJ/kg; thus to supply the 1.7 kJ in one hot flash, we need $(1.7/30) = 60$ g of wax. The cost of this amount of wax is under $1. We suggest putting these waxes in a porous sponge which has air gaps every 0.5 cm, so that we can handle both cooling and sweating. The resulting composite is about 60% wax. We recognize that this design lacks important details, but it does seem feasible.

Risk. The analysis above suggests both the Peltier cooler and the controlled melter will work. The next step is to choose at least one of these alternatives and to build at least one prototype. Which one is chosen will depend on the strengths and weaknesses of the manufacturer and on any additional marketing data. For example, we might be interested in how much more women are willing to pay to control the cooling, which is possible with the Peltier cooler. We would be interested in how much women will value being able to machine wash the product, which is probably easier with the controlled melter.

Before we build a prototype, we must also decide where the product implies risk and how we may mitigate this risk. Some of the risks involved are summarized in Table 7.5–6. Sweating, which jeopardizes the entire project, might be reduced with reversible dessicants. The other big risk for the Peltier cooler is the required cooling, assumed to be 17 kJ/day. If this is ten times too small, then both the battery weight and the cooler cost will miss our specifications. To reduce this risk, we need a much more complete review of the existing medical literature. For the controlled melter, this is not as severe because we can more easily meet the cooling demands with this self-regenerating system. We may be able to reduce the cost of Peltier devices by talking to manufacturers about which devices are

Table 7.5–6 *Risks in hot-flashes-relief products.*

Risk	Probability	Consequence	Risk level
Both alternatives			
Topical cooling fails	0.5	0.9	0.45
Sweating continues	0.5	0.7	0.35
Device poorly washable	0.1	0.9	0.09
Peltier cooler			
Requires heat larger than assumed	0.5	0.9	0.45
Cost excessive	0.5	0.3	0.15
Control system ineffective	0.1	0.1	0.01
Controlled melter			
Requires heat larger than assumed	0.5	0.3	0.15
Effective over wrong temperature range	0.3	0.3	0.09

easiest to make, and we expect thermostat manufacturers are likely to have an off-the-shelf control system which is suitable.

The largest risk for the controlled melter may be the choice of wax. While we have not investigated this in detail, we are sure that some wax or chemical combination can be found which will work. We suggest adding solvents to the wax or checking eutectic mixtures of inorganic hydrates. The estimates here show that we can easily be in the correct range.

Both the Peltier cooler and the controlled melter should give women relief from the discomfort caused by menopause-induced hot flashes. We suggest building prototypes of both.

7.6   Conclusions for Chemical Devices

The design of chemical devices uses the same design template as the design of commodity chemical processes, but the steps have different importance. The first "needs" step for devices depends on convenience, on how easily the device is used. This use depends strongly on start up and shut down and less strongly on cost minimization.

The second step, "ideas," uses similar concepts of reaction engineering and unit operations, but the choices made among options are different. The process flow diagram is now more often steps in time, like a recipe for a cake. It is less frequently a chart in space, with the steps representing flows. The reactors are usually batch, not continuous, and the separations depend on adsorption more often than distillation.

The third design step, "selection," is the most important and the most difficult. This is because the selection often involves comparing radically different options on the basis of both technical and non-technical factors. The basis for this comparison is not only cost, but also simplicity. Whenever possible, we will avoid recycles; we will choose compact over extended processes, even when these are more expensive. We will make this selection step without worrying as much about manufacture. This is a sharp contrast to commodity manufacture, where our effort will always be driven by minimizing cost.

Device design is challenging because convenience is a vaguer objective than minimum cost. It is interesting for this reason as well, because it always requires judgment.

Problems

1. *Needle-free injection.* Accidental injection or scratching with syringe needles is a significant cause of injury for health professionals, with the added risk of transmission of blood-borne infections. Your medical devices company is interested in producing a needle-free injection system, probably based on blasting nano-droplets through the skin. Write a list of needs for such a device and list alternative ways of meeting these needs.

2. *Lab-on-a-chip.* Lab-on-a-chip technology can be defined as the miniaturization of chemical and physical processes and their integration on a microchip. Recent years have seen rapid advances in this technology: micron-sized features now are routine for pipes, reactors, mixers, and separators. Your company has experience in the fabrication of such chips for research applications, and is interested in commercializing this technology. Suggest possible applications for lab-on-a-chip technology. Sort and screen these to determine the most attractive few for further research and prototyping.

3. *Waiting to exhale.* Measuring key indicators of health, like blood cholesterol, hematocrit, and white-blood-cell count, is an important aspect of health management. However, this requires analyzing a blood sample, which is slow and expensive. Your company is interested in quickly measuring compounds in exhaled breath and using these as indicators of health. For example, some studies have shown that exhaled breath can be used to monitor efforts to lose weight.

   Choose diseases and conditions where such an analysis could be useful. Select compounds in the breath that could correlate with these diseases.

4. *Bronze vs. teflon extrusion dies for pasta.* Traditionally pasta is made by extrusion of durum wheat paste through a bronze die. This gives the pasta both its characteristic shape and a surface roughened on a sub-millimetre scale. This roughness, a result of instabilities caused by the high friction between the die and extrudate, is desirable because it helps the pasta retain sauce. However, extrusion through a bronze die is slow. Die corrosion and cleaning are also problems. Because of this, pasta is mainly produced by extrusion through a Teflon die, giving the same shape with reduced production costs. However, the surface of the pasta is smoother, less able to retain sauces.

   Explore ideas for combining the advantages of the Teflon and bronze dies.

5. *Partial de-alcoholization of wine.* There is increasing awareness of damage to health caused by alcohol, but also increasingly widespread appreciation of the pleasures of a good bottle of wine. We seek to maintain the flavor of wine but with significantly reduced alcohol content. Such a product would be taxed at a lower rate. Four devices are under consideration for reducing alcohol in wine before bottling:
   (1) Pervaporation through hollow fibers of a membrane preferentially permeable to alcohol.
   (2) Liquid–liquid extraction of the alcohol into a water-insoluble organic phase.
   (3) Absorption of the alcohol into a swellable gel.
   (4) Adsorption onto activated carbon or another adsorbent.
   Flesh out details of how each of these processes could work and select the best one.

6. *Nanoreactors.* Progress in microelectronics has included development of etching techniques capable of making very small structures. These small structures have simulated interest in nano-technology: nanovalves, nanoturbines, etc. This example is concerned with nano-sized chemical reactors. These reactors can provide rapid mixing. However, turbulent mixing can produce eddies around 30 μm. In gases, these mix in microseconds; even in liquids, they mix in seconds. Thus, fast mixing alone is not a good reason for developing these nanoreactors.

   The reactors also offer better selectivity. If their walls are coated with catalyst, they may provide a single, well-defined residence time at a single temperature. Dispersion due to diffusion within porous catalysts is eliminated. This should be equally true for reactions in gases or in liquids.

   Suggest reactions where these reactors might be valuable.

7. *A modular nuclear plant.* Babcock and Wilcox have announced a 125 MW nuclear reactor which should ease concerns surrounding scale-up, grid connectivity, and safety. The device, based on an advanced light water reactor, lets customers scale electricity requirements in 125 MW increments. This allows a distributed nuclear-power network, attracting industrial users, municipal suppliers, and developing countries with poorly developed national grids. The reactors are designed to operate below ground with storage space for 60 years of fuel and spent fuel. The design will be operated for five years on a single batch of fuel.

   Consider the commercial, legal, and political risks of the project. Assume all technical safety issues have been addressed in the design. For each risk you identify, make an estimate of probability and consequence, and consider ways of mitigating the risks.

8. *Keeping drinks carbonated.* Colas and tonic water are usually sold in 1.5 L bottles. Any remaining beverage is often stored for long periods and loses

---

Table 7.8A *Ideas for keeping drinks carbonated.*

A. Mechanical
  (i) *Add CO_2 to the bottle*
    1. A soda stream
    2. Pressure-sensitive CO_2 reservoir
    3. Soda siphon
    4. Widget
  (ii) *Improve bottle*
    1. Use less-permeable plastic
    2. Use smaller containers
      a. individually
      b. sub-divide larger bottle
    3. Use a plunger or siphon to force liquid out of bottle
    4. Valve on bottle to keep pressure up
    5. Improve cap re-seal
  (iii) *Larger devices*
    1. Beer pump
    2. Tonic-on-tap
    3. CO_2 cylinder in the refridgerator
  (iv) *CO_2 addition to glass*
    1. Put CO_2 in a cocktail stick
    2. Put CO_2 in ice
    3. Add a small widget
B. Chemical
  (i) Increase CO_2 capacity of liquid
    1. Control pH
    2. Use a complexing agent like heme or diethylamine
    3. Store colder
  (ii) *Add CO_2 as solid (6)* – to glass or bottle (possibly via lid)
    1. (Sugar) encapsulated CO_2
    2. $NaHCO_3$ (with acid trigger?)
  (iii) *Don't use CO_2*
    Use N_2 bubbles (see also A(i)4 and A(iv)3)
  (iv) *Control nucleation*
    1. Don't disturb while pouring
    2. Make inside of bottle smoother
C. Other
  (i) Include color change to indicate adequate CO_2 content
  (ii) Make portions in variable or various sizes

---

its carbonation. The alternative, buying the drinks in 150 mL cans or bottles, is twice as expensive.

Your company is interested in finding better ways to keep colas carbonated after opening, i.e. in sustaining the 5 bar pressure when the bottle is resealed. Studies to date have generated the list in Table 7.8A. Select the best ideas.

9. *Glucose sensor.* Current optical sensors for glucose use hologram technology: in response to glucose, the hologram will change color or display a

specific pattern. This elegant technology works well. However, its manufacturing is complex, expensive, and requires flat surfaces, so that holograms are only suitable for high-value applications.

Your company has invented a method of evenly applying onto surfaces layers of polymer which are around the wavelength of light in thickness. By laying down layers of two polymers with different optical densities, a diffraction grating can be produced, which will reflect light of different wavelengths as a function of angle, according to Bragg's law. The method is cheap, quick, and can be applied to surfaces of any shape. Your company has also developed a probe molecule that, when incorporated into one of the polymer layers, reacts with glucose to cause a swelling of the layer. Thus, the color of the surface coating will change in the presence of glucose.

While the probe is currently specific to glucose, your company believes that it will not be difficult to extend it to other biologically and medically important small molecules. Generate ideas for the application of this new technology. Sort and screen these ideas. Which are the best few, worthy of more detailed examination?

10. *Hand-held spectroscopic analyzers.* Consumers are often worried about trace chemicals in imported products, including lead paint in children's toys. In some cases, the paint can be quickly analyzed with a hand-held X-ray fluorescence (XRF) analyzer, which looks like the bar-code reader used in grocery stores. Similar detectors may also be useful in identifying phthalates and other compounds. Such hand-held detectors are not intended to give accurate chemical analyses, but to screen products, identifying those meriting further analysis.

Our company is interested in identifying areas where such hand-held analyzers would be useful. We are not interested in XRF specifically, but rather in broad market areas where any type of fast chemical screening would be useful. Generate ideas for such applications, sort and screen these ideas, and determine three or four good ideas to take forward for further investigation.

11. *Desiccant wheel specifications.* Air conditioning takes warm, humid air, cools it, and pumps it into a house to increase comfort. Because the outside air is humid, cooling it often produces condensation. Even after the liquid is removed, the remaining air is saturated at 100% relative humidity (RH), which is uncomfortable. As a result, air conditioners normally overcool the air, remove the condensed water, and reheat the air so it is comfortable.

To make this more specific, assume we have 35 °C outside air at 50% RH, giving a water vapor pressure around 21.0 mm Hg. This air is to be cooled to 9 °C and 100% RH, where the saturation vapor pressure is 8.6 mm Hg. The difference between vapor pressures yields liquid water, which must be removed. The remaining air is reheated to 20 °C, where its vapor pressure is 8.6 mm Hg but its RH is a comfortable 50%.

This process works well but uses a lot of energy. As energy costs escalate, we seek to design houses that are more energy efficient. Old houses leak so much that air remains in the house only about 40 minutes. New houses leak so little that air remains in the house for a day. This allows CO, cooking smells, and CH_2O to accumulate, which can cause "sick-house syndrome," a condition in which house inhabitants develop acute or chronic negative health effects as a result of these accumulating chemicals. As a result, the Society of Air-Conditioning Engineers (ASHRAE) now suggests that house air be exchanged 8.4 times per day. The US government suggests six times per day, or once every four hours. We will use this four-hour standard here.

The air-conditioning system used now is different to the older design outlined above. The warm inlet air (35 °C; 50% RH) is dried isothermally with a desiccant to give air with a water partial pressure of 8.6 mm Hg. This air then flows countercurrently with exhausted air from the house, which enters at 25 °C and 11.9 mm Hg (50% RH at 25 °C). The inlet air exits the heat exchanger at 27 °C and enters the air conditioner to be cooled. The exhaust air exits the heat exchanger at 33 °C, is heated, and flows through the desiccant to regenerate it. This system is more expensive to install but much cheaper to run.

The key to this process is the desiccant, which is mounted on a wheel. The wheel rotates so it is exposed first to the wet inlet air and then to the hot exhausted air. Many websites illustrate the process, including some cleverly animated ones. However, we are not interested in the process; we are interested in the desiccant.

We want to write specifications for the desiccant. To do so, we assume we want a circular bed with 400 cm^2 area, 1 cm deep. We want to size this bed for a 250 m^2 house with 3 m ceilings, the air in which is exchanged every four hours, with outside and inside air as given above. We want specifications in each of four areas:

(1) Identify an adsorbent that adsorbs only water and whose adsorption changes sharply with humidity.
(2) Identify a manufacturing procedure for this adsorption unit. Packed beds, monoliths, and coating are possibilities.
(3) Set a maximum pressure drop across the adsorbent.
(4) Specify the number of cycles that the adsorbent must achieve.

While all four specifications are important, #2 is essential.

REFERENCES AND FURTHER READING

Bronzino, J. D. (ed.) (2006) *Tissue Engineering and Artificial Organs*, 3rd Edition. CRC Press, Boca Raton, FL.

Burban, J. H., Spearman, M. R., Thundyil, M., and **Zia, M.** (2003) Oil Dehydrator. US Patent 6,517,725 issued February 11.

Cussler, E. L. (2009) *Diffusion*, 3rd Edition. Cambridge University Press, Cambridge, ch. 15, 19.

Douglas, J. M. (1988) *Conceptual Design of Chemical Processes*. McGraw-Hill, New York, NY.

Kale, **K. M.**, **Cussler**, **E. L.**, and **Evans**, **D. F.** (1980) Characterization of micellar solutions using surfactant ion electrodes. *Journal of Phyical Chemistry* **84**, 593.

Koryta, **J.** and **Stulik**, **K.** (2009) *Ion Selective Electrodes*. Cambridge University Press, Cambridge.

Li, **X.** and **Jasti**, **B. R.** (2005) *Design of Controlled Release Drug Delivery Systems*. McGraw Hill, New York, NY.

Nauman, **E. B.** (2008) *Chemical Reactor Design, Optimization, and Scaleup*. Wiley, New York, NY.

Turton, **R.**, **Bailie**, **R. C.**, **Whiting**, **W. B.**, and **Shaeiwitz**, **J. A.** (1998) *Analysis, Synthesis, and Design of Chemical Processes*. Prentice-Hall, Upper Saddle River, NJ.

Wilson, **E. E.** (1915) A basis of rational design of heat transfer apparatus. *ASME Journal of Heat Transfer* **37**, 47–70.

8

Molecular Products

Molecular products, the subject of this chapter, are exemplified by pharmaceuticals. These products, which sell for much more than the cost of their ingredients, are sold to perform a particular task, like curing a disease. They differ dramatically from the commodity chemicals summarized in Chapter 6 and explored in detail in courses on chemical process design. Commodities sell for only a slight premium over the cost of their ingredients. They are sold into bulk markets like those for polymers, lubricants, and fertilizers. Patents on commodity products normally describe processes for their manufacture, not their application to a specific function. Not surprisingly, the presidents of commodity chemical companies tend to be engineers. The presidents of pharmaceutical companies, and of other molecular product companies, are usually chemists or physicians.

Molecular products depend on two keys: their discovery and their time to market. Drug discovery is remarkably inefficient. In justifying the high prices for drugs, company executives sometimes assert that it takes 10,000 candidates to find one successful drug. If this is true, it makes Napoleon's invasion of Russia, shown in Figure 8.0–1(a), look like a success. After all, Napoleon began with 472,000 men, and returned with 4900, a success rate one hundred times better than the drug industry.

We can imagine drug development as a similar campaign, shown lightheartedly in Figure 8.0–1(b). We begin by identifying the target disease, and if possible, what we wish to manipulate (e.g. a particular protein). We then spend perhaps four years and over half a billion dollars seeking drugs which influence this disease. This huge, expensive, inefficient search is where so many compounds are identified, synthesized, and abandoned. At the end of this saga, we begin animal testing, where our success rate is only about 10%. This is much worse than Napoleon's Battle of Borodino, where two thirds of his soldiers emerged unscathed. The animal testing will normally involve a sequence of mice, rats, and dogs. The dog tests are the most demanding, leading to the quip that "Drugs kill dogs, and dogs kill drugs." At the end of this ordeal, which is beyond the scope of this book, we will have identified a possible target molecule. This is the point where engineering starts to become involved.

(a) Napoleon's 1812 Russian Campaign

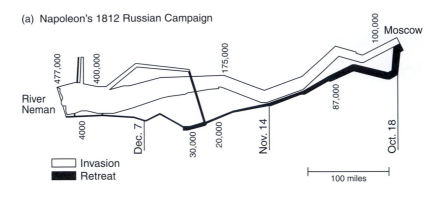

(b) A Typical New Drug Campaign

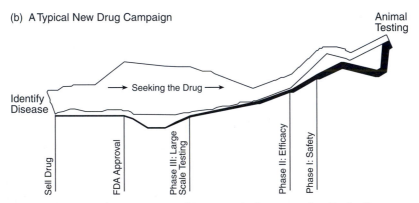

Figure 8.0–1 *Napoleon's invasion of Russia vs. finding a new drug* Both efforts start with many resources but finish with few survivors.

This engineering involvement centers on the second key aspect of molecular products, the speed of their development or their time to market. This is important because the first molecular product to be sold for a specific task normally garners two thirds of the sales in this area. In this engineering-based development we will use the same design template of needs, ideas, selection, and manufacture to bring the chemical product to market. We will decide what amounts and purity of the target molecule are needed. We will generate ideas to make and purify this molecule, and we will use generic equipment to manufacture batches of our product.

In this chapter, we show how to apply this design template to minimize time to market. We emphasize the selection step because it tends to be critical to the engineering, as it was for chemical devices. In Section 8.1, we summarize the chemistry of these products, stressing how they differ from commodities. We also discuss their discovery, the amounts needed, and the disciplinary frictions in their development. In Section 8.2, we describe how to start development, including reaction path synthesis and separation selection. Sections 8.3 and 8.4 review the reactors and separators available, and the choices which are usually made.

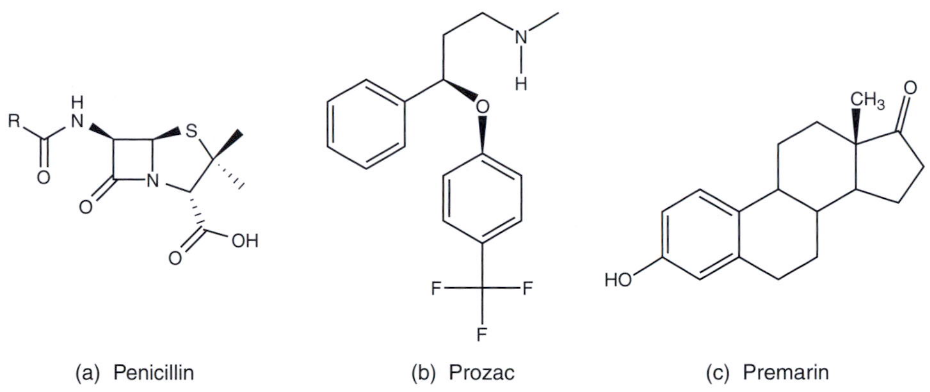

(a) Penicillin (b) Prozac (c) Premarin

Figure 8.1–1 *Three drugs from three different sources* Penicillin is made by fermentation, Prozac is synthesized chemically, and Premarin is an extract from a natural source.

Finally, Section 8.5 gives some more complex examples in which these ideas are used.

8.1 Characteristics of Molecular Products

Molecular products are high-value molecules, such as pharmaceuticals, pigments, flavors, and electronics chemicals. These molecules usually have molecular weights of 500 to 3000 daltons, though some antibodies can have molecular weights of several million. These species, exemplified by the three molecules in Figure 8.1–1, are obtained in three different ways. Antibiotics, like penicillin shown in Figure 8.1–1(a), are examples of molecules produced by fermentation. Prozac (fluoxetine), an antidepressant shown in Figure 8.1–1(b), is an example made by chemical synthesis. In most cases, the products are chemically well defined and of exceptionally high purity. Occasionally, products are mixtures of pharmaceutically active species prepared by a specific process, often from a biological feedstock. The hormone-replacement treatment Premarin, which contains large amounts of the species shown in Figure 8.1–1(c) as well as many other compounds, is an example of this type.

For molecular products, we are interested in molecular structure, including details like chiral centers. The key size scale is nanometers, much smaller than the size scale of centimeters used for chemical devices, or even the size scale of micrometers important to microstructures. Of course, commodity chemicals also have well-defined molecular structures which are nanometers in scale, but these structures are relatively simple and not central to their function. Because of this simplicity, we rarely draw these structures again and again as we develop processes for commodity manufacture.

Molecular products are typically made in small quantities, often less than ten tons per year. They sell for high prices, often over $100/kg. One good example is Zoladex (goserelin acetate), a decapeptide used to treat both breast and prostate cancers and shown in Figure 8.1–2. Forty-six kilograms of the drug is made each

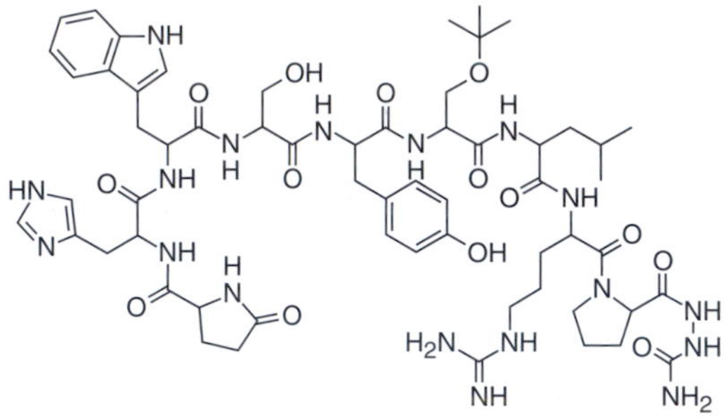

Figure 8.1–2 *Zoladex (goserelin acetate)* This decapeptide, sold as the acetate, is produced synthetically.

year, selling for $800 million. Compounds like this are not synthesized in optimized, dedicated equipment, but are made periodically in generic equipment. Each period is often called a "campaign," an echo of the parallel with Napoleon. Moreover, molecular products like these are often a single stereoisomer, typically having several asymmetric centers.

These properties are the antithesis of commodity products. Commodities are normally made in quantities greater than 10,000 tons per year. They are cheap, selling for less than $5/kg. Because their ingredients often cost a large percentage of the selling price, they must be made very efficiently, continuously, in dedicated equipment. Commodities are chemically simple, usually without stereoisomers.

As described in the introduction, the key to molecular products is discovery. Once a target molecule is identified, the core team needs to make enough for clinical trials to see if the molecule has the desired pharmacological benefit. Such clinical trials are often discussed in terms of three phases mandated by the Food and Drug Administration (FDA). The core team must choose the reaction path, that is, chemical reactions which will be used to make the target molecule. It must choose the purification procedure. Moreover, the choice of reactions and purifications are binding because the FDA will normally not approve dramatic changes without further testing, which is slow and expensive.

We will discuss the drug discovery, clinical trials, and common development problems in the following paragraphs. While these issues involve more than product chemistry, they occur frequently in molecular product design.

8.1–1 THE RULE OF FIVE

We first consider drug discovery. The details are beyond the scope of this book, but we can see some general characteristics. First, we should not be surprised that it is difficult to discover which chemical compounds are best for one particular disease, because over thirty million such compounds are known. We should ask what if any general guidelines are available. Surprisingly little is known, but there are some clues.

One such clue, suggested by C. A. Lipinski of Pfizer, is called "the Rule of Five." This rule, which does not apply to natural products, estimates the chance of clinical success for a drug to be administered orally. It suggests that successful drugs must meet four criteria:

(1) *The drug must have five or fewer hydrogen-based donors.* This is the sum of –OH and –NH groups.
(2) *The drug must have fewer than ten hydrogen-based acceptors.* This is the sum of the molecule's nitrogen and oxygen atoms.
(3) *The drug's molecular weight must be under 500.* This does not include species like HCl or NaOH added to enhance solubility.
(4) *The logarithm of the drug's partition coefficient K_{ow} between octanol and water must be less than five.* In other words, its octanol solubility must be less than 100,000 times its water solubility.

Note that each criterion contains a number which is a multiple of five: hence "the rule of five."

The first and second rules describe the drug's chemistry; the third is a measure of size; and the fourth suggests that to reach human tissue, the drug must pass through water. This "Rule of Five" has spawned many extensions, some of which seem more effective. But in some ways, the simplicity of this rule underscores the absence of other guides to drug discovery. If such guides exist, they are presumably closely held commercial secrets.

8.1–2 CLINICAL TRIALS

Once a target molecule is identified, those responsible for its development must make more fast. Typically, this scale-up will take place in three stages, often as required by Phase I, Phase II, and Phase III clinical trials. We need to estimate how much drug is involved, and know what the different phase trials involve.

The amount of drug to be made varies widely, so that generalizations may be inaccurate. Typical amounts may be one gram, one hundred grams, or ten kilograms. However, the amounts involved could easily be ten times smaller for a polypeptide, or one hundred times larger for an antibiotic. Still, the amount is obviously tiny compared with the amounts routinely produced for any commodity product.

The three levels of clinical trials refer to the use of the drug in humans. Phase I is a small study of around twenty healthy, paid volunteers. The study accesses the molecular product's toxicity and pharmacokinetics, usually in an outpatient clinic. Once the safety of the drug is established, the Phase II trial starts, typically with 50 volunteers, some of whom have the target disease. Phase II includes studies of how much drug should be given, and how well the drug works. Phase II is the step where most drugs fail, often because side effects are too serious. The Phase III trial, involving perhaps 1000 patients at several different locations, aims to collect statistically significant data for final submission, seeking approval for the drug from the FDA. These trials are tedious, averaging eight years for a new drug. They are a major reason why drugs are expensive, for the final price reflects not only the costs of the successful molecule but also the cost of all the failures.

8.1–3 CULTURE WARS

The development of molecular products often involves friction between those trained in different disciplines. Some of this is inevitable because the skills of those involved in drug discovery are so different to those involved in drug development. In many traditionally organized drug companies, the development chemist will rough out the process, and then the engineer will take over to work out details. The inefficiency and hostility which result can sometimes be reduced by forming core teams of both chemists and engineers right from the start of the project.

However, even within core teams, the relation between team members is often fractious, filled with unintended and intended insults in both directions. To illustrate this, consider the following conversation to make a steroid, like that in birth-control pills:

Chemist This is an easy reaction which anyone intelligent should be able to run. I just dissolve the crude steroid in methylene chloride and then add *n*-butyl lithium. The reaction is … Wait, let me put it in terms you'll understand. At –40 °C

$$A + B \rightarrow AB$$

You can't run too long because there's a side reaction:

$$AB + B \rightarrow AB_2$$

I then add acetone, which knocks out the product (i.e. causes it to precipitate). I decant the solvents and add DMF (dimethylformamide) to redissolve it. Then I add water to make the alcohol:

$$AB + H_2O \rightarrow AOH + BH$$

All these reactions are pretty exothermic. Still, they run easily, though the overall selectivity is often low, around 40%. You shouldn't have any trouble getting that higher.

Engineer Why is the selectivity so low?
Chemist I don't know. It often is in reactions like these.
Engineer How much does the temperature increase?
Chemist Quite a lot. Even at –40°C, you can see the temperature jump when you add the *n*-butyl lithium. However, I've kept the temperature rise small by running in an acetone–CO_2 bath. Sometimes, I've kept it from jumping too much by turning off the stirrer for a while.
Engineer Can you use any different solvents?
Chemist I don't know. You probably can't replace methylene chloride; it really is the best for these reactions.
Engineer You remember that it's viewed as a dangerous carcinogen.
Chemist Yeah, but lots of chemicals are dangerous.
Engineer Could methylene chloride be replaced with butyl acetate?
Chemist I don't know. Look, I really like methylene chloride. It works really well, and I think you'll have trouble replacing it.
Engineer Did you ever check for the maximum temperature rise in this reaction?

Chemist No, but it could be big enough to boil the solvent. But you can slow the reaction by shutting down the stirring.

Engineer Does that work if the reaction mixture starts to boil?

Chemist I don't know. My experiments never boiled.

Engineer Why do you always run in a round-bottom flask? You could get faster conversion in a tubular reactor.

Chemist Look, I need to slow the reaction down, not speed it up. When it runs too fast, it makes too much by-product. Then the product goes brown, not white, like it probably should be.

Engineer How can you remove the color?

Chemist I don't know. Sometimes activated carbon works on problems like these.

Engineer Can you try to get any purification when you make the acetone knock-out?

Chemist You mean add the acetone slowly so that you get purer crystals? That's a good direction to go, though it's hard at –40 °C. I didn't do it, because I was just trying to rough out the chemistry.

Engineer Did you measure the purity of that intermediate precipitate?

Chemist No. I don't think it is important.

Engineer How do you separate the product? The one after hydrolysis.

Chemist Actually I didn't. I just ran the solids that were knocked out and hydrolyzed through the HPLC (high-pressure liquid chromatography). I knew where the peaks should be because of earlier experiments using combinatorial chemistry.

Engineer Do you know how to purify the product?

Chemist Sure.

Engineer I mean at large scale.

Chemist But that's your job. I finished this one, and I did it right. I've got other reactions to run. Come back and see me if you need help. This isn't hard. See you later.

So ended the discussion. Parenthetically, when we show this fictional conversation to those involved in the drug industry, they laugh and say "Actually, you know, it's worse than that ... "

So molecular products are different from commodity chemicals because their chemistry is harder and their value is higher. Such products are subject to difficult, inefficient discovery and the need to get to market quickly. How these factors affect their design is described in the next section. Before describing this design, we review the properties of molecular products via some examples.

EXAMPLE 8.1–1   TESTING THE "RULE OF FIVE"

Prozac, penicillin, and Premarin were given as three illustrations of drugs in Figure 8.1–1. Prozac is made by in vitro synthesis; penicillin is produced by fermentation; and Premarin, an extract of horse urine, is a mixture, the principal component of which is estrone sulfate. Because only Prozac is not a natural product, only it should be covered by the "Rule of Five." Nevertheless, see if this rule works.

Table 8.1–1 *Testing the "Rule of Five"* These rules are listed in the text of this section; M_r is the molecular weight, in daltons.

Compound	$(OH + NH_2) < 5$	$N + O < 10$	$M_r < 500$	$\log_{10} K_{ow} < 5$
Prozac	1	2	309	3.8
Penicillin	2	6	334	2.3
Premarin (estrone sulfate)	1	2	372	3.1

SOLUTION

As shown in Table 8.1–1, the "Rule of Five" works for all three molecules. This result is interesting partly because it gives such a clear comparison for the properties of three very different drugs.

EXAMPLE 8.1–2 CONCLUSIONS FROM A CONVERSATION

What should the engineer have learned from talking to the chemist, as described in the conversation above?

SOLUTION

The engineer must scale up a highly exothermic reaction whose selectivity is strongly temperature dependent. The reaction is possibly mass transfer controlled, because its rate depends on stirring. The mixture from which the reaction products are separated will include raw materials and the results of side reactions. Separation by adsorption – the basis of chromatography – works, at least on a small scale. Solvents are important, but largely uninvestigated. Still, note that at least one of the engineer's suggestions – using a tubular reactor – is almost certainly stupid given the amounts involved.

In a case like this, the engineer must first check the chemist's results. He must repeat the reactions in a round-bottom flask, carefully watching the temperature vs. time. He should imitate the way that the chemist combines the reagents, and he should use the same solvents, even the methylene chloride. The engineer should first separate the products by HPLC. In most cases, he will not initially get results that are as good as the chemist's, a result of the chemist's greater laboratory skill and of the inadvertent omission in the chemist's description of nuances of chemical technique. Eventually, he should equal or surpass the chemist's laboratory results. He is then ready to design, as described next.

8.2   Getting Started

We want to design a small process which produces the target molecule. We will assume that the discovery stage is over; we know which molecule we want to make. We will normally know the purity and the amount we need because we will know which stage clinical trial we are targeting. Thus, our objective for a

---

Table 8.2–1 *Synthesis of material flow for molecular vs. commodity products* The sequence of decisions is the same, but the alternatives chosen are different.

Decisions	Molecular product	Commodity product
Process type	Batch	Continuous
Flow chart	In time	In space
Reactions	Often complete; few recycles	Partial; requires recycles
Separations	Extraction, precipitation ("knock out"), chromatography, crystallization	Distillation, gas absorption

---

molecular product is a process like that for a commodity chemical, but at perhaps one billion times smaller scale.

We will develop the process for such a product using our normal design sequence of "needs," "ideas," "selection," and "manufacture." Our "needs" step is easy: it is just the amount and purity of the target molecule to be made fast to meet the demands of the clinical trial. Needs may also include secondary constraints, such as discharge streams and available raw materials. The "ideas" step is a plethora of possible processes drawn from laboratory experiments and published patents. Each process idea will include inventing a reaction-path sequence and a separation train. Our "selection" step is the hard one because we will need to simplify our process ideas quickly, and we will not have time to do the obvious but tedious experiments which would assure our success. The "manufacturing" step will again be easy because its scale isn't so different to the bench process. Thus we need to work on the "ideas" and "selection" steps.

For both "ideas" and "selection," we will use a synthesis of material flow much like that used for commodity products. However, while the synthesis steps are parallel, the details of the steps are usually very different, as shown in Table 8.2–1. In both cases, we must first decide on what type of reactor we will use. For molecular products, this will normally be batches in a stirred round-bottomed flask. We will frequently start with one reagent in the flask and add another, but we will rarely remove material until we are ready for the next step.

The flow chart will be a series of these steps, like a gourmet recipe, not a collection of pipes. One feature of the reaction is that it rarely involves recycles; instead, the desired product will just be precipitated, or "knocked out," by adding a non-solvent. Unreacted species and the mixed solvents will be decanted, leaving the solid precipitate. This precipitate is then redissolved; another reagent is added; and a new reaction occurs, the next step in the sequence. Such a process, which makes good sense in laboratory explorations, may unnecessarily complicate the eventual process chosen. Finally, the separation train for molecular products may be more elaborate than that used for commodities because the desired product purity is so much higher.

The "ideas" and "selection" steps of molecular product design are thus straightforward parallels of those for commodity products. While different tools are usually adopted, its basic intellectual thrust is the same. However, it does have two aspects which merit more detailed discussion: reaction-path synthesis, and common separations.

8.2–1 REACTION-PATH SYNTHESIS

Once the target molecule is identified, the synthetic chemists on the team will start to think of possible routes for the synthesis. Going backwards from the target molecule to simple precursors, called "the disconnection approach," is common. This approach, outlined by Warren and Wyatt (2009), makes successive "disconnections" to reduce the target molecule to simple, available precursors. Each disconnection involves imagining breaking the structure of the target molecule: this breakage is the reverse of a synthetic step. A disconnection should be related to a well-established synthetic method but go in the opposite direction. Thus, a disconnection will be closely connected with the functional groups available in the target molecule.

Usually, several different disconnections are possible for any target molecule, and many successive disconnections are necessary before simple precursors are reached. Thus, many alternative synthetic routes can easily be deduced. An experienced organic chemist will usually be able to eliminate most of the routes as impractical, leaving a handful of reasonable alternatives. Sometimes, none of the potential routes will look viable. In this case, the target molecule may be discarded as a useful idea; it may be highly efficacious, but if it cannot be commercially made it will be no use as our product. Alternatively, it might be extracted from a natural product or made via fermentation.

To illustrate the type of approach required to fill in the gaps in our information, we will give two examples.

EXAMPLE 8.2–1   SYNTHESIS OF THE TRANQUILIZER PHENOGLYCODOL

The structure of this species is given in the upper left-hand corner of Figure 8.2–1. Suggest several routes by which it may be synthesized.

SOLUTION

This is a fairly complex molecule and so many paths are possible. The most obvious paths, also shown in Figure 8.2–1, use commercially available precursors. Which synthetic route we prefer will depend on other factors, such as cost, safety, etc.

EXAMPLE 8.2–2   TAXOL SYNTHESIS

As a result of biological screening of natural products by the National Cancer Institute in the USA, taxol was isolated from the bark of the Pacific yew (*Taxus brevifolia*) and identified as a promising anti-cancer compound. It has a unique mode of anti-cancer action, promoting the formation and stabilization of microtubules and so inhibiting cell mitosis. This implies a whole new class of anti-cancer

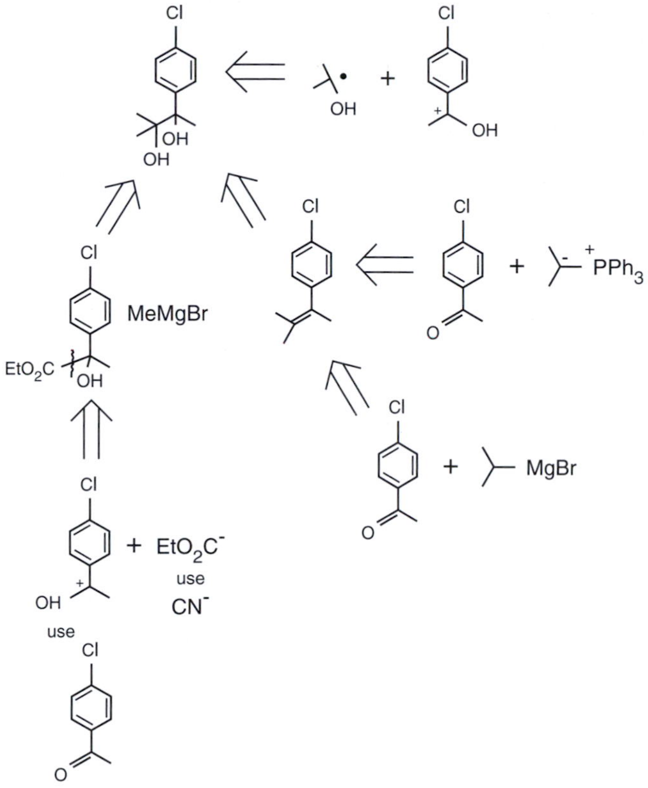

Figure 8.2–1 *Phenoglycodol synthesis* The drug, shown in the top left of the figure, could be synthesized by many different routes. Three of the most promising possibilities are sketched here. Each "disconnection" step involves going backwards along one of the arrows shown to simpler precursor molecules.

drugs. To evaluate this class, large quantities of taxol were required. However, the natural source of taxol, the Pacific yew, is one of the world's slowest growing trees and a protected species; moreover it would require six such yew trees to produce enough taxol to treat a single patient. Clearly direct use of the natural product was not possible. Suggest how greater amounts of taxol could be obtained.

SOLUTION

The obvious place to start is to look at other similar species. Intensive screening found an analog of taxol, 10-deacetylbaccatin III, in the leaves of the European yew, widely used as a hedging plant in gardens, and so easily available as clippings. 10-Deacetylbaccatin III can be converted into taxol by a relatively easy chemical transformation, given in Figure 8.2–2(a). Although the extraction and subsequent chemical elaboration of 10-deacetlybaccatin III was laborious, this synthetic route allowed clinical trials to proceed. In 1992, taxol was approved by

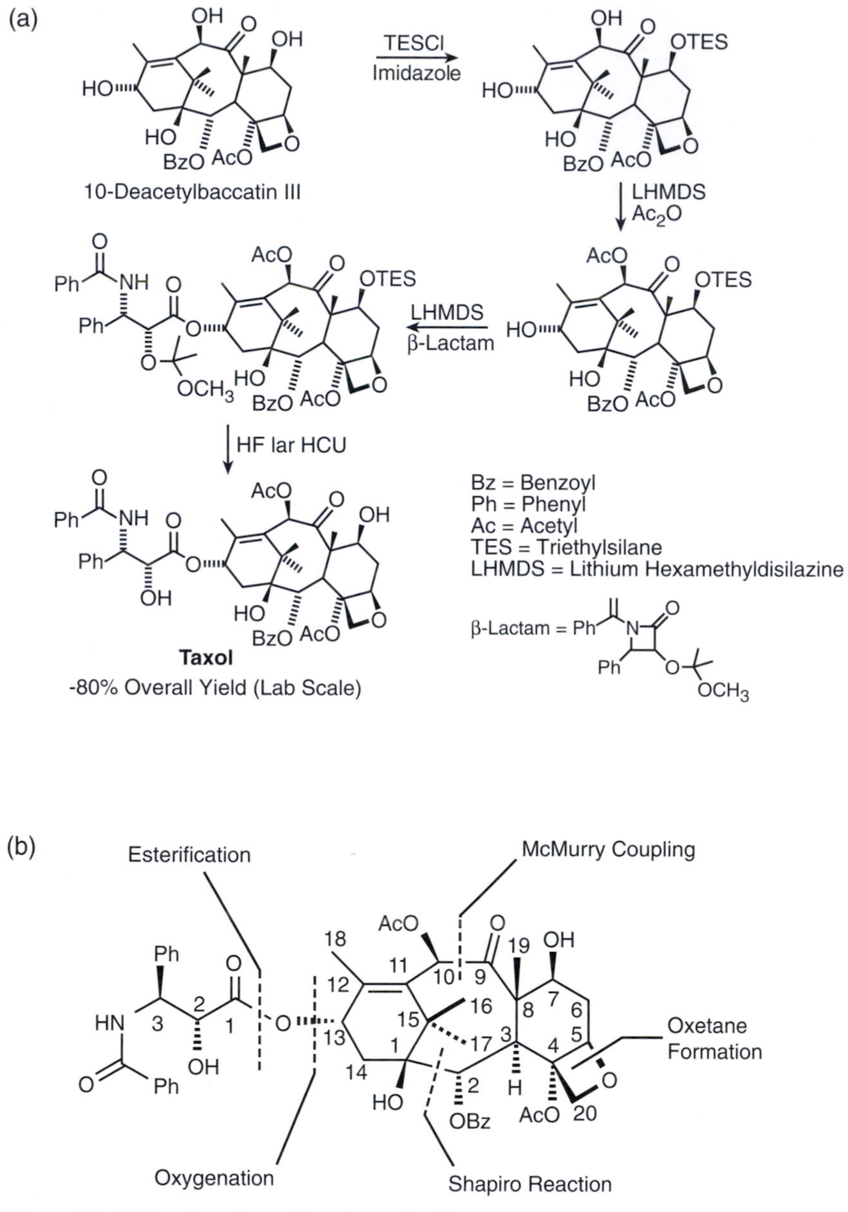

(a)

10-Deacetylbaccatin III

TESCl
Imidazole

LHMDS
Ac₂O

LHMDS
β-Lactam

HF lar HCU

Taxol
-80% Overall Yield (Lab Scale)

Bz = Benzoyl
Ph = Phenyl
Ac = Acetyl
TES = Triethylsilane
LHMDS = Lithium Hexamethyldisilazine

β-Lactam =

(b)

Esterification

McMurry Coupling

Oxetane
Formation

Oxygenation

Shapiro Reaction

Figure 8.2–2 *Taxol synthesis* That shown in (a) starts from the needles of the European yew. That sketched in (b) is entirely synthetic.

the FDA for treatment of ovarian cancer and in 1994 for breast cancer. Its relatively non-toxic nature made taxol a particularly attractive anti-cancer drug.

However, because of the high cost of this semi-synthetic route, major effort was put into the total chemical synthesis, including that shown as a series of disconnections in Figure 8.2–2(b). This path is a good example of convergent synthesis, since three fairly complex smaller molecules are synthesized independently and then assembled into the full molecule. The key creative steps are identifying the appropriate "disconnections" in the target molecule, leading to manageable sub-targets which can be more easily managed. Work on this topic continues, with exciting results (Ajikumar *et al.*, 2010).

8.2–2 SEPARATION SYNTHESIS

Just as the core team's chemists will seek alternative syntheses, so the team's engineers will be thinking about separations. In so doing, they will be reviewing general separation strategies. These strategies are complicated by the tendency of specialty chemicals to be produced at high dilution. For example, therapeutic proteins may be synthesized by genetically modified mammalian cells at concentrations around 20 μg/L. We need to produce pure protein from this solution, a task analogous to finding in the entire world's population a hundred persons who have a single, uncommon genetic defect.

Geveral Rules

Separating molecular products will normally begin with the contents of the batch reactor. These contents, fed to generic separation equipment, will contain perhaps one gram to one kilogram of product. While the separations involved vary widely, the following heuristics can guide how we proceed. These are given in descending order of importance:

(1)   Concentrate the product before purification.
(2)   Remove the most plentiful products early.
(3)   Do the hardest separations last.
(4)   Remove any hazardous materials early.
(5)   Avoid adding new species during the separation. If they must be added, remove them promptly.
(6)   Avoid extreme temperatures by using different solvents.

These important guidelines merit additional discussion.

(1)   *Concentrate before purification.*

This heuristic suggests that the first step in any separation train should focus on concentrating both the product and the principal impurities from the dilute feed. This heuristic gains support from a "Sherwood plot," a graph of product concentration in the feed vs. product selling price shown in Figure 8.2–3. The implication of this figure is that concentrating the product is more important than purifying it. To see why this may be true, we first note that the slope of the line in the figure is about minus one. Thus, if the product's concentration in the feed can be improved ten times, then its cost should drop ten times.

Now imagine that for a particular antibiotic we have developed a highly selective separation. This new separation is inexpensive as well. Then this antibiotic should appear as a point well below the line in Figure 8.2–3. Like other antibiotics, it would have a low feed concentration, but, unlike the others, it could be cheaply separated and so have a low price. The fact that no points occur well off the line in the figure suggests that whether a selective separation exists will not

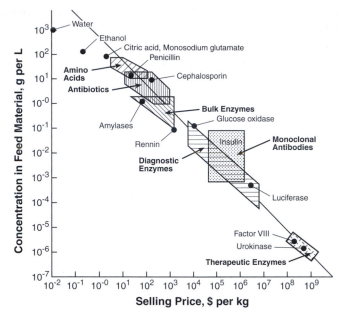

Figure 8.2–3 *Feed chemical concentration vs. product selling price*
This correlation, called a "Sherwood plot," is valid over 12 orders
of magnitude of concentration.

matter as much as the original concentration at which the antibiotic was manufactured.

We recognize that this generalization is imperfect, that in a few cases we may
do better to separate selectively before we concentrate. This could cut the cost
of making our product, and still not make a dramatic difference to the Sherwood
plot. Still, the common message in most cases is clear: concentrate first.

(2)   *Separate the more plentiful products early.*
(3)   *Do the difficult separations last.*
(4)   *Remove any hazardous materials early.*

These three heuristics, often given for any chemical separation sequence, are best
understood by imagining we want to sort the tableware removed from a big dish-
washer. The tableware has all sorts of utensils: forks, spoons, spatulas, knives,
and so forth. We want to sort them and put them away. We will first remove any
sharp knives. Then we will quickly sort the forks and spoons. Finally, we will separate Aunt Evetta's antique coffee spoons, used only on special occasions, from
the other teaspoons that we use every day.

What we have done is an illustration of these three heurstics. We separate the
sharp knives first because they can cut us: they are a potential hazard. We sort the
forks and spoons early, because there are a lot of each and because the separation
is easy. We separate Aunt Evetta's spoons last because they look similar to some
of the other spoons – the separation is difficult.

These heuristics are certainly valuable for commodity separations, where they
are widely useful. We believe that heuristics (3) and (4) are also applicable in

molecular products, but we are less sure that heuristic (2) is as reliable. The difference is a reflection of the value of by-products. For commodities, we will seek markets for any by-products. For example, if we are making ethylene, we will certainly find uses for any by-product hydrogen. However, for molecular products, we may not have any by-products which have significant value. In these cases, heuristic (2) will not be useful.

(5)   *Avoid adding new species. If you must add them, remove them promptly.*

We will need to add new species in many specialty chemical separations. We will add solvents to extract many fine chemicals from the original extraction mixture. We will use adsorbents, especially ion exchangers, for purification. We will add detergents to lyse cell walls and hence release intracellular proteins which are of therapeutic value. We do so because adding these species gives us a more effective separation.

However, the caution that we remove these added species quickly is the real message of the heuristic. In some cases, we will do so almost automatically. For example, we will normally evaporate an extraction solvent or further concentrate our product back into water at a different pH. But in some cases, these added species will be much harder to remove. The detergents used to lyse cells are an excellent example. Those detergents which rupture cell walls most effectively are also those which just keep moving on through the entire separation process until they are hardest to get out of the final product.

(6)   *Avoid extreme temperatures by using different solvents.*

High temperatures can decompose many specialty chemicals, and low temperatures can be expensive. After all, we are not going to want to put our filter press, used for several different products, into a walk-in freezer. In many cases, we can avoid these challenges by switching solvents. In doing so, we will have trouble getting help from the synthetic chemists, who will normally have identified a few which work well and will not be that sympathetic with our efforts to change reaction conditions in order to make the purification easier. While we may get some help from the solubility parameters described in Section 4.1, we are going to depend mostly on new experiments.

Fermentation Rules
The heuristics given above are valuable for any type of fine chemical process. They presume that the product is valuable, that it has a high price per kilogram. They also presume that it is being made in small amounts, in generic equipment. However, many such molecules will be made by fermentation, as suggested by the example of penicillin, whose structure was given in Figure 8.1–1. In this case, the reaction will yield a dilute broth containing suspended microbes. Often, the product will be in very dilute solution in the broth: this is the case for most antibiotics. In some cases, the product will be inside the microbes: this is the case for some genetically modified protein products.

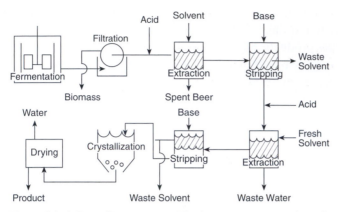

Figure 8.2–4 *Penicillin separation* The fermentor's output is purified by the sequence shown.

There are four other rules for the separations from fermentation broths which may be more useful than the more general ones given above. This four-step prescription can be remembered by the acronym *RIPP*:

(1) *Removal of insolubles.* The fermentation broth or other natural feedstock is filtered.
(2) *Isolation.* The product in the filtrate or the cake is isolated, that is, concentrated, often by extraction.
(3) *Purification.* The concentrate is partially purified, most often by chromatography.
(4) *Polishing.* The purified product is purified again, most often by crystallization.

Typical product mole fractions for these four steps are 10^{-5} for removal of insolubles and 0.99999 for polishing. This scheme works especially well for antibiotics, but also works for many vegetable extracts.

To see how this four-step scheme can simplify our planning, we consider the process for making penicillin shown in Figure 8.2–4. The process begins with a fermentor, shown in the upper left-hand corner of the figure. The output from the fermentor goes into the separation processes which follow, including filtration, extraction, and crystallization. It seems complex, hard to understand without a lot of effort.

But if we apply the *RIPP* sequence, we can understand the process much more easily, as shown in Figure 8.2–5. Each step in the *RIPP* sequence is outlined by shading. The first step, removal of insolubles, is accomplished in this case by filtration. The next step, isolation, is accomplished by extraction. Purification, the first *P* in *RIPP*, is by two more extractions. Finally, the second *P*, polishing, is by crystallization. This acronym often makes the design of separation sequences of fermentation products simpler. For products made synthetically, however, the six heuristics seen earlier are often better.

These strategies are illustrated by the examples which follow.

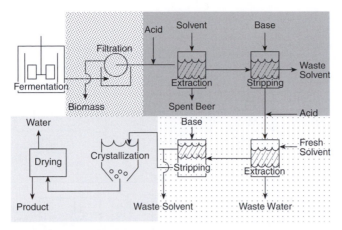

Figure 8.2–5 *Penicillin separation organized* The separation can be summarized by the acronym *RIPP*.

EXAMPLE 8.2–3   TETRACYCLINE ANALOG PURIFICATION

We are producing a modified tetracycline from a fermentation which yields the target molecule in a dilute broth containing bacteria. How should we separate and purify this antibiotic?

SOLUTION

The structure of tetracycline, shown below, has a phenolic ring, an amide, and an amine. If any of these groups is ionized, the target molecule will be more soluble in water than in organic. If none is ionized, the molecule will be less soluble in water than in organic.

This suggests a four-step process. In the first step, we filter out the solid biomass, often with the help of a solid adsorbent. Diatomaceous earth, often called "Filter-Aid," is a common choice. We then adjust the pH of the clarified aqueous solution so that the antibiotic is easily extracted into an organic solvent like butyl acetate. This will significantly concentrate the antibiotic. After we have again changed the pH to recover the product in water, we purify this concentrate in an adsorption bed, which can be activated carbon particles or ion-exchange beads. When the bed is "loaded," that is, largely saturated with antibiotic, we stop the flow of extract and elute the bed, probably with an aqueous solution whose pH is chosen to dissolve the product. We collect this eluent and crystallize the product from it,

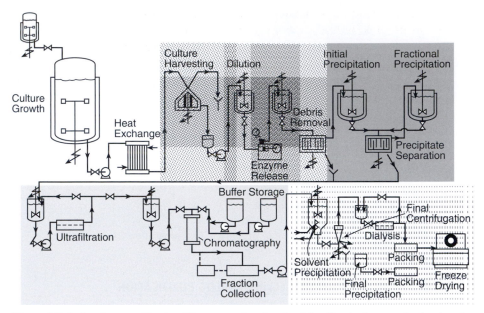

Figure 8.2–6 *Amylase purification* The separation involved is effectively organized using the *RIPP* sequence.

less frequently by chilling and more often by adding a water-miscible solvent in which the tetracycline is not soluble.

This example illustrates the rules given above. It concentrates the product by extraction, without much thought of purification (rule #1). It removes biomass easily and early (rule #2). It gives the greatest purification last, by means of crystallization (rule #3). Extraction and elution solvents are used, contrary to rule #5; these should be chosen to be non-hazardous and to facilitate crystallization (rules #4 and #6). The process also follows the *RIPP* sequence: removal of insolubles by filtration, isolation by extraction, purification by adsorption, and polishing by crystallization. Without even thinking, we have followed the rules.

EXAMPLE 8.2–4 AMYLASE ISOLATION

Amylase enzymes are used to convert cornstarch to simple sugars as a first step in the manufacture of high fructose corn syrup. Enzymes are proteins with catalytic activity, controlled by an active site, in this case involving calcium and chloride ions. The enzymes are produced intracellularly in a fermentation using genetically engineered *E. Coli*. How can these enzymes be separated?

SOLUTION

This process, shown in Figure 8.2–6, follows the *RIPP* sequence. However, it differs from that in the penicillin example because the product is inside the cells in the fermentor, not outside the cells in the surrounding broth. Thus, the whole broth is first fed to a centrifuge, and the cells are separated; this step corresponds to the removal of insolubles. The resulting paste of cells is subjected to high shear to rupture the cells, and the enzyme is precipitated by a pH change; this precipitation is the isolation step. After the precipitate is filtered and dissolved, the solution is fed to an adsorption column. The eluent is again precipitated to make the product. This process is interesting because the enzyme is precipitated twice. Such precipitation, which produces non-crystalline, amorphous material, often changes the structure of the protein and hence destroys the active site. Such structurally changed proteins are called "denatured." Here, the protein is robust and the precipitation fast, so the active site remains intact and the amylase is still a catalyst.

8.2–3 SEQUENTIAL REACTION AND SEPARATION

The examples given above are idealized in many ways. One serious idealization is the division suggested between reaction and separation. In many cases, we will carry out a reaction, followed by a separation, followed by another reaction, followed by another separation, and so forth. Often, we will have around six sequential steps; sometimes, we can have twenty.

In cases like these, the yield of each reaction plus separation becomes a major issue. For example, if we have a 60% yield for each of six steps, then the overall process yield will be only

$$(0.6)^6 = 0.046$$

less than 5%. Thus, we will often run the reaction for unusually long times, or run an extraction for what normally would seem a ridiculous number of steps, just to try to keep the overall yield high.

At the same time, we must examine the conditions found by the product's discoverers to see if these conditions are risky for the environment or for the health of those making the drug. In particular, the solvent methylene chloride is very useful in drug synthesis but is a carcinogen. Thus, during idea generation, we will often need to question whether such a solvent is the only viable choice. Moreover, the FDA in the USA insists that the drug produced commercially must be made by the same process used in the clinical trials. If we use methylene chloride as a solvent in our original scale up, we must use it in our commercial process. We cannot arbitrarily replace methylene chloride with toluene, even if we later discover that it gives a higher yield of a purer product. If we want to make this replacement, we must first petition the FDA to do so, a slow and sometimes capricious procedure. While we want to do the scale-up fast, we must remember that we must later live within our original process. These constraints are illustrated in the examples which follow.

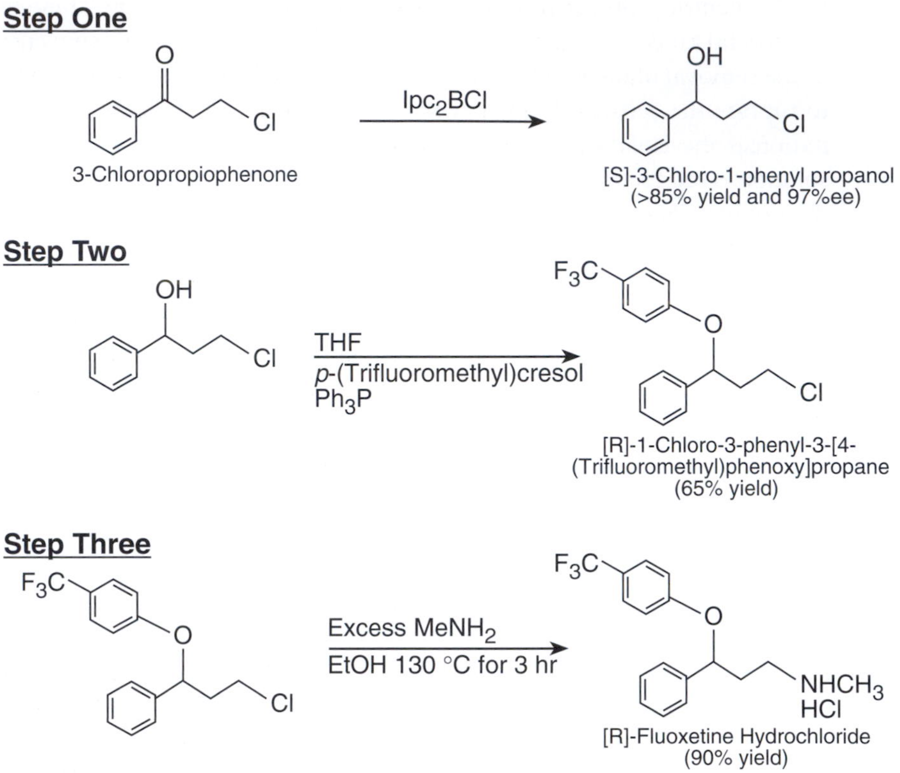

Step One

3-Chloropropiophenone

Ipc$_2$BCl

OH

[S]-3-Chloro-1-phenyl propanol
(>85% yield and 97%ee)

Step Two

OH

THF
p-(Trifluoromethyl)cresol
Ph$_3$P

F$_3$C

O

[R]-1-Chloro-3-phenyl-3-[4-
(Trifluoromethyl)phenoxy]propane
(65% yield)

Step Three

F$_3$C

O

Cl

Excess MeNH$_2$
EtOH 130 °C for 3 hr

F$_3$C

O

NHCH$_3$
HCl

[R]-Fluoxetine Hydrochloride
(90% yield)

Figure 8.2–7 *Synthesis of Prozac (fluoxetine hydrochloride)* We seek to use the patents to choose the details for the reactions.

EXAMPLE 8.2–5 PROZAC SYNTHESIS

Because the patents on the commonly prescribed antidepressant Prozac have expired, we are looking at the prospects for a generic equivalent. The synthesis involved centers on the three steps shown in Figure 8.2–7. In the first step, the keto group on a commercially available starting material is converted to an alcohol using a borane, di-isopinocamphenylchloroborane. Next, the hydroxyl is combined with trifluoromethylcresol to make an ether. Finally, the terminal chloride group is replaced by a methylamine, yielding the final product.

What details can we glean from patents about these three steps? How should we consider modifying these?

SOLUTION

The patents provide considerable details. In the first step, the starting ketone is dissolved in methylene chloride, and the borane is added drop-wise at room temperature to give an 85% yield. The alcohol product is precipitated with methyl ether. The mixed solvent, presumably containing unreacted borane, is decanted, and the precipitate is washed.

This first step almost certainly works well. The only questions are the solvents. We should consider replacing methylene chloride with less dangerous toluene, a substitution which is often successful. We could also consider replacing the explosive ether with a solvent with a similar solubility parameter, like butanediol. However, the ease of the reaction and the high conversion may make us reticent to change much.

In the second step, the precipitated alcohol is dissolved in tetrahydrofuran (THF). The cresol is added, giving a 65% conversion. In the third step, the solvent is switched to ethanol: the THF is slowly evaporated while ethanol is slowly added. The system is then refluxed under pressure and excess methylamine for three hours at 130 °C. Yield is 90%. The only difficulty with these steps is the solvent switch from THF to ethanol, a procedure which is both tedious and common in syntheses like these. We will want to explore avoiding this, but the chances of doing so are not high. Overall, the process looks good, better imitated than improved.

8.3   The Molecular Toolbox: Chemical Reactors

We are now ready to select the best of the ideas which we have generated. We are past the most difficult step of drug discovery, dominated by organic chemists or by microbiologists. We have identified ideas to make enough of the target molecule, most often for clinical trials. We want to choose the best idea and to develop a commercial process as quickly as possible. The need is speed: any patent protection is limited and the first to market with a drug offering a specific benefit tends to get two thirds of the sales. Thus the goal is selecting the best idea fast.

The reaction engineering of molecular products is much more primitive than that for commodities. Because for molecular products we are using generic equipment, we will normally want to know how much we can make in an existing reactor for each of our best ideas for synthesis. The reactors and the kinetic analysis will normally be as simple as we can make them in order to be able to select between ideas.

The form of the reactors depends most importantly on the source of the target molecules given in Figure 8.1–1. If the source is natural, the process has no reactor, but begins with separations. If the source is synthetic, then the reactors are just large versions of the round-bottom flasks used in the laboratory.

If the source of the drug is a fermentation, then the reaction uses a fermentor, a reactor with unusual and remarkably constant properties. A typical fermentor, shown in Figure 8.3–1, is a tank with a volume typically between 1000 and 100,000 liters. Made of stainless steel, it is steam jacketed, so that its contents are easily sterilized. It is run at modest temperature, typically less than 40 °C; efforts to run hotter with thermophilic microorganisms continue, because these could give greater productivity. The fermentor is usually run under slightly positive pressure, perhaps a few psi, so that if it leaks, it leaks outwards, and so the contents of the tank remain uninfected.

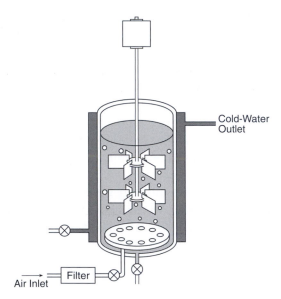

Figure 8.3–1 *A typical fermentor* Equipment like this, with a height-to-diameter ratio of 3:1, is the workhorse for manufacture of a wide range of products.

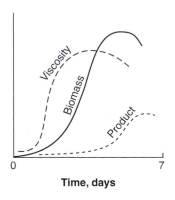

Figure 8.3–2 *Fermentor concentrations vs. time* The concentration of microbes increases sooner than the concentration of product.

Fermentations are typically run aerobically, so oxygen must constantly be supplied. The oxygen, fed as air, is first filtered through a loose bed of glass wool or a similar substance. The air is then forced through a sparger, that is, a shower nozzle mounted upside down near the bottom of the fermentor. The air from the sparger is injected as small bubbles which rise through the fermentor, supplying the oxygen and removing the carbon dioxide produced by the fermentation. Very often, the rate of reaction in the fermentor is controlled by the rate of oxygen mass transfer. Thus fermentation rates often involve discussions of $k_L a$ values, the mass transfer coefficient times the area per volume, which is the key rate constant for effective aeration.

The rate of the manufacture of the target molecule is complex, as suggested by the data in Figure 8.3–2. Fermentations typically take about a week. The fermentor is first charged with the reagents, which often include sugars and sources of nitrogen and phosphorous. In many cases, these reagents will include agricultural waste products: in the classic case of penicillin, the original reagent was corn steep liquor, a by-product of making corn starch. The contents of the fermentor are sterilized, and the tank is inoculated with the microorganism developed to make our product. Typically, nothing much happens for a day or so. Then the microbes grow exponentially for a couple of days, so that the biomass in the tank increases rapidly. During this logarithmic growth, the concentration of the target molecule does not increase much, but remains at very low levels. After the growth of biomass has stopped or gone through a maximum, the growth of the target molecule rises. This too reaches a maximum after typically six days, at which point the fermentation is stopped and the contents of the fermentor are harvested for separation, often called "downstream processing."

This overview of a fermentation often makes reaction engineers eager to start a program of improvements. Such efforts have rarely been effective because the fermentations are much more complex than the simple description suggests. For example, the viscosity of the fermentor's contents varies dramatically during the fermentation, as suggested in Figure 8.3–2. The viscosity is usually non-Newtonian, often with a yield stress. This means that the effectiveness of the stirring is changing, as well as the $k_L a$ and the aeration. Efforts to make the fermentation more scientifically specified have often not been as successful as expected, perhaps because developers move on to new products before their understanding of the old products is complete.

These generalities are illustrated in the examples which follow.

EXAMPLE 8.3–I   PENICILLIN MODIFICATION

We are adding a phenyl group to a β-lactam ring using an enzyme catalyst, supported on ion exchange beads. Kinetic data taken in a stirred flask are as follows:

Time, sec	Concentration, mol/dm³
0	0.0110
200	0.0072
400	0.0047
600	0.0032

The inexpensive reagent with the phenyl group is present in excess. We want at least 98% conversion and have enzyme particles available both in a larger stirred round-bottom flask and a small packed bed. Which should we use?

SOLUTION

The data given let us identify the reaction mechanism because a plot of the logarithm of concentration vs. time is linear. For a first-order reaction

$$\frac{dc_1}{dt} = -kc_1$$

If the initial concentration is c_{10}, we can integrate this to obtain

$$\frac{c_1}{c_{10}} = e^{-kt}$$

which is consistent with the data. Note that this is not always the answer: many enzyme-catalyzed reactions are zero order in the limiting β-lactam reagent, so c_1 should be linear with time. Here, where c_1 drops exponentially, the reaction is first order, with k equal to 2×10^{-3} per sec.

The stirred tank and the packed-bed reactors will give the same conversion vs. time or space time (bed volume/flow through the bed). Because the tank is simpler, it seems the obvious choice. For 99% conversion, we need a reaction time of about 2300 sec, or 40 min. If filling, starting, stopping, and cleaning the reactor takes another 40 min, we should be able to complete about four runs per shift.

EXAMPLE 8.3–2 PIGMENTS FOR FARMED SALMON

Salmon grown in fish farms are fed on an artificial diet which normally does not contain algae or small crustaceans. As a result, the flesh of farm-raised salmon is white, not the striking pink of the wild fish who do eat crustaceans. To overcome this, farm-raised fish are fed supplements of astaxanthin, whose structure is

This compound is like carotene, responsible for the color of carrots, but has added hydroxyl groups.

The current worldwide market for astaxanthin, about $200 million, is made in three ways. Two of the ways are synthetic, practiced by BASF (US Patent 5,210,314) and by DSM (purchased from Hoffman-Laroche, US Patent 4,245,109). Material made in this way costs about $2000/kg. Astaxanthin can also be made by fermentation using yeast. Because this method can be claimed to produce a "natural" color, its product sells for $7000/kg. Companies using this route include Igene Biotechnology, ADM, and Cyanotech (e.g. U.S. Patent 5,356,809.)

Our core team is investigating producing this material, not only as a colorant but also as a nutritional supplement. We want to compare the three syntheses and to recommend one for further development. Which should we choose?

SOLUTION

We begin by questioning whether the pink color has any value. The color does not change the nutritional value of the fish but is viewed by consumers as an

Table 8.3–1 *Characteristics of three oil extraction processes* Each process produces 1 kg astaxanthin per day.

Feed	Reactor size	Batch time	Feed concentration	Oil required	Extraction time	Extract concentration	Remarks
Shellfish	(1500 L/day waste)	–	750 ppm	0.14 kg/kg feed	1 hr at 88 °C	650 ppm	Feed variable, depending on seasonal fishing
Yeast	40,000 L	4 days	110 ppm	1 kg/kg feed	1 hr at 65 °C	1000 ppm	Has been operated commercially
Algae	180,000 L	9 days	50 ppm	1 kg/kg feed	24 hr at 25 °C	80 ppm	Operated only at laboratory scale

indication of quality. Thus, in the market, consumers pay a premium for pink fish, which is why farmed fish are fed astaxanthin. In addition, there is demand for nutritional supplements which are antioxidants and free-radical scavengers. Astaxanthin is such a molecule, as are β-carotene and lycopene. Thus, the core team believes that future demand for astaxanthin will be strong.

Our core team also wonders if "natural" astaxanthin has added value over synthetic. There is a subtle difference in the chemical structure. The natural methods produce only one enantiometer, but the synthetic methods produce a mixture of all possible enantiometers. While there is no known difference in the behavior of the enantiometers in the human body, the natural product could have a perceived difference for consumers. At present, however, any perceived difference does not justify the three-fold difference in selling price; and the synthetic material from BASF is driving the natural material out of the marketplace.

As a result, our core team decides to examine three alternatives: a new natural source, a different chemical synthesis, and a different pigment. Each is explored below.

A new natural route for astaxanthin will most likely involve the extraction of the compound from one of three potential sources into soybean or mackerel oil. The three sources to be considered are the shells and tissues produced as waste when harvesting the edible meat from shellfish; the yeast *Phaffia rhodozyma;* and the algae *Hematococcus*. The two microorganisms are fermented to produce cells containing astaxanthin. The extraction process for all three feeds follows essentially the same scheme. A set amount of oil is mixed with the feed in a large pot, heated, and stirred. The mixture is then allowed to separate by gravity.

Some characteristics differentiating the three processes are shown in Table 8.3–1. All are calculated on the basis of 1 kg/day astaxanthin produced. The feed concentration of the astaxanthin assumes a shell feed which is half water; those of the fermentations are based on the broth produced. If this broth were dried, the

Table 8.3–2 *Matrix scaling of three extraction processes.*

Faster	Weight	Shellfish	Yeast	Algae
Feedstock reliability	0.3	2	7	7
Feed concentration	0.3	7	3	2
Volume equipment	0.2	7	3	1
Volume oil	0.1	7	4	4
Maturity	0.1	9	8	2
TOTAL	1.0	5.7	4.8	3.5

concentrations would be 6000 ppm and 25,000 ppm for yeast and algae, respectively. The oil required for the shellfish is significantly less; the extraction time for the algae is significantly longer. Because the shellfish-based process depends on a natural feedstock, it may be seasonal.

Selection between these processes is aided by the matrix scaling shown in Table 8.3–2. The important criteria are the feedstock's reliability and concentration. The volumes of equipment and oil, which are less heavily weighted, are essentially indications of capital investment and operating expense, respectively. The fifth criterion, process maturity, is less heavily weighted, retained just to remind us of this risk.

The results in Table 8.3–2 suggest that shellfish extraction is somewhat better than yeast fermentation. However, we recognize that this conclusion would be changed if the yeast or the algae could be inexpensively dewatered. If these options are to be examined further, we urge coagulation and filtration studies of the fermentation broth as a route to increased feed concentration. However, in our opinion, none of the processes in Table 8.3–1 looks strong.

We are impressed that BASF's synthesis, which includes 14 steps, produces astaxanthin over three times less expensively than the existing fermentation. We have no striking new ideas for synthesizing this molecule, and we feel that we are unlikely to outperform BASF's chemists in extending their process in significant ways. As a result, we recommend not selecting a process producing astaxanthin. Instead, our core team believes in exploring what other less-expensive pigments can be fed to salmon to produce the pink color. While we understand that such pigments are not the "natural" color, the current color in farmed fish is from BASF's synthetic pigment. If our new, cheaper pigments were in themselves natural products, we might blunt this criticism. However, in suggesting this route, we are abandoning the nutritional supplement market.

8.4   The Molecular Toolbox: Separations

We now turn to the separation processes used to purify molecular products. These unit operations are similar to those used to purify commodity chemicals. However, their relative importance and their process engineering are different because the products are so different. Here, we are dealing with small amounts of valuable materials which we want to purify fast, using generic equipment. For commodities, we are processing large amounts of inexpensive compounds which we want to purify at minimum cost, using dedicated equipment specifically designed for that particular separation. For molecular products, the key separations are extraction, adsorption, crystallization, and steam distillation.

8.4–1 EXTRACTION

In extractions, we normally begin with the dilute solution containing the molecular product produced by our batch reactor. We contact the solution with a different solvent, immiscible with the solvent used in the reactor, in which the product is more soluble. We thus concentrate the product, not necessarily selectively.

Three topics merit discussion for extraction: how we choose the solvent; how much separation we can get in one batch; and how we can purify by using repeated extractions. The first topic, choosing a solvent, seeks a large partition coefficient m:

$$m = \frac{y_1}{x_1} \tag{8.4–1}$$

where y_1 and x_1 are the mole fractions of the product in the feed and solvent respectively, at equilibrium. In the following analysis we assume that m is not a function of feed concentration.

The second topic is to estimate how much separation we can get in a single stage. We start with a feed of solute concentration y_{10} and amount G, and a pure solvent with solute concentration $x_{10} = 0$ and amount L. Then from a mass balance,

$$Gy_{10} = Gy_1 + Lx_1 \tag{8.4–2}$$

Combining with the partition coefficient m, we find

$$y_1 = \frac{y_{10}}{1 + \frac{L}{mG}} \tag{8.4–3}$$

The fraction recovered in one extraction is $(1 - y_1/y_{10})$.

The third topic for extraction is how much we can get with multiple stages. When we wash the raffinate of one extraction with a second batch of initially pure solvent, we get

$$y_2 = \frac{y_1}{1 + \frac{L}{mG}} = \frac{y_{10}}{\left(1 + \frac{L}{mG}\right)^2} \tag{8.4–4}$$

After N washings,

$$y_N = \frac{y_{10}}{\left(1 + \frac{L}{mG}\right)^N} \tag{8.4–5}$$

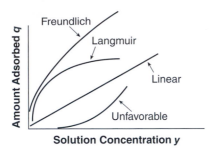

Figure 8.4–1 *Adsorbent isotherms* Most are favorable, with downward curvature.

When we pool all the extracts, we will discover that the fraction θ of the molecular product we have extracted is

$$\theta = 1 - \frac{y_N}{y_{10}} = 1 - \frac{1}{\left(1 + \frac{L}{mG}\right)^N} \tag{8.4-6}$$

This is normally the figure of merit for our extraction.

Those experienced in separations will recognize this as a staged, cross-flow extraction. This is less efficient than a staged countercurrent extraction, such as that practiced for the large-scale dewaxing of lubricating oils. The cross-flow scheme is used for molecular products because it is simpler, much easier to apply to a batch feed. It is the norm.

8.4–2 ADSORPTION

The second important separation is adsorption, in which the partially purified extract solution is pumped through a packed bed of small particles of a solid adsorbent. Because the adsorbent is usually microporous, it has a large surface area on which it can adsorb the product. Solute–surface interactions are frequently more selective than the solute–solvent interactions which occur in extraction. Thus, adsorption is especially effective for product purification, though it can also be used for product concentration.

Like extraction, adsorption is conveniently discussed as three topics: how we choose the adsorbent, how it will work in batch, and how we will use it to purify the product. The choice of the adsorbent depends on experimental measurements of the equilibrium between product adsorbed vs. product in solution. These experimental results, called isotherms, are often presented graphically, as in Figure 8.4–1. Isotherms are usually non-linear, implying that the thermodynamics is more complex than that responsible for the partition coefficient used in extraction. In many cases, for a plot of the amount adsorbed vs. the concentration in solution, the slope decreases as the concentration increases, so that adsorption is most effective in dilute solution. Such an isotherm is common, and is termed "favorable."

Adsorbents are often expensive so we seek to minimize the amount we need. Three classes of adsorbents are common: carbons, inorganics, and synthetic polymers. The carbons have non-polar surfaces which adsorb non-polar solutes. They

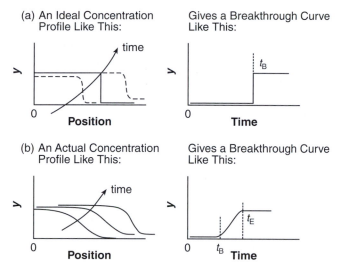

Figure 8.4–2 *Breakthrough curves for adsorption* We suggest starting designs by assuming the breakthrough is ideal.

are manufactured from a variety of sources, including coke, wood, and coconut shells. Carbons made from a mixture of sawdust and pumice are often used to remove color from fine chemical solutions. Inorganic adsorbents center on activated alumina and silica gel, both of which are used as dessicants. These materials have polar surfaces, and so tend to be more effective for polar solutes. Synthetic polymers include ion exchangers. While ion exchangers were originally designed to capture multivalent ions in exchange for monovalent ones, they are often remarkably effective for selectively adsorbing high value-added non-polar solutes like drugs and pigments.

Once we have chosen an adsorbent and measured its isotherm, we will normally measure its breakthrough curve. In this measurement, we pump a solution into a packed bed of adsorbent, and measure the concentration coming out. We hope that the exiting concentration will be a step function, that it will be zero until the bed is completely saturated, and then jump to the feed concentration. For such an "ideal breakthrough curve," like that in Figure 8.4–2(a), we can estimate the mass of adsorbent needed M as

$$Qy_{10} = Mq_{10} \qquad\qquad\qquad (8.4\text{--}7)$$

where Q is the total volume of solution fed, y_{10} is the molecular product's concentration in this solution, and q_{10} is the concentration on the adsorbent. This equation neglects the product in the solution remaining in the spaces between the adsorbent particles. Such neglect is often justified because q_{10} is normally much larger than y_{10}, which is why we chose this particular adsorbent.

However, such an estimate usually seriously underestimates the amount of adsorbent needed. This is because the actual breakthrough curve is normally far from a step function, as shown in Figure 8.4–2(b). This actual breakthrough is due to four major causes. Three tend to blur the breakthrough: axial diffusion, dispersion, and adsorption kinetics. Axial diffusion, due to product diffusing ahead of the solution, is almost always minor. Dispersion, caused by uneven flow in the packed bed, is normally the chief cause of blurring. Adsorption kinetics, blurring due to slow uptake of product by adsorbent, is also significant.

The fourth major cause of non-ideal behavior is due to the isotherm itself. If the isotherm is favorable, the front will get sharper, more like a step function. Because the sharpening caused by the isotherm tends to balance the blurring caused by dispersion and kinetics, the amount of blurring tends to move through the bed without changing. This is the key to our analysis.

In practice, we are anxious to capture and concentrate most of our valuable product. Thus we will stop the adsorption when the exiting concentration from the bed y_1 is a small fraction, perhaps less than 1%, of the feed concentration y_{10}. We will make a quick test experiment to determine how much the bed will hold, found from

$$\begin{bmatrix} \text{bed} \\ \text{capacity} \end{bmatrix} = \int_0^\infty Q(y_{10} - y_1)\,dt \tag{8.4--8}$$

where Q is the flow which we intend to use in the large bed. We can also find the amount of the bed actually saturated from

$$\begin{bmatrix} \text{bed} \\ \text{used} \end{bmatrix} = \int_0^{t_B} Q(y_{10} - y_1)\,dt \tag{8.4--9}$$

where t_B is the "breakthrough time," when we stop the experiment to avoid losing more material. The fraction of the bed used θ is

$$\theta = \frac{\int_0^{t_B} Q(y_{10} - y_1)\,dt}{\int_0^\infty Q(y_{10} - y_1)\,dt} \tag{8.4--10}$$

The meaning of an experimentally measured θ can easily be clarified by considering its value for an ideal breakthrough curve. There, y_1 is zero until t_B, and equal to y_{10} thereafter. Thus, θ is one: for an ideal breakthrough, the bed is saturated, and all its capacity is used.

We may now estimate the amount of adsorbent needed for a real bed. We begin by defining the length of unused bed l' as

$$l' = l(1 - \theta) \tag{8.4--11}$$

where l is the actual bed length. We now assume that l' is constant, regardless of the bed length, because the sharpening due to a favorable isotherm balances the blurring due to dispersion and kinetics. We can then parallel Equation 8.4–7 to get a relation for the mass of adsorbent needed in a real bed

$$y_{10}Q = q_{10}\frac{M}{1 - \frac{l'}{l}} \qquad (8.4\text{–}12)$$

This lets us use a breakthrough curve like that in Figure 8.4–2 to estimate the amount of adsorbent needed in a larger bed.

This approximate analysis is surprisingly reliable. It is prey to several hidden constraints. Three merit mention. First, the analysis suggests that because l' is constant, we can get a higher fraction of the bed used by designing very long beds. While this is true, the higher pressure drop required by such beds often makes them inoperable. Second, the value of l' does depend on operating conditions:

$$l' \propto \frac{v}{ka} \qquad (8.4\text{–}13)$$

where v is the superficial velocity in the bed, k is the mass transfer coefficient into the particles, and a is the particle area per volume. Thus, doubling the flow through the bed can double the length of unused bed. Similarly, because both k and a often depend inversely on adsorbent particle size, doubling that size will increase l' by a factor of four. Third, removing the product often involves washing the bed with a different solvent, perhaps at a different temperature or pH. This elution can supply additional purification.

8.4–3 CRYSTALLIZATION

The third important separation process for molecular products is crystallization and its bastard cousin, precipitation. Both these separations involve making solids form in solution. Precipitation is usually a poorly controlled process, done quickly to concentrate the product, to facilitate its isolation. Crystallization is done much more slowly, and aims at dramatic purification. It is often the penultimate step in specialty separation, followed only by drying. Both are reviewed below.

In most cases, we will trigger precipitation by adding a non-solvent to the solution produced by the chemical reactors. The non-solvent is miscible in the solution, but causes solutes – including the product – to precipitate. Such a precipitation occurs because the free energy of the product in solution is increased above that of the solid product.

Non-solvents normally have a very different polarity to that of the product. For example, if the feed is aqueous, the non-solvent may be acetone or *t*-butanol. If the feed is in a solvent like ethanol, the non-solvent is usually water. If the target solute in the feed is ionic, the precipitation can be effected by excess salt. Ammonium sulfate and potassium citrate are two common choices. Beyond these generalities, we suggest four heuristics:

(1) Precipitation increases as temperature decreases.
(2) Precipitation of high-molecular-weight products is easier than low-molecular-weight ones.

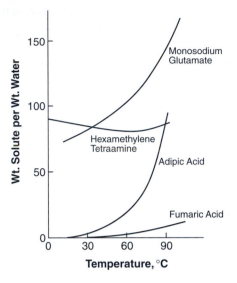

Figure 8.4–3 *Solubility vs. temperature* Solubility normally rises with temperature, but not always.

(3) Precipitation tends to be easier if many solutes are present.
(4) Precipitation from water is easier when the ionic strength is around 0.1 M.

Beyond this, we must depend on experiment.

Crystallization is one of the most important separation processes for molecular products. Like precipitation, crystallization produces solids from solution, known as the "mother liquor." Crystallization is a more ambitious process than precipitation because it tries to purify the product, not just concentrate it. Crystallization usually also aims at large crystals, perhaps around 300 μm. Large crystals are easier to wash and filter, normally the next steps in the separation.

Crystallization depends on three key factors. First, it depends on solubility variation with temperature and solvent composition, an equilibrium factor parallel to the partition coefficient for extraction and the isotherm for adsorption. Second, crystallization depends on the crystal growth rate. Third, crystallization depends on the "cooling curve." Each of these three key factors merits more explanation.

Solubility changes with temperature and solvent composition, as suggested in Figure 8.4–3. Usually, the solubility increases as temperature increases. By reducing the temperature or changing the solvent concentration, we can potentially initiate crystal formation. The difficulty with this simple picture is that solutions can often contain more solute than that present at saturation. Such supersaturated solutions are not thermodynamically stable. However, they can be metastable, that is, they can remain unaltered indefinitely. This metastability, the result of the surface energy of small crystals, can often be overcome by adding seed crystals. Ideally, these seeds will be of pure product and their number will equal the number of large crystals eventually grown. In fact, in industrial crystallizers, the seeds dominate crystal formation only at the start of the process. At later times, secondary nucleation may become important, especially in larger crystallizers. We will return to this point later. For now, remember that nucleation and supersaturation complicate the effect of equilibrium solubility, still the basic requirement for this separation.

In addition to the first key factor of solubility, crystallization depends on a second key factor, crystal growth rate. Crystal growth is, in many instances, controlled by diffusion, and described by the equation

$$\frac{dM}{dt} = k A (c_1 - c_1^*) \tag{8.4–14}$$

where M is the crystal mass, A is the total crystal area, k is a mass transfer coefficient, and c_1 and c_1^* are the solute concentration actually in the solution and at saturation, respectively. Other mechanisms of crystal growth also occur, but the focus on the diffusion-controlled case illustrates the ideas involved.

This rate equation is complicated because the crystal area A varies with the crystal mass M. For simplicity, we will assume cubic crystals. For such a crystal of side l,

$$M = l^3 \rho \tag{8.4–15}$$
$$A = 6l^2 \tag{8.4–16}$$

and

$$\frac{dl}{dt} = \frac{2k}{\rho} (c_1 - c_1^*) = G \tag{8.4–17}$$

where G is the growth rate of a single crystal. Note that this rate is independent of crystal size (but note that k can be a function of crystal size), and linearly dependent on the degree of supersaturation.

We now have two of the three key factors for crystallization. From experiment, we will establish the saturation concentration, the first factor. Also from experiment, we will determine the growth rate, the second key factor. This growth rate should give us crystals large enough to filter.

We next turn to the operation of our batch crystallizer. Normally, we will effect crystallization either by cooling the crystallizer or by adding a second solvent, miscible in the mother liquor but not a good a solvent for the product. Adding this non-solvent is frequently called a "knockout."

Whether crystallization is induced by temperature or knockout, we most often get the best crystals if we keep the degree of supersaturation small but constant. Such constant supersaturation reduces secondary nucleation and favors growth of large, pure crystals of uniform size. Keeping supersaturation constant is not easy because it is affected both by crystal growth rate and by crystal size.

While the details are beyond the scope of this book, we can illustrate the ideas involved by considering the result for non-solvent addition. We begin by generating supersaturation and then adding seed crystals of uniform size l_s to start the growth. The non-solvent concentration c_2 should then vary with time t as

$$c_2 = \frac{M_s / V}{(-dc_1^*/dc_2)} \frac{3Gt}{l_s} \left(1 + \frac{Gt}{l_s} + \frac{1}{3} \left(\frac{Gt}{l_s} \right)^2 \right) \tag{8.4–18}$$

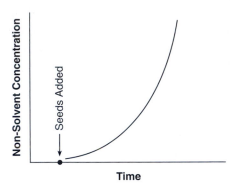

Figure 8.4–4 *Non-solvent addition for crystal-lization* The non-solvent should be added more rapidly as crystallization continues.

where $(-dc_1^*/dc_2)$ is the rate of change in solubility with non-solvent concentration, V is the volume of the batch crystallizer, and M_s is the mass of seeds. In deriving this equation we have assumed that the supersaturation is sufficiently low that there is no nucleation other than that caused by seeds, that all growth occurs on the added seeds, and that $(-dc_1^*/dc_2)$ is constant over the range of non-solvent concentrations involved. Note that this equation teaches that the non-solvent should first be added slowly and then added faster and faster to ensure constant supersaturation. A typical plot of this addition is shown in Figure 8.4–4.

8.4–4 STEAM DISTILLATION

Finally, we mention distillation. While this separation is by far the most useful for commodity products, it is not often valuable for molecular products, which are normally only slightly volatile and which often degrade when heated. The molecular products which can be distilled are flavors and fragrances, which are often separated by steam distillation. In steam distillation, we are concerned with the evaporation of a single volatile product, which is normally not miscible with water. In a two-component single-phase liquid mixture, a first approximation to the total pressure of the mixture is given by Raoult's law:

$$p = x_1 p_1 + x_2 p_2 \tag{8.4–19}$$

where x_i is the mole fraction of component i in the liquid mixture and p_i is the vapor pressure of pure i at the temperature being considered. The total pressure is always lower than that of the more volatile component. However, when the two components are immiscible and a two-phase liquid mixture forms, the vapor pressures are approximately additive:

$$p \approx p_1 + p_2 \tag{8.4–20}$$

The total pressure is now always higher than that of the more volatile component on its own. Since the boiling point is the temperature at which the total vapor pressure equals the external applied pressure, this implies that the boiling point of the two-phase mixture is lower than both the boiling points of the individual components. Thus, even relatively non-volatile species can be steam distilled at a temperature lower than the boiling point of water. Moreover, once the distillate is condensed, the two liquids separate, so the removal of most of the added water is easy. We discuss these molecular separations in the examples which follow.

EXAMPLE 8.4–1 STREPTOMYCIN ANALOG EXTRACTION

We want to extract a product from 4500 L of an unfiltered fermentation broth with amylacetate, for which the partition coefficient m equals 0.018. What percent recovery do we get with one 400 L extraction? How about two 200 L extractions? What total volume of acetate is needed for three extractions giving 99% recovery? Which extraction sequence should we use?

SOLUTION

The answer depends on Equation 8.4–6. For the one 400 L wash,

$$\theta = 1 - \frac{1}{1 + \dfrac{400}{0.018(4500)}} = 0.83$$

For the two 200 L washes,

$$\theta = 1 - \frac{1}{\left(1 + \dfrac{200}{0.018(4500)}\right)^2} = 0.92$$

For the three washes giving a 99% recovery,

$$0.99 = 1 - \frac{1}{\left(1 + \dfrac{L}{0.018(4500)}\right)^3} \Rightarrow L = 295\,\text{L}$$

A total volume of three times this, or around 900 L, will be required.

We cannot decide which choice is best without first deciding on the percent recovery we can tolerate. Once this is known, we can decide on the number and size of washes needed. If our estimates suggest more than three washes, we may want to consider other solvents.

EXAMPLE 8.4–2 PRODUCTIVITY OF A POLYPEPTIDE ADSORPTION

We have a working adsorption process for concentrating a polypeptide with promise for rheumatoid arthritis. Because Phase I trials were promising, we need more polypeptide as soon as possible for Phase II trials. At present, we use a 1 m bed 24 cm in diameter, with an unused bed length of 16 cm. Because our pressure drop is currently modest, we can triple the flow through the bed. How much could this increase our productivity?

SOLUTION

When the velocity triples, the length of unused bed will also triple, to about 48 cm. The capacity in the original bed at some breakthrough time t_B is

$$\text{capacity} \propto q_{10}(100-16) + \frac{q_{10}}{2}16 = 92q_{10}$$

Under the new conditions at $t_B/3$, it is about

$$\text{capacity} \propto q_{10}(100-48) + \frac{q_{10}}{2}48 = 76q_{10}$$

Thus the increased productivity is approximately

$$\frac{3 \times 76q_{10}}{92q_{10}} = 2.5$$

We can get about two and a half times the productivity by running three times as fast. These estimates deserve experimental checks.

EXAMPLE 8.4–3   SCALING UP A LINCOMYCIN ADSORPTION

We are adsorbing two isomers of a lincomycin analog from a clarified fermentation broth onto a modified dextran resin. The resin, which can stand pressure drops up to 1000 kPa, shows a highly favorable isotherm for these products. In the laboratory, we have run the beads in a 1.6 cm diameter tube packed to a depth of 34 cm. With a pressure drop of only 60 kPa, we get breakthrough at 46 min, and an exhausted bed at 62 min.

We want to run this system in an existing pilot plant adsorption bed which is 30 cm in diameter. We have already operated this bed using a pressure drop of 410 kPa. How much can we scale up this process? How should we operate to scale up 5000 times?

SOLUTION

To begin, imagine that we operate the 30 cm diameter bed under exactly the same conditions as the laboratory bed, including a bed length of 34 cm. Because the pressure drop and the bed length are unchanged, the velocity, the breakthrough time, and the exhaustion time are all the same. Thus, the gain in capacity is solely due to the gain in cross-sectional area:

$$\left[\begin{array}{c}\text{gain in}\\\text{scale up}\end{array}\right] = \frac{\text{cross section of pilot bed}}{\text{cross section of laboratory bed}}$$

$$\frac{\frac{\pi}{4}(30\,\text{cm})^2}{\frac{\pi}{4}(1.6\,\text{cm})^2} = 350$$

This reasonable increase is smaller than we seek.

As a second alternative, we can operate at the same velocity but in a deeper bed. Because we know that we can increase the pressure drop to 410 kPa, we can use a deeper bed:

$$\begin{bmatrix} \text{pilot bed} \\ \text{depth} \end{bmatrix} = \begin{bmatrix} \text{laboratory bed} \\ \text{depth of 34 cm} \end{bmatrix} = \frac{410\,\text{kPa}}{60\,\text{kPa}} = 230\,\text{cm}$$

Because the isotherm is favorable, the length of unused bed is constant. For the laboratory bed, the length of unused bed l' is proportional to the difference between the exhaustion time t_E and the breakthrough time t_B:

$$l' = l \left(\frac{t_E - t_B}{2t_B} \right)$$

$$= 34\,\text{cm} \left(\frac{62\,\text{min} - 46\,\text{min}}{2\,(46\,\text{min})} \right) = 6\,\text{cm}$$

Thus the increased capacity is now

$$\text{gain in scale up} = \frac{\text{volume used in pilot bed}}{\text{volume used in laboratory bed}}$$

$$= \frac{\dfrac{\pi}{4}\,(30\,\text{cm})^2\,(230\,\text{cm} - 6\,\text{cm})}{\dfrac{\pi}{4}\,(1.6\,\text{cm})^2\,(34\,\text{cm} - 6\,\text{cm})}$$

$$= 2800$$

This is a substantial increase, but still less than the factor of 5000 we had hoped for.

To go still higher, we will need to take risks tempered by additional experiments. If possible, we still want to use the same bed velocity. Since the adsorbent is said to stand 1000 kPa, we could investigate running at the same velocity in a still deeper bed. To do so, we must make sure that our pilot column can also stand the higher pressure. Alternatively, we can increase the adsorbent diameter and hence the velocity through the bed. Doing so will almost certainly increase the length of unused bed, perhaps dramatically. Before we make this more radical change, we will need more laboratory experiments using bigger velocities past bigger particles.

8.5 Using the Molecular Toolbox

This chapter is reassuring because so many of the ideas useful for inexpensive, large-scale commodity processes can be used to make small amounts of high value-added molecular products. The key ideas again center around reaction engineering and separation processes. To be sure, reactors for molecular products are normally batch, while reactors for commodity products are continuous; separations for molecular products center on adsorption and rarely involve distillation. Nevertheless, the concepts involved are similar.

At the same time, these concepts tend to be applied towards somewhat different ends. For molecular products, we stress simplicity and speed because the time to market dominates who gets the most sales. We will usually avoid recycles and run reactions close to completion whenever we can. We may discard or burn spent solvents, rather than fret over their fate. We won't spend much time on optimization, though we will be concerned with scale-up from a tiny scale to a small scale.

We normally will not be heavily involved in molecular discovery. Instead, we will center our attention on selecting our best idea for product manufacture. Normally, we will have a lot of ideas and will narrow them down quickly. Selecting between the last three or fewer ideas will be our chief challenge, as the examples below indicate.

EXAMPLE 8.5–1   BUTTER-FLAVOR SYNTHESIS

Our company is considering the manufacture of ten tons per year of 2,3-butanedione (commonly known as diacetyl), whose chemical structure is

This compound, which sells for $120/kg, tastes like butter, and is used as an inexpensive cholesterol-free flavoring for margarine, popcorn, and other products. Regular exposure to high concentrations of diacetyl vapor can cause asthma and other breathing problems, but these seem successfully managed with good housekeeping.

The company has formed a core team to evaluate the feasibility of this synthesis by three chemical routes and by fermentation. The first chemical route (US Patent 2,455,631) is dehydration of a diol:

These reagents are fed to a reboiler connected to a distillation column. The column removes the products, leaving the catalyst behind.

The second chemical route (US Patent 2,799,707) uses acetol and formaldehyde

The process configuration is similar. The third chemical route (Anunziata *et al.*, *Cat. Lett.* (2004) **75**, 87–91) oxidizes methyl ethyl ketone (MEK) directly

This gas-phase catalysis may reduce the separation problems encountered by the other chemical routes.

The fermentations (US patents 4,867,992 and 5,075,226) use lactobacillus organisms, sometimes immobilized on an inert support. The feed varies: while some suggest using solids remaining after the removal of coffee flavors, most use glucose, yielding a solution of 2 g diacetyl per liter and an efficiency of 0.04 g diacetyl per gram glucose metabolized.

Which process looks most promising?

SOLUTION

We will refer to the four alternatives as the diol, the acetol, the MEK, and the fermentation processes. While little kinetic information is given in the patents, our estimates for the diol process are 50% conversion per hour in a reaction mixture that is 10% diol, and a 60% selectivity. Because 10 ton/year is 20 g/min, this implies a reactor volume V of

$$\frac{100\,g}{L}\frac{0.5 \times 0.6}{60\,min}V = \frac{20\,g}{min}$$
$$V = 40\,L$$

More likely, assuming that the reactant concentration of 100 g/L can be achieved, we will want to run a generic 400 L stirred tank reactor for one month to get the desired amount per year.

The kinetics of the acetol process are more rapid. In two hours, we can get complete conversion and an 83% selectivity from 78.6 g acetol diluted with formaldehyde to 120 mL. The reactor volume V is now

$$\frac{0.83 \times 78.6\,g}{120 \times 10^{-3}\,L \times 120\,min}V = \frac{20\,g}{min}$$
$$V = 4\,L$$

The reactor is ten times smaller than for the diol process. The kinetics of the third chemical route, the MEK process, are not clear, though the best operating conditions give only a 5% MEK conversion and a 32% diacetyl selectivity. This

Table 8.5.1 *Matrix scaling for diacetyl manufacture* The advantages of selling a "natural flavor" are not enough to overcome the cheaper synthetic processes which produce the same chemical.

Factor	Weight	Diol	Acetol	MEK	Fermentation
Capital cost	0.4	5	7	5	1
Scientific maturity	0.3	5	7	1	3
Operating ease	0.3	5	6	3	10
Score		5	6.7	3.2	4.3

means that this process must separate and recycle the unreacted MEK, a significant complication for this gas-phase reaction.

The volume needed for fermentation is much larger because the best results yield 10 g/m^3 hr of diacetyl. Thus

$$\frac{10\,g}{m^3\,hr}\left(\frac{m^3}{10^3\,L}\right)\frac{hr}{60\,min}V = \frac{20\,g}{min}$$
$$V = 120{,}000\,L$$

While it is true that fermentors are not expensive, this is an increase in reactor volume of more than three hundred times, a significant disadvantage.

Other engineering problems center on the separation of the diacetyl and water, which form an azeotrope. However, this problem is common to all the methods proposed and so is not a basis for choosing among them.

These four possible processes are compared in Table 8.5.1. The diol process, which was developed first, is used as a benchmark. The three standards of comparison are the capital cost, roughly proportional to the size of the reactor; the scientific maturity, an indication of the extent to which the chemical kinetics are known; and the operating ease, an estimate of chemical complexity. On this basis, we judge that the MEK and fermentation processes are sharply inferior to the other two. The kinetics of the MEK process are not known in enough detail to be useful. Moreover, while the gas-phase reaction at moderate temperature can offer effective processing, the small conversion implies extensive recycle, which probably means liquification and distillation of the product gases. Similarly, the fermentation is just not productive enough to be competitive. While modern advances in genetic engineering could make the fermentation more productive, these gains are a research dream, not close to commercialization. Even the hope that we could sell the fermentation product at a higher price because under current regulations it could be sold as "natural" is not enough to make this attractive.

The two remaining processes, based on the diol and on acetol, are more attractive. The top-ranked acetol process should be selected for further development, with the diol process as a backup, in case of unexpected problems.

EXAMPLE 8.5–2 LUTEIN PURIFICATION

Our company is interested in expanding its base in specialty chemicals into dietary supplements, or "nutraceuticals." One possible target is lutein, whose chemical structure is shown below:

If the –OH groups are removed, the result is carotene, the color of carrots; if the terminal rings are removed, the result is lycopene, the color of tomatoes; if a keto group is added to each ring, the result is astaxanthin, the color of shrimp and salmon.

Carotenoids like these have significant medical effects, especially as antioxidants. These effects are complex: for example, carotene is said to prevent a decline in cogitation with age but to increase the risk of lung cancer. Lutein is believed to prevent oxidative stress in the eyes and to decrease the risk of age-related macular degeneration. A second market for lutein is more pedestrian: it is added to chicken feed to give chicken meat a yellow tint. It is the application to improved vision which is of greater interest to our company.

Carotenoids like lutein can be made by three routes: by fermentation, by chemical synthesis, and by extraction from natural sources. Our company is most interested in natural sources, and, in particular, in marigolds. It already gets other impure products from India, where marigolds grow easily. Thus the company plans to produce crude lutein in India and to purify it further in this country.

We want to design a simple process for producing 10 kg/day lutein from marigold flowers. In marigolds, each hydroxide of the lutein molecule is esterified to a fatty acid. Our process must remove these fatty acids and purify the lutein. The process should require a small amount of capital and should be able to be operated seasonally by workers with the equivalent of a high-school education. Our process will have three steps:

(1) *Hexane extraction of the flowers.* Current use is 300 kg hexane per kg flowers, which seems high. The extract can then be concentrated.
(2) *Reaction with base.* Current practice uses roughly stoichiometric KOH. While reported reaction times vary widely, four hours at 80 °C is probably sufficient to give 90% saponification. This is true when the lutein concentration is about 1% by weight.
(3) *Soap removal.* The reaction mixture will contain soaps, which are the product of the reaction with hydroxide. These are removed by washing with water, and the organic is dried for shipment.

Our design seeks 99% lutein recovery. Because the waste flowers in the raffinate will yield about 2 kg hexane for each kg dry flowers, we may want to dry the flowers to recover the residual hexane. We certainly want to reduce

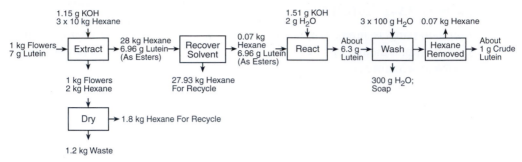

Figure 8.5–1 *Purifying lutein* The process uses extraction and saponification to recover lutein from marigold flowers.

the hexane used. The lutein concentration in the living flowers is around 8×10^{-4} g/g; that in dried flowers is typically 0.007 g/g.

SOLUTION

We are interested in preparing 10 kg of crude lutein per day from about 1.5 tons of dried marigold flowers. Our process is described below in four parts. The first part gives a superficial description, including the two critical steps of hexane extraction and lutein ester saponification. The second and third parts describe these critical steps in more detail. The fourth part outlines where we expect the greatest risks to realizing this process, and how we would plan to mitigate these risks.

Overall description. The overall process we suggest is shown in Figure 8.5–1. The process washes each kilogram of flowers with three 10 kg batches of hexane to recover 99.4% of the lutein present. This is ten times less hexane than the 300 kg hexane/kg flowers which is currently practiced and so represents a major improvement. After this step, the extract is concentrated by evaporating most of the hexane. The resulting 10% lutein solution is then ready for saponification.

The solution is fed to a reactor where it is treated with a 10 percent excess of aqueous potassium hydroxide to remove fatty acids from the two terminal rings on the lutein molecule. The kinetics of this two-phase reaction are probably controlled by mass transfer of the aqueous hydroxide to the interface between the aqueous phase and the hexane phase. This means that the reaction rates would be a function of interfacial area and hence of stirring, a variable which does not seem to have been carefully considered. Here, we assume 90% saponification in a stirred tank reactor run four times per day. We will need a reactor volume of about 50 L.

Finally, we will repeatedly wash the reactor contents to remove soaps and any excess hydroxide. We then will dry the product for shipping. Details of the extraction and reaction steps are given next.

Extraction details. Current practice is to use 300 kg solvent per kg flowers, which does give a recovery of 99% of the lutein esters. We propose instead using three washes, each of 10 kg hexane. This crossflow extraction should give about

Table 8.5.2 *Risk analysis for lutein process* The most serious risk, from the reactor kinetics, should be mitigated by experiments.

Risk	Probability	Consequence	Level
1. Economics	–	–	–
2. Rapid lutein analysis	0.1	0.7	0.07
3. Extraction			
Lutein doesn't dissolve	0.2	0.2	0.04
4. Reaction			
Hexane removal excessive	0.3	0.1	0.03
Lutein saponification too slow	0.5	0.2	0.10
5. Scheduling	0.2	0.2	0.04

the same recovery of 99% of the lutein esters. Such a process change reduces the amount of hexane needed ten times and cuts the size of the equipment ten times. While we could reduce the hexane needed even more by using more, smaller hexane washes, or by running a countercurrent extraction, we believe that the three hexane washes proposed are a good balance between simplicity and efficiency.

However, this saving will be realized only if all of the lutein esters dissolve quickly in the hexane, and if the solution formed is not saturated. This is reasonable if hexane is a good choice of solvent. "Dissolve quickly" means that the residence time in the extraction vessel is longer than the time to dissolve the esters. While literature references suggest that with moderate stirring, this will not be a problem, it is a point which must be checked by experiment.

Reactor details. We propose using a simple stirred tank reactor, operated in four-hour runs, with two hours per run for filling, emptying, and cleaning. The reactor feed is the pooled hexane extracts but with over 99% of the hexane evaporated. This presumes that the lutein saponification will occur at least as fast if the lutein concentration is increased. This would normally be true if the reaction were homogeneous. In the present case, the reaction involving potassium hydroxide in water and lutein in hexane is probably mass transfer controlled, and so depends on the interfacial area between the liquid phases. This in turn will depend on stirring to generate this area. We worry that this area will be harder to generate for the much greater concentration in the planned extract. In particular, the extract may contain other cell lipids which could increase the extract viscosity, or consume some of the hydroxide. This point will also require experimental study.

Risk analysis. We have identified five major risks to the technology described here, as shown in Table 8.5–2. The most serious risk is economic: we are not certain that we can make money. Our brief did not involve considering costs of supply or manufacture, so no estimates are in the table. We see no technical challenges which we expect will cause major trouble, so the economics should be further investigated.

The other risks are lutein analysis, extraction, reaction, and scheduling. We should seek a lutein analysis which is easy, fast, and approximate. The analysis should be able to measure concentrations of lutein and lutein esters to within 10% and within one hour. Reliable, accurate chromatographic analyses do exist, but these may be tedious and expensive. One possible alternative is spectropho-tometry, especially if it can distinguish between lutein and its esters.

A third risk is our use of crossflow extraction, which can so dramatically reduce the use of hexane. This reduction depends on the assumption that all the lutein dissolves. To mitigate this risk, we suggest experiments which measure extract concentrations and compare these with estimates based on the assumption of total dissolution. If experiments and estimates agree, we can proceed with confi-dence. If not, we should investigate extraction times, seeking a model which does work.

The next serious risk is that the reaction will not proceed as expected. We note that the literature on this point is confusing, suggesting times between 40 minutes and 10 hours (although at different temperatures). We must deter-mine by experiment what the reaction time will be under our conditions. If we need a longer time or if we cannot evaporate the amount of hexane planned, then we will need a larger reactor. Nonetheless, we urge sticking to a stirred tank to keep the process as simple as possible.

Finally, we are concerned about scheduling our process. We expect that marigold flowers will be periodically harvested, and hence our feedstock will be available seasonally. We would like to store this feedstock and to process it as steadily as possible. This will allow smaller equipment, and hence minimize cap-ital investment. At the same time, we want the extractors and reactors to run with the same productivity. These issues should be our focus after the risks given earlier are reduced.

We see no technical obstacles to making lutein that we expect to be insur-mountable. If the economics look positive, we recommend that the project should proceed.

EXAMPLE 8.5–3 SYNTHESIS OF PROGESTERONE

Progesterone, for birth control pills, is central to population control. Our non-profit company is interested in producing this material at low cost in Asia. We want to review the current synthesis, well described in expired patents; see how this process is currently accomplished; and decide how it might be improved.

This example illustrates a situation which we will commonly face. We know our target molecule: progesterone. We know manufacturing procedures, which offer a large number of options which will work to varying degrees. We will have plenty of ideas. We will need to select the best ideas using the toolbox summa-rized previously.

The chemical synthesis of progesterone is shown in Figure 8.5–2. This begins with the natural product stigmasterol, obtained by first extracting soybeans with hexane to remove the oil, and then by extracting the soybean husks with acetone

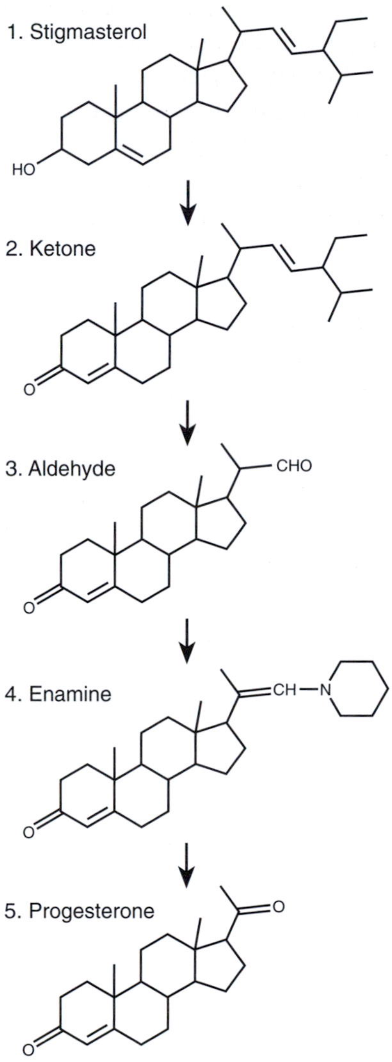

1. Stigmasterol

2. Ketone

3. Aldehyde

Figure 8.5–2 *Synthesis of progesterone* We seek reaction conditions for the four key steps shown.

4. Enamine

5. Progesterone

to remove the sterols. It is then separated from the closely related β-sitosterol, a by-product which can be separately fermented to produce more stigmasterol. In the first step of the process of interest, stigmasterol is coverted to a ketone. The ketone is converted to an aldehyde, which is then coverted to an enamine. This is reacted to make the final product.

What do the patents say about this transformation? How should we proceed in our efforts?

SOLUTION

The relevant patents include 2,601,287 ("Partial Synthesis of Progesterone") and 2,752,369 ("Oxidation of Steroid-Enamines"). These teach that the first step, stigmasterol to the ketone, can be effected by an Oppenauer oxidation. In this process a solution of 4% stigmasterol, 8% cyclohexanone (the reagent), and 88% toluene should be refluxed with a small amount of aluminum propionate catalyst at 50 °C for five hours under argon. After the reaction, the catalyst is easily removed by washing with water. Presumably, the product is dried to remove the solvents, but no details are given. We will need to explore this step further, including ways by which the reaction can be accelerated.

In the second reaction, making the aldehyde, the patents teach making a 1% solution of the ketone in a mixed solvent of 5 percent pyridine and 94% chloroform. A mixture of ozone in oxygen is bubbled through this solution at −20 °C to cause an immediate reaction. Other possible solvents are said to include carbon tetrachloride, methylene chloride, methanol, diethyl ether, and glacial acetic acid. The solvents are switched to a glacial acetic acid–ether mixture by vacuum distillation, and then mixed with zinc dust. The resulting suspension is washed with acetic acid and dissolved in ether. After filtration, the filtrate is dried to yield the aldehyde.

This step offers a number of opportunities for rationalization, exorcising the ghosts of the original laboratory experiment. While the ozone treatment seems reasonable, the solvent choices give a chance of simplification. This will require experiments to check how much the kinetics are compromised.

The third step, making the enamine, starts with making a 7% mixture of the aldehyde with 3% barium oxide and 90% benzene. This suspension is then combined with a 9% solution of piperidine in 91% benzene. The temperature should be raised so the reaction can be completed in two to four hours. The enamine product can be crystallized from the reaction mixture. In our efforts, we will almost certainly want to replace benzene with toluene. We will want to explore the yield and purity vs. temperature, anticipating that higher temperatures will give faster, less-selective reaction. Finally, we will want to consider recrystallizing the enamine either from toluene or, as the patents suggest, from methanol and pyridine.

In the fourth and final step, we make the progesterone by preparing a 10% solution of the enamine in pyridine. We add this slowly at 5 °C to a suspension of chromic anhydride. The reaction takes one to four hours. The patents then teach extraction and crystallization with benzene; we will want to try toluene. This gives us our product. We can be optimistic that this scheme will work.

8.6   Conclusions for Molecular Product Design

The design of molecular products is both difficult and familiar. The most difficult step is active-ingredient discovery, the province of synthetic chemists and microbiologists. This difficult step of discovery is beyond the scope of this book.

The familiar step, which may also be difficult, is the manufacture of the product. For a pharmaceutical, this involves rapid production of small quantities for clinical trials. The need to be rapid reflects the importance of a short time to market, which means that low cost and optimum operation are often forgotten. In this process invention, we can effectively use the design template of needs, ideas, selection, and manufacture. Again, the key step is selection, as it was for chemical devices. Again, this selection rests on reaction engineering and separation processes. The key separations are extraction, adsorption, and crystallization.

The result for us is reassuring: we see the same types of thinking which are familiar from commodity-chemical production. It seems the same language, with a new unit operations vocabulary. By following the steps given in this chapter, our chances for success are excellent.

Problems

1. *Xenon recovery.* Xenon is an excellent general anesthetic: patients go under and emerge quickly due to the very low solubility of xenon in blood and tissues. However, the gas is expensive, approximately $10 per liter, or roughly 2000 times the cost of nitrous oxide (N_2O). To be economically viable as a practical anesthetic, an efficient xenon recycling system is essential. Write specifications for such a system and suggest ways in which it might be achieved.

2. *Hormone-replacement therapy.* Premarin is a compound drug used to treat symptoms of menopause. Although there are several active ingredients, the principal one is estrone sulfate. As its name implies, Premarin is extracted from the urine of pregnant mares: the horse is the chemical reactor, producing a liquid mixture including the product at low concentration along with a large number of impurities. Using the *RIPP* heuristic, outline a separation scheme to purify Premarin.

3. *N_2O degradation.* Nitrous oxide was first made by Joseph Priestley in 1772 and later investigated by Humphrey Davy, who records that breathing the gas helped relieve his toothache. While it was the first anesthetic used, it was more valued in the nineteenth century for recreation than in medicine. A strong analgesic and mild anesthetic, it now has a number of uses in hospitals and dental surgeries. It is commonly supplied as a 50/50 mixture with oxygen and is typically used at a rate of 10–20 L/min by patients.

   There are two concerns about its use in open hospital wards. The first is that long-term exposure has deleterious effects on health, so exposure levels for clinical staff are normally restricted to 100 ppm or below. Second, nitrous oxide is an extremely potent greenhouse gas, 300 times as potent as carbon dioxide and accounting for around 6% of the warming effect of greenhouse gases in the atmosphere. Generate, sort, and screen ideas for allowing the safe use of nitrous oxide in open wards.

4. *Glass restoration.* King's College Chapel at Cambridge University has the largest surviving area of fifteenth-century Flemish glass in the world. The traditional instructions for repairing this glass use human urine as an ingredient. Urine is not a standard chemical preparation. Suggest what should be used as a substitute.

5. *Better baby food*. Most dietary fats are triglycerides, consisting of a glycerol molecule esterified with three fatty acids. Common fatty acids are stearic acid and palmitic acid, saturated 18-carbon and 16-carbon straight-chain acids respectively; and oleic acid, an 18-carbon straight-chain fatty acid containing a single double bond. Triglyceride names are often abbreviated, for example as POP, signifying a triglyceride with palmitic acid esterified in the 1- and 3-positions and oleic acid in the 2-position. Human breast milk is rich in OPO. Vegetable fats, which have traditionally been used in infant feed supplements, are rich in POO. Such synthetic infant feeds increase diarrhea and decrease the absorption of calcium and palmitic acid, particularly in pre-term babies. This is believed to be because the lipases present in the gut are 1,3-selective, i.e. they selectively hydrolyze the ester bonds only in the 1- and 3-positions of triglycerides. Thus, significant amounts of palmitic acid are released into the gut from vegetable fats, but almost exclusively oleic acid from human milk. Calcium oleate is soluble, but calcium palmitate is much less soluble. Hence, when vegetable fats are put into the guts of infants, calcium palmitate precipitates, cannot be absorbed through the gut wall, and cause diarrhea.

   We want to improve infant formulations by including only symmetric fats with palmitate at the 2-position. Propose synthetic routes to produce such fats. Note that 1,3-selective lipases are readily available from genetically modified yeasts, and also operate as 1,3-selective trans-esterifiers for triglycerides.

6. *Fructose purification*. Fructose has the same dietary calories as glucose but is twice as sweet. By using fructose rather than glucose, the same food product can be the same sweetness, but have fewer calories and lower cost. Our company seeks to produce 99.9% fructose from corn syrup. The existing process for making high fructose corn syrup involves three enzyme treatments of corn starch. First, α−amylase is added to break the starch down into sugars. Then glucoamylase is used to further hydrolyze the carbohydrate into simple sugars, mainly glucose. Finally, xylose isomerase interconverts glucose and fructose, giving an equilibrium yield of just over 40% fructose. The first two enzymes are cheap and can be disposed of after each use. Xylose isomerase is more expensive and must be re-used many times.

   Design a separation scheme for purifying fructose.

7. *Generic Prozac (fluoxetine)*. Depression is a real illness with real causes. These include severe psychological stress such as death in the family, divorce, financial difficulties; physical illnesses like cancer, diabetes, and stroke; medications, including some cardiovascular drugs and hormones; and alcohol or drug abuse. One out of eight people needs treatment for depression at some point. While depression is not fully understood, it is believed to result from an imbalance of the brain's neurotransmitters, those chemicals which allow nerve cells in the brain to communicate with each

other. The imbalance of one of the neurotransmitters, serotonin, may be an important factor in the development and severity of depression.

A new class of drugs, called selective serotonin re-uptake inhibitors (SSRIs), may help patients with depression by increasing the availability of serotonin in the brain. Fluoxetine hydrochloride is one active ingredient of SSRIs. The registered trademark of fluoxetine is Prozac, made by Eli Lilly. Fluoxetine has been used by more than 60 million people worldwide, and its market share of drugs for treating depression is about 30%. The patent of Prozac expired in 2003.

Our core team seeks to design a new process for fluoxetine as a generic. It should be made at lower cost and be available to the market as soon as possible. Develop such a process.

8. *Degumming soybean oil.* Soybeans supply the standard cooking oil in North America. The beans are crushed and extracted with hexane. The spent beans are then leached with base, producing an aqueous protein solution and a high-fiber residue. The proteins are separated from water and used as a food supplement. For example, soy protein is an inexpensive, low-fat, low-cholesterol egg substitute in commercial baking.

The oil-containing hexane extract is currently washed with water to remove "gums" and then washed again with aqueous caustic to remove fatty acids. These two steps produce a dilute, low-value waste stream which can be difficult to sewer. As a result, there is significant interest in alternative methods to "degum" this extract. One possible technology is the membrane process of ultrafiltration, which could remove the gums to produce an oil–hexane solution easily separated by distillation. However, ultrafiltration must overcome two major barriers:
(1) The ultrafiltration membranes must remain viable in the hexane, which is not now the case.
(2) The gums must not severely foul the membrane.
Note that ultrafiltration is currently successful in separating proteins from cheese whey, a roughly similar separation.

Using cheese whey separation as a benchmark, set tentative specifications for the ultrafiltration of soybean oil extracts.

9. *Isopropyl alcohol recycle.* Our core team is interested in manufacturing ultrapure isopropyl alcohol (IPA) for cleaning microelectronic parts. These parts were formerly cleaned with freons, which were then allowed to evaporate. Since these freons damage the ozone layer in the stratosphere, they may no longer be used for this purpose. Many alternative solvents have been suggested; the one most frequently chosen for these cleaning operations is IPA.

The basic market for IPA is shown in Table 8.9A. Obviously, there is a substantial premium for IPA containing low concentrations of metals and of water. You are interested in one of the four technologies shown in Figure 8.9A. Choose the best of the four and justify your answer.

Table 8.9A *IPA Market.*

	Metals concentration	Water concentration	Market size ($M)	Value ($/gal)
Bulk	–	1 wt%	500	2.24
Technical grade		1000 ppm		3.75
Electronics grade	1 to 10 ppm	1000 ppm	6	4.90
	<1 ppm	300 ppm	12	6.80
	<100 ppb	30 to 50 ppm	6	11.70

Technology	Schematic	Skid Cost (Capacity)	System Capabilities
Distillation	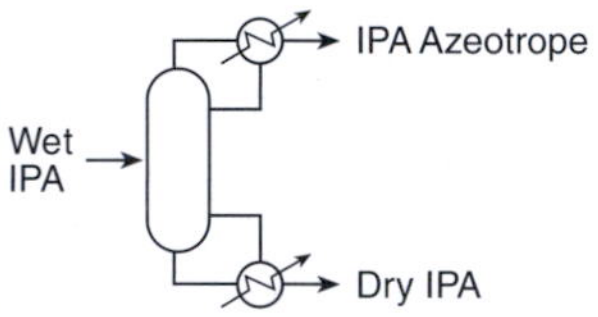	$100 K (9 gph)	Product quality — fair • 1000 ppm water • <2 ppb metals IPA recovery = 40%
Zeolite	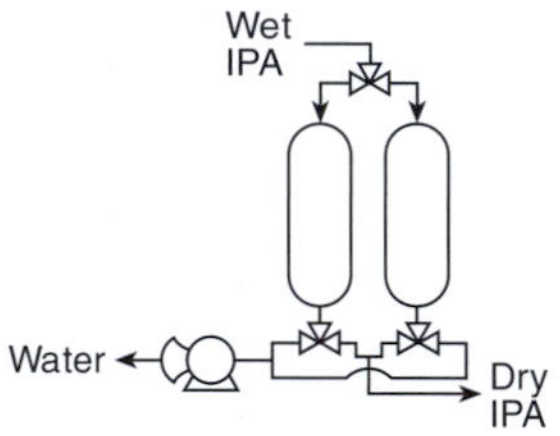	$92 K (4 gph)	Product quality — good • 100 ppm water • 3 ppb metals IPA recovery = 85%
Pervaporation	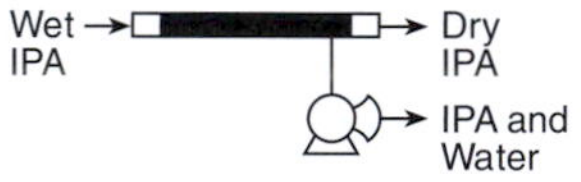	$600 K (7.5 gph)	Product quality — good • 1000 ppm water • 4 ppb metals IPA recovery = 95%
Vapor Permeation	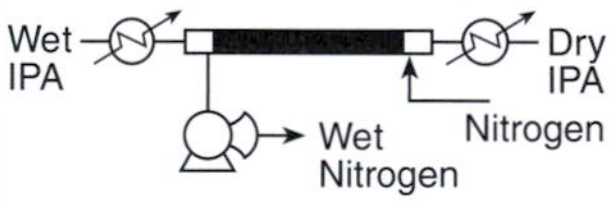	$150 K (9 gph)	Product quality — excellent • 10 ppm water • <2 ppb metals • Low particulates IPA recovery = 98%

Figure 8.09A

10. *Polypeptides to order.* Your laboratory instrument company is interested in making machines to effect chemical synthesis. As an initial target in this market, you decide to explore building a machine to make a decapeptide, a polymer of ten amino acids bound together in a specified sequence. Such a machine is sometimes called a Merrifield synthesizer.

   The benchmark chosen for comparison can be idealized as a packed bed 30 cm high and 1 cm in diameter, filled with 100 μm particles which already have 10^{-3} mol/g of a binder attached. The bed uses fluoroenylmethoxycarbonyl (FMOC) chemistry for adding the amino acids. With this chemistry, each amino acid addition involves the following steps, each of which adds 15 ml of reagent per gram of solid phase:

(1)   Wash the column with dichloromethane (DCM) (2 × 3 min).
(2)   Wash with dimethylformamide (DMF) (2 × 3 min).
(3)   Deprotect with DMF:piperidine (1:1) (20 min).
(4)   Wash with DMF (2 × 3 min).
(5)   Wash with dioxane:water (2:1) (2 × 10 min).
(6)   Wash with DMF (3 × 5 min).
(7)   Wash with DCM (3 × 5 min).
(8)   Add derivitized FMOC amino acid in DCM:DMF solution (typically 3 moles of amino acid per mole of bound linker) (30 min).
(9)   Wash with DCM (5 × 8 min).
(10) Wash with DCM(5 × 8 min).
(11) Wash with isopropyl alcohol (5 × 8 min).
(12) Repeat sequence to add the next amino acid.

The times given are for a synthesis carried out by hand, so the total time for amino acid addition is four hours. When this production is automated, as it is in commercial machines, the total time per amino acid is one hour. Even so, the automated machine produces only 25 μmol of the decapeptide in a 10–12 hour cycle. This corresponds to perhaps 30 mg per cycle, or 60 mg per day.

Your company wants to make 100 g per day, a scale-up of about 1500 times. While you could obviously increase bed diameter $(1500)^{1/2}$ times, you would like to invent a more efficient process. However, your efforts at generating new chemistries in new phases have not been that promising. You have been driven back to the established packed-bed geometry.

Your challenge is thus to speed up the existing process. When you look at the sequence above, you are struck by the fact that three quarters of the total time is washing, i.e. mass transfer. You decide to accelerate this washing by using smaller particles. One quarter of the total time is chemical change, i.e. reactions. You decide to accelerate these reactions with higher temperatures.

Design a new column and operating procedure based on these ideas.

11.   *Lipitor synthesis*. Lipitor (atorvastatin) is the most successful of a drug class known as statins, which when taken orally can lower blood cholesterol. Statins also stabilize vascular plaque and thereby reduce the risk of stroke. Lipitor is enormously successful, with sales of almost $13 billion per year. Your company is interested in making generic Lipitor using a newly opened facility in China, because Lipitor comes off patent in June 2011.

The open literature on Lipitor is extensive. US patents 4,681,893 and 5,003,080 describe Pfizer's early manufacturing routes. Pfizer also identified the only optical isomer of Lipitor that is biologically relevant, so now synthesizes Lipitor using a process outlined by Baumann *et al.* (1992). The

key to this process is combining two ingredients, a diketone and a stereoiso-
mer, to produce atorvastatin Three references, Li *et al.* (2004), Baumann
et al. (1992), and Brewer *et al.* (1992) give the basic structure of atorvastatin
and details of the synthetic strategy used by Pfizer.

Design an overall process for making this drug, summarized as a detailed
process diagram. For the purposes of your design, you may assume that –
in addition to common laboratory reagents – *only* compounds **3** and **5a–c** in
Baumann *et al.* and **8** in Brewer *et al.* are commercially available. Include a
choice of solvents, reaction conditions and times, and purification and separation methods. You may find it helpful to summarize these choices in tables.
Explain your selection of solvents, conditions, and separations used. If you
can, identify the steps which you believe will cause the most difficulty in
making your company's generic atorvastatin.

REFERENCES AND FURTHER READING

Ajikumar, **P. K.**, **Xiao**, **W.**, **Tyo**, **K. E. J.**, *et al.* (2010) Isoprenoid pathway optimization for
taxol precursor overproduction in Escherichia coli. *Science* **330**, 70–74.

Bailey, **J. E.** and **Ollis**, **D. E.** (1986) *Biochemical Engineering Fundamentals*, 2nd Edition,
McGraw Hill, New York, NY.

Baumann, **K.**, **Butler**, **D. E.**, **Deering**, **C. F.** *et al.* (1992) The convergent synthesis of CI-
981, an optically active, highly potent, tissue selective inhibitor of HMG-CoA reductase.
Tetrahedron Letters **33**(17), 2283–2284.

Blanch, **H. W.** and **Clark**, **D. S.** (1997) *Biochemical Engineering*. CRC Press, Boca Raton,
FL.

Brewer, **P. L.**, **Butler**, **D. E.**, **Deering**, **C. F.**, *et al.* (1992). The synthesis of (4R = cis)-l, l-
dimethylethyl, a key intermediate for the preparation of CI-981, a highly potent, tissue
selective inhibitor of HMG-CoA reductase. *Tetrahedron Letters* **33**(17), 2279–2282.

Harrison, **R. G.**, **Todd**, **P. W.**, **Rudge**, **S. R.**, and **Petrides**, **D.** (2002), *Bioseparations Science
and Engineering*. Oxford University Press, Oxford.

Li, **J. J.**, **Johnson**, **D. S.**, **Sliskovic**, **D. R.**, and **Roth**, **B. D.** (2004) Atorvastatin Calcium
(Lipitor®). In *Contemporary Drug Synthesis*. John Wiley & Sons, Hoboken, NJ, ch. 9,
p. 113.

Lipinski, **C. A.**, **Lombardo**, **F.**, **Dominy**, **B. W.**, and **Feeney P. J.** (2001) Experimental and
computational approaches to estimate solubility and permeability in drug discovery and
development settings. *Advanced Drug Delivery Reviews* **46**, 3–26.

Ng, **R.** (2008) *Drugs: From Discovery to Approval*, 2nd Edition. Wiley-Blackwell, New
York, NY.

Shuler, **M. L.** and **Kargi**, **F.** (2001) *Bioprocess Modeling*, 2nd Edition. Prentice Hall, Englewood Cliffs, NJ.

Warren, **S**. and **Wyatt**, **P.** (2009) *Organic Synthesis: The Disconnection Approach*, 2nd
Edition. Wiley, New York, NY.

Wong, **D. T.**, **Bymaster**, **F. P.**, and **Engleman**, **E. A.** (1995) Prozac (fluoxetine), the first
selective serotonin uptake inhibitor and an antidepressant drug: twenty years since its
first publication. *Life Science* **57**, 411–441.

9

Microstructures

In this chapter we turn to chemical products whose value is strongly connected to their microstructure. By "microstructure," we mean chemical organization on the scale of micrometers. Some microstructured products are listed in Table 9.0–1. To understand how large such structures are, we remember that hair and beards grow about one millimeter in three days, or 300 μm per day. Thus, beard growth in eight hours – a "five o'clock shadow" – is about 100 μm. We can just feel this size. On the other hand, blood cells are discs 8 μm across and 3 μm thick, and we cannot feel individual blood cells. In many cases, the microstructures which we make will be between these sizes. Because they will be smaller than 100 μm, they will feel like continua, but because they are often the size of blood cells, they are not in reality homogeneous.

Another way to judge these products is their size on a logarithmic scale, shown earlier in Figure 1.5–1. On such a scale, the relative size of microstructures lies midway between the sizes of us and of individual molecules. Our little fingers are about 1 cm, or 10,000 μm, across. A microstructure of 1 μm, or 10,000 Å, is much larger than a molecule 3 Å in diameter. Microstructures are around the wavelength of visible light.

Products with microstructure supply added value because of the physical and chemical properties of the microstructure. Skin creams make skin soft and smooth, etchants remove photoresists to produce electronic devices, polyacrylics keep the wearer warm and comfortable, and polishes give metal a shine. The microstructures help the customer satisfy a specific need.

The value of microstructured products has a different basis to the other product types covered in the three previous chapters. In Chapter 6, we discussed commodity chemicals, whose success depends on the cost of making the products. The key step in controlling cost is the manufacturing process, which is the subject of chemical process design. The synopsis in Chapter 6 briefly reviews this subject, covered in a variety of well-established process design texts. In Chapter 7, we discussed the design of chemical devices, whose value depends both on cost and on convenience. This subject, which involves process scale-down as often as process scale-up, includes devices such as separators to enrich oxygen for

363

Table 9.0–1 *Microstructures by industrial sector or end use* The properties of these materials are often evaluated in non-scientific terms.

Product form	Aerosols	Liquids and liquid mixtures	Creams and pastes	Structured solids
Specialty chemicals	Bug sprays, spray paints	De-greasers, lube additives	Adhesives, anti-icers	Polymer composites
Agrochemicals	Herbicides, insecticides	Liquid fertilizers	Dairy, cleaners	Controlled-release chemicals
Pharmaceuticals and healthcare products	Inhalants	Syrups	Athlete's foot, anti-itch creams	Injectable powders, dietary supplements
Foods, flavors, and fragrances	Perfume, cooking spray (PAM)	Juices, extracts, cooking oils	Ice cream, ketchup	Margarines, candies
Personal care products, cosmetics	Hair sprays, deodorants	Shampoos, liquid detergents	Sunscreen, toothpaste	Solid detergents, soaps, diapers
Home and office products	Air fresheners, bug sprays	Inks, adhesives	Waxes, paints	Paper, packaging

emphysema patients and reactors for destroying small amounts of flammable radioactive waste.

In Chapter 8, we turned to the production of molecular products, where the key is speed, that is, the time to market. This need for speed has two components. The more important is the discovery of the candidate molecule itself, the active ingredient. This discovery, which is difficult and inefficient, depends mainly on chemistry and microbiology. The second, less-important component is rapid scale-up to produce larger amounts of any chemical identified as promising. This component is where engineering is important. As Chapter 8 describes, production is at a much smaller scale than practised for commodity chemicals or for many chemical devices. Production often uses unit operations which are uncommon in commodity manufacture, like filtration and crystallization. It rarely depends on more familiar chemical engineering tools like plug flow reactors or staged distillation columns.

The microstructured products reviewed in this chapter are different to these more familiar types of products. This difference is broader and more fundamental than differences between commodities, chemical devices, and molecular products. Developing these earlier types of products is often limited by making the choice between alternatives, all of which will work. In other words, the hard step in the design template is selection. Development of commodities, devices, and molecules uses the familiar tools of reaction engineering and unit operations.

Thus, while the targets may be different to those familiar from traditional chemistry and chemical engineering, the tools remain pretty much the same.

For microstructured products, this is not true. The most difficult step in design is no longer selection but needs. Converting these needs into specifications may require measuring consumer reactions and hence can include psychological factors. Making a microstructure involves selection of ingredients and process conditions, usually necessitating considerable empiricism. Sometimes large areas of science are missing, largely unknown. Reaction engineering and unit operations aren't as important as for other product types.

Thus, our discussion of microstructured products is different to, and less complete than, those for other classes of chemical products. We begin in Section 9.1 by summarizing the physical properties of materials in general and of microstructured products in particular. We also review the major complexities which these products show during their use. In Section 9.2, we outline a strategy for beginning the design of microstructured products. This strategy includes outlining a general model. While this model is necessarily complicated, it can often be simplified as two special cases. In the first special case, the product's specifications can be written in physical and chemical terms. In the second case, the product's specifications are only available as consumer reactions, which need to be rewritten as physical and chemical specifications.

Sections 9.3 and 9.4 describe the tools needed to design these products. These include both physical methods, like mixing; and chemical issues, like phase equilibria and colloid stability. The necessary tools include rate processes, like time–temperature superposition; and measures of product performance, like rheology. Note that the tools required are not restricted to reaction engineering and unit operations. The chapter ends with Section 9.5 which includes more complex examples.

9.1   Properties of Microstructures

The physical and chemical properties of microstructures vary widely, partially because their ingredients vary widely. The properties of some common ingredients, shown in Table 9.1–1, give some indication of the scope of materials involved. Fluids are characterized by their viscosity, and solids by their Young's modulus and yield stress. While the properties of liquids vary by a factor of 100, those of metals, semiconductors, synthetic polymers, and ceramics can vary by more. We will choose among all these possible ingredients in making microstructures.

The physical properties of the microstructures themselves are often very different to the properties of the ingredients from which they are made, as shown in Table 9.1–2. This is, of course, a consequence of their structure. This structure is often anisotropic; for example, the Young's modulus of cotton cloth is different in the direction of the warp than in the direction of the weft, a fact which was exploited by Vionnet in 1922 in developing the "bias cut." The microstructure often involves several phases which may not be in equilibrium. For example,

Table 9.1–1 *Properties of ingredients for microstructures* Most of these properties are available in handbooks.

Fluids	Hydrogen	Air	Water	Gasoline	Olive oil
Cost, $/kg	4	0	0.01	1	10
Specific gravity	10^{-4}	10^{-3}	1	0.8	0.9
Viscosity, 10^{-3} kg/m sec	0.008	0.02	1	0.6	100
Kinematic viscosity, 10^{-6} m^2/sec	90	15	1	1	100
Thermal conductivity, W/mK	0.2	0.02	0.6	0.2	0.2
Thermal diffusivity, 10^{-6} m^2/sec	150	20	0.2	0.1	0.1
Diffusion coefficient, 10^{-6} m^2/sec	100	20	10^{-3}	10^{-3}	10^{-5}

Solids	Steel	Silicon	Polypropylene	Concrete	Glass
Cost, $/kg	0.5	100	1.1	0.05	1.4
Specific gravity	8	2.3	0.9	2,4	2.5
Thermal conductivity, W/mK	40	150	0.1	1	1
Thermal diffusivity, 10^{-6} m^2/sec	10	10	0.1	1	0.3
Diffusion coefficient, 10^{-6} m^2/sec	10^{-6}	10^{-20}	10^{-11}	10^{-15}	10^{-20}
Young's modulus, 10^9 kg/m sec^2	200	200	1	30	60
Yield stress, 10^6 kg/m sec^2	500	700	30	40[a]	50[a]

[a] In compression.

Table 9.1–2 *Properties of microstructured products* The properties of these materials, normally dominated by the continuous phase, are approximations, because they often do not obey rules like Hooke's law of elasticity or Newton's law of viscosity.

Fluids	Paint	Shampoo	Toothpaste	Milk	Beer
Cost, $/kg	10	5	20	0.7	3
Specific gravity	1.1	1.1	1	1	1
Viscosity, 10^{-3} kg/m sec	1000	100	1000	2	2
Kinematic viscosity, 10^{-6} m^2/sec	1000	100	1000	2	2
Thermal conductivity, W/mK	1	0.6	2	0.6	0.6
Thermal diffusivity, 10^{-6} m^2/sec	0.1	0.6	1	0.14	0.1
Diffusion coefficient 10^{-6} m^2/sec	0.1	1	0.01	1	1

Solids	Cotton fabric	Wool fabric	Steak	Bread	Breakfast cereal
Cost, $/kg	40	80	20	4	10
Specific gravity	0.5	0.3	1.0	0.16	0.2
Thermal conductivity, W/mK	0.07	0.04	0.4	0.16	0.1
Thermal diffusivity, 10^{-6} m^2/sec	0.1	0.06	0.2	0.05	0.1
Diffusion coefficient, 10^{-6} m^2/sec	20	20	10^{-4}	10^{-2}	10^{-5}
Young's modulus, 10^9 kg/m sec^2	8	3	0.5	0.2	0.04
Yield stress, 10^6 kg/m sec^2	10	10	100	0.1	50

beer on the tongue bubbles as dissolved gases are released; and butter melts in the mouth as fatty liquid crystals lose their structure. In many cases, the phases present may not be in thermodynamic equilibrium, but may be trapped in a metastable state. This is true for latex paint, for some shampoos, and for cosmetic creams.

In addition, the physical properties of microstructured properties are often complex. At low shear, toothpaste doesn't leak out of the tube because it has a yield stress. At the higher shear caused by squeezing the tube, the paste flows easily, the result of non-constant, non-Newtonian viscosity. Such non-Newtonian effects are the rule, not the exception. Moreover, the physical properties which are important may depend on the application. For example, the thermal conductivity is a key factor in the perceived comfort of polyacrylic textiles, which regulate heat loss from the human body. However, the thermal diffusivity, including a contribution from melting colloidal fats, is key for the assessment of perceived creaminess of whole milk.

The complex properties of microstructured products are often simplified because the continuous phase dominates many of the physical properties. For example, if we rub a sun cream on our arms, the cream originally feels thin – of low viscosity – because the cream is an emulsion of viscous oil droplets in much less viscous water. As the water evaporates, the cream suddenly becomes much thicker – much more viscous – because it inverts or "breaks" to become an emulsion of water droplets in the more viscous oil. The continuous phase of the emulsion controls the ease with which it can be spread. One phase dominates even when the product is bicontinuous, as in bread. Diffusion of aromas in fresh bread occurs largely through the air channels and not through the starch phase. The Young's modulus, on the other hand, results from the continuous starch matrix, and is relatively unaffected by the gas channels key to aroma diffusion.

Many microstructured products change their structures during use. These include steak, cut by the teeth; ice cream, melting in the mouth; and paint, drying on the wall. Many other products, like textiles, should not undergo these changes. Almost all microstructured products are chemically ill-defined. All of this complexity might drive us to despair until we remember two key points:

(1)   Microstructured products often sell for considerably more than the cost of their ingredients.
(2)   Microstructured products are a major opportunity for growth in the chemical industry.

These two points will sustain our interest.

EXAMPLE 9.1–1   LIQUID BANDAGES

Many first-aid kits now contain "liquid bandages," polymer solutions which, when applied to cuts, act just like traditional adhesive bandages. After drying, liquid bandages stick better than conventional solid ones. They survive washing or swimming without falling off. Most of these liquid bandages are water-based solutions of polyvinylpyrolidone or alcohol solutions of poly(methylacrylate-isolutene-monoisopropylmaleate). Some are even based on octylcyanoacrylate, a less toxic alternative to the ethylcyanoacrylates used in "superglue."

Our core team wants to develop a product like this. Any liquid bandages must meet three chief needs: they must keep out bacteria and debris; they must maintain a proper moisture balance, allowing the cut to dry; and they must reduce pain by covering nerve endings. What microstructure should they have?

SOLUTION

Because the first goal of the liquid bandage is keeping the cut clean, it must not have any pores larger than the size of a bacterium, around 1 μm. To allow water evaporation, it should have a resistance to water transport which is less than the resistance in the air surrounding the wound. This resistance is the thickness of the boundary layer in the air, perhaps 0.1 cm, divided by the diffusion coefficient of water vapor in air, about 0.3 cm^2/sec. Finally, the liquid bandage solution should contain an approved disinfectant, like iodine. Parenthetically, we note that one folk remedy used as a liquid bandage was willow sap, which includes salicylic acid, the active part of aspirin.

EXAMPLE 9.1–2   BIODEGRADABLE PACKAGING

Polystyrene foam is an effective packing material for many objects, especially for fragile products of complicated shape. It is the familiar "popcorn" in every garbage can after Christmas. It does its job well. However, after its use, it is an environmental nightmare for the same reasons that it is a successful product. It has low density, so it blows all over the place. It is not wet by water, so it floats on lakes and streams. Most importantly, it doesn't biodegrade: it is immortal.

Our core team sees a need for a biodegradable substitute for polystyrene packaging. The team members want to identify what specifications this substitute product must meet. They wish to support any qualitative specifications with calculations. They want to avoid suggesting actual products, like real popcorn made from corn, but instead to state the properties desired in the new material.

SOLUTION

After considerable discussion, the core team decides that the new product must meet three critical needs. First, it must be biodegradable, defined as losing its microstructure within one month after being released into the environment. Further, the basic polymer should degrade into carbon, water, and carbon dioxide within ten years of its release. Second, the product should have mechanical properties within a factor of five of the existing polystyrene foam. Third, it should be stable at 50 °C so that it can be used much as the current product is. The obvious benchmark for our new product is a foam based on the biodegradable polymer polylactic acid (PLA).

Table 9.2–1 *Microstructured products with differing types of specifications* Different microstructures are evaluated in different ways.

Physical science	Psychophysics	Physical science and psychophysics
Controlled-atmosphere packaging	Tender meat	Spreadable, effective sunscreen
Anti-icing chemicals	Warm, comfortable	Paint color and hiding power
Optical films that sharpen visual	clothes	Soap that cleans and freshens
imaging	Silky coatings	

Biodegradation could involve hydrolysis, ultraviolet decomposition, and enzymatic degradation. The team suggests focusing ideas on composites of a natural waste material, like sawdust, and a more expensive binder. We suggest seeking binders which decay within one month of release. The tensile strength of polystyrene foam of a density of 2 lb/ft^3 is 60 psi, and such a foam also has a thermal conductivity of 0.02 Btu/ft/hr/°F. These seem sensible target values for preliminary screening.

9.2   Getting Started

As explained above, we expect that microstructured product design involves two key steps. In the first, the need for a product must be expressed as quantitative specifications. Often, these will be physical properties of the product, like its rheology, its permeability, or its Young's modulus. In other cases, the need will be expressed as consumer desires, like creamier beer or silkier fabric.

Examples of such products are given in Table 9.2–1. The first group has specifications based solely on physical properties. For example, controlled-atmosphere packaging is used for vegetables which, even after harvest, continue to produce carbon dioxide and to consume oxygen. The permeability of the packaging is adjusted to provide the concentrations of gases which prolong freshness most. This concentration is known and easily specified for each vegetable. In the same sense, anti-icing chemicals, sprayed onto planes while they wait in a snowstorm to take off, depend on freezing-point depression. Because these chemicals coat the wings, the planes would fly poorly. The anti-icing chemicals must peel off during take-off. They must have a yield stress so they stick to the wings of a stationary plane, but they must have a sufficiently low viscosity at high shear to peel off before take-off speed is reached. The needs required of this group of microstructured products are easily specified.

The characteristics of the second group of products in Table 9.2–1 are much harder to quantify. Each of us may understand what he feels is "tender meat," but efforts to relate this attribute to properties like Young's modulus and ultimate tensile strength have had mixed success. Similarly, a specification for clothes that are warm and comfortable probably involves fabric thermal conductivity and water vapor permeability, but in ways that are not fully understood. Still other products may involve combinations of physically definite properties and consumer assessments. For example, the sunscreen mentioned in the third product group in Table 9.2–1 must screen a certain percentage of short ultraviolet light (UVB; cf. Example 2.4–4). It must also be "spreadable," a consumer assessment probably related to viscosity.

Thus, for many microstructured products, we will first evaluate consumer needs. How we attempt to do so is reviewed in Subsection 9.2–1. Once we have done this evaluation, we must relate these needs to particular physical properties, which is the subject of Subsection 9.2–2. Third, we must specify the desired values of these physical properties, as discussed in Subsection 9.2–3. In all these steps, we may be forced to take shortcuts and make rough approximations. Making these approximations will usually increase the risk that our product will fail but reduce the time required to bring the product to market.

9.2–1 CONSUMER NEEDS

In establishing needs, we will speak to actual or potential customers, probably focusing on lead users. We may also make use of focus groups and expert test panels. All of these groups will describe the desired product attributes in qualitative language, using non-scientific terms like "crunchiness," "body," "softness," and "redness." Our first task is to convert these descriptors into some sort of numerical scale. To do so effectively requires that we formulate appropriate questions before speaking to the consumers. We might want to ask them "Is product A 'creamier' than B?"; or "How many times is A 'creamier' than B?"; or if they prefer A or B, and by how much. How we pose the questions will determine into what sort of quantitative scale we convert the answers. This was mentioned in Section 2.2, where the concepts of "hedonic" and "intensity" scales were introduced. The former describes consumers' preferences, and the latter measures their perceptions. Whether we use a "hedonic" or "intensity" evaluation, we want to emerge from this exercise with our assessment of the consumers' needs placed on some type of quantitative scale.

9.2–2 NEEDS VS. PHYSICAL PROPERTIES

We now want to convert what the consumers desire into specifications of products with particular physical properties. Doing so involves answering three questions. First, we must decide how to tell when two properties are related. Second, we must choose the best descriptors, like "creamy," from the considerable vocabulary that is available. Third, we must discover relations between the best descriptors and relevant combinations of physical properties. These three challenging questions are discussed next.

The best way to test for relations between two variables is by means of correlation coefficients. To define these, we imagine assessments of two variables, x

and y. We then define relations between these variable in terms of the following averages:

$$s_{xx} = \sum_{i=1}^{N} (x_i - \overline{x}_i)^2 \qquad (9.2\text{--}1)$$

$$s_{yy} = \sum_{i=1}^{N} (y_i - \overline{y}_i)^2 \qquad (9.2\text{--}2)$$

$$s_{xy} = \sum_{i=1}^{N} (x_i - \overline{x}_i)(y_i - \overline{y}_i) \qquad (9.2\text{--}3)$$

where the overbar represents the arithmetic average over each assessment "i" for a total of N samples. The correlation coefficient r is then given by

$$r^2 = \left[\frac{s_{xy}^2}{s_{xx} s_{yy}} \right] \Big/ N \qquad (9.2\text{--}4)$$

As an example, consider the assessments given below:

x	y
1	2
2	4
3	6

The correlation coefficient r between these values is one. If the relation were inverse (e.g. $y = 1$, $\tfrac{1}{2}$, and $\tfrac{1}{3}$), the correlation coefficient r would be -1. If x and y were unrelated, the correlation coefficient r would be zero.

To illustrate these ideas more completely, we consider data for class attendance vs. grade, shown in Table 9.2–2. We may expect that attending class helps earn high grades. However, the correlation between attendance and grades is less than might be expected, as shown in Figure 9.2–1. The r^2 is 0.56. Class attendance may help, but factors like intelligence are also important. We will use this tool of correlation coefficients to answer the first question about needs vs. properties: we will assume high correlation means a close relation.

The second question which we must address is the choice of which descriptors are best for characterizing the product. In many cases, we will guess because identifying the best descriptors requires excessive work. Making this guess means accepting the risk that our descriptors are not optimal. If we decide not to take this risk, we can measure the attributes of a variety of products and see if customer assessments are correlated. In this effort, we will discover synonyms and antonyms. For example, for liquids in the mouth, "thickness" and "thinness" are inversely related, as shown in Figure 9.2–2. They have an r of -0.92. "Thickness" and "smoothness" are not as closely related, as also shown in Figure 9.2–2. Some published efforts have tried to define the entire vocabulary for a given product area, with results like those in Figure 9.2–3 for liquids in the mouth. In this figure, points "1" and "2" are "hard" and "soft," and points "6" and "7" derive from

Table 9.2–2 *Class attendance data* The expectation that attending class will lead to better grades can be tested by these data.

Attendance %	Points earned %	Attendance %	Points earned %
35	55	85	83
40	21	85	64
45	62	85	80
50	60	85	57
65	43	85	76
65	65	85	82
70	75	85	86
70	59	90	75
70	53	90	79
70	57	90	80
75	70	90	73
75	64	90	67
75	65	95	89
75	64	95	85
75	80	95	85
80	69	95	77
80	80	95	71
80	60	100	86
80	81	100	90
85	78	100	83

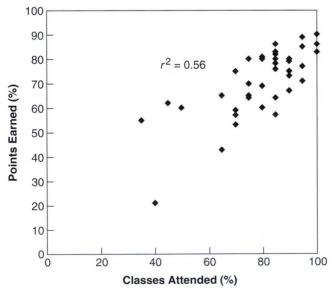

Figure 9.2–1 *Class attendance vs. grades earned* While attending class tends to give higher grades, factors like intelligence are also important.

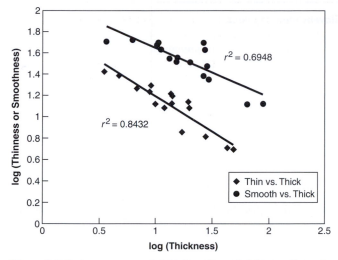

Figure 9.2–2 *Assessments of "thickness" and "thinness" in the mouth* These descriptors of soups and sauces are antonyms.

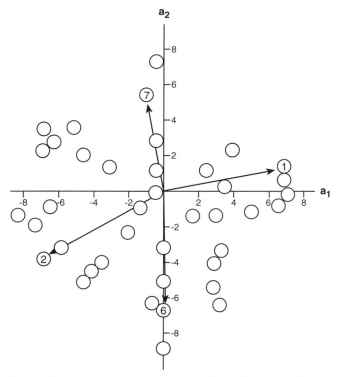

Figure 9.2–3 *An effort to map all words describing mouthfeel* The data suggest the abscissa is associated with viscous forces, and the ordinate is related to heat transfer. More complete descriptions will require more dimensions.

Table 9.2–3 *Stevens's law exponents.*

Attribute	Stimulus	Exponent
Length	Visual	1.0
Area	Visual	0.7
Temperature	Cold	1.0
Brightness	Light	0.3
Electric shock	Current in fingers	3.5
Thickness	Viscosity	0.4
Spreadability	Viscosity	0.5

"warm" and "cold." Other points are other words. We are unsure how useful such analyses are in practice.

The third question which we need to answer is the relation between a consumer descriptor and a particular combination of physical properties. We usually seek these relations empirically, though these empiricisms are often called "laws." We could assume a linear law:

$$[\text{response}] = a[\text{stimulus}] \qquad (9.2\text{–}5)$$

where a is an empirical constant. For example, the assessments of "how long" are proportional to the actual physical length, as measured with a ruler. Alternatively, we can assume a semi-logarithmic relation, called the "Weber–Fechner law":

$$\log[\text{response}] = a[\text{stimulus}] \qquad (9.2\text{–}6)$$

where a is a different constant. This relation, suggested on the basis of experiments with weights, is often quoted but rarely fits experimental data. The third possibility, associated with the physiologist S. S. Stevens, assumes a power-law relationship:

$$[\text{response}] = a[\text{stimulus}]^b \qquad (9.2\text{–}7)$$

where a and b are empirical constants. This "Stevens law" is often more successful than the linear and semi-logarithmic ones, which is not surprising because it has two empirical constants, rather than just one. Table 9.2–3 gives some values of b.

9.2–3 SPECIFICATIONS

At this point, we will normally have some idea of what our customers want. We will understand which physical properties are important, and we will have good estimates of the range of properties we seek for our product. We will be ready to develop products which meet the customer needs.

However, at this point, product design often falters because our description is so completely empirical, without much basis in science. In particular, we must recognize that the science of human perception is incomplete. In this assertion, we do not refer to neural processing of well-defined stimuli, which has been carefully studied. Rather, we are uncertain how different products produce these stimuli. We can explore this further by considering how perception occurs. The best understood perception is vision. In this case, the stimulus is of a particular spectrum of colors, a distribution of wavelengths. We perceive this distribution as an average over three types of receptors in the eye, which are sometimes approximated as being responsive to blue, green, and red light. This integrated average is our perception of color. Notice that the spectrum of a new product does not have to be the same as the benchmark to generate the same color; only the integrated average does.

There is a vivid example of this in every hardware store. Imagine we want to paint our bedroom the specific color of yellow that van Gogh used so successfully in his "yellow period." We could of course paint our room using the same pigments in van Gogh's paintings, which are known. But using those pigments would both be toxic and expensive. Instead, we simply toddle off to the hardware store to match the yellow in van Gogh's pictures against one of the small color chips available in the store. The colors available in the store are not made with the same pigments used by van Gogh; they do not give off the same spectrum of light as van Gogh's pigments do. They do give the same integrated average over the same three receptors in our eye, and so they give the same perception of color. Such an understanding of perception is available not only for color but also for sound.

Such an understanding of perception is not available for touch, taste, or smell. The last few years have produced major advances in the molecular biology of smell, but not in the simulation of smell. In other words, many details of the chemical changes involved when neural receptors are challenged with specific molecules are now understood; but the way in which chemical challenges diffuse to and competitively react with the receptors remains a mystery. The chemical senses seem for the moment incomplete; the sense of touch also has remained out of reach. For example, the attribute "creamy" is highly desirable in food and cosmetics, but in spite of a century of focused scientific work, no one knows what "creamy" is.

We can speculate about how such an understanding may develop, as shown in Table 9.2–4. The first column in this table is the human sense involved, and the second suggests products evaluated with this sense. The third column gives the physical quantities assessed during perception; this column is speculation. As already explained, vision assesses color as an integrated average over wavelengths. Our speculation is that touch assesses forces. The fourth and fifth columns give the phenomena giving rise to sensory perception and the branches of science describing these. We will explore the implications of this table in some of the examples given next.

Table 9.2–4 *An overall schematic of human perception*

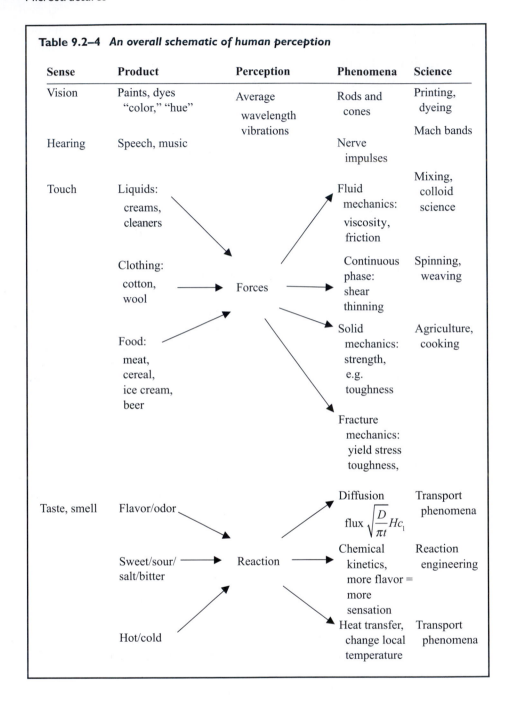

Sense	Product	Perception	Phenomena	Science
Vision	Paints, dyes "color," "hue"	Average wavelength vibrations	Rods and cones	Printing, dyeing
				Mach bands
Hearing	Speech, music		Nerve impulses	
Touch	Liquids: creams, cleaners		Fluid mechanics: viscosity, friction	Mixing, colloid science
	Clothing: cotton, wool	Forces	Continuous phase: shear thinning	Spinning, weaving
	Food: meat, cereal, ice cream, beer		Solid mechanics: strength, e.g. toughness	Agriculture, cooking
			Fracture mechanics: yield stress toughness,	
Taste, smell	Flavor/odor		Diffusion flux $\sqrt{\dfrac{D}{\pi t}} Hc_1$	Transport phenomena
	Sweet/sour/ salt/bitter	Reaction	Chemical kinetics, more flavor = more sensation	Reaction engineering
	Hot/cold		Heat transfer, change local temperature	Transport phenomena

EXAMPLE 9.2–1 UNDERSTANDING SOFTNESS

We are interested in developing cushions with specific degrees of "softness" and "firmness." To do so, we seek a better understanding of how "softness" is perceived, either by squeezing the cushions with our fingers or by sitting on them. How can we develop this understanding?

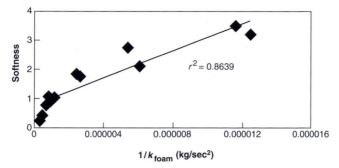

Figure 9.2–4 "*Softness*" *of cushions* The assessment of "softness" is related to the Hookean spring constant of the cushions.

SOLUTION

To begin, we ask customers to evaluate a range of cushions using ratio scaling. We also measure the spring constant of these cushions, that is, the deformation per applied force. We find a good correlation between the customers' assessments and the spring constants. To explain this correlation, we assume that what is actually assessed is the deformation of the fingers, probably with Merkel or Ruffini receptors. What is perceived may be either the vertical displacement with a given force or the force required for a given displacement. In either case, the perception of softness will be:

$$\text{softness} \propto \frac{1}{k_{\text{flesh}}} + \frac{1}{k_{\text{cushion}}}$$

where the k values are the Hookian spring constants of the materials involved. Thus, a plot of perceived "softness" should vary linearly with the reciprocal of the cushion's spring constant. It does, as shown in Figure 9.2–4. This correlation lets us predict softness for other cushions.

EXAMPLE 9.2–2 KEY FACTORS FOR CONTROLLED RELEASE

Our company has developed a microencapsulation technology for controlling the humidity in microelectronic devices during shipping. The technology consists of small, spherical bubbles made of a water-permeable polymer. Each bubble contains a solid desiccant which could damage the microelectronics if it were not separated by the polymer.

You are a member of the core team charged with deciding whether this technology can be adapted for controlled release of a drug taken orally. The drug is solid, within the polymer bubble. Once in the gut, the bubbles hydrates, producing a saturated drug solution within each bubble. The drug then diffuses out steadily over time, so that the total drug ingested can be smaller, giving the same improvements in health but with smaller side effects.

What key parameters of the polymer bubbles need to be specified?

SOLUTION

Any polymer used must be non-toxic and cheaper than the drug. It must be able to be processed using the existing microelectronics-driven technology. The key additional property depends on the desired dose per time J_1 given in, for example, micrograms per hour. If the polymer is non-porous, this rate is given by

$$J_1 = nAj_1 = \frac{nADH(c_1(\text{sat}) - 0)}{l}$$

where n is the number of microcapsules; A is the surface area of one microcapsule; l is the thickness of the bubble wall; DH is the wall's permeability, the product of the drug's diffusion coefficient D and its solubility in the wall H; and $(c_1(\text{sat}) - 0)$ is the concentration difference across the microcapsule's wall. When solid drug is present, the concentration inside will be saturated, $c_1(\text{sat})$; and that outside will be relatively small, and hence effectively zero. Note that n, A, and l are functions of the microcapsule's geometry; and D and H vary with microcapsule chemistry. We should have some latitude in trying to meet the desired specifications.

EXAMPLE 9.2–3   SEEKING EQUIVALENT "THICKNESS"

We want to reformulate a variety of soups and sauces to reduce the concentration of saturated fats. To do so successfully, we must retain the same product properties. Surveys of our customers show one key attribute of these products is their "thickness." Thus, we need a theory of "thickness." How can we develop such a theory?

SOLUTION

We suspect "thickness" is probably assessed as shear forces on the tongue. When we collect published data for these forces, we find a strong correlation between this force and "thickness," as shown in Figure 9.2–5. This correlation includes not only assessments in the mouth but also assessments with the fingers. It includes not only Newtonian fluids but also non-Newtonian fluids.

This correlation implies that we do not need to duplicate rheology completely. For example, if our current recipes use Newtonian fluids, we can potentially use new recipes with non-Newtonian fluids as long as the new fluids cause the same forces in the mouth. Similar correlations can be found for other descriptors, as shown for *spreadability* vs. *force* in Figure 9.2–6. Our theory for the perception has given us a means of rationally changing the ingredients from which we make our products.

9.3   The Microstructure Toolbox: Reactions

Because we have rewritten the customer needs as specifications, we are now ready to generate ideas that can fulfill these specifications and to select the best of these ideas. These steps imply a toolbox of processes, including both reactions

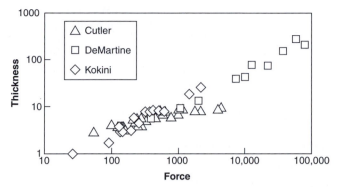

Figure 9.2–5 *"Thickness" vs. measured viscosity* The correlations shown are for both force on the tongue and on the fingers as a function of viscosity. The data for DeMartine are on the fingers; the other data are in the mouth. These correlations have a strong basis in fluid mechanics, and result from the same physical effects.

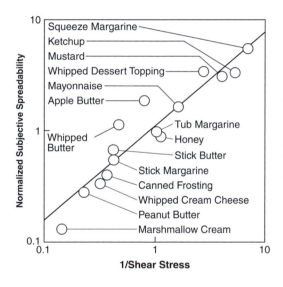

Figure 9.2–6 *"Spreadability" vs. force on the fingers* The results show that the same spreadability can be achieved with any ingredients giving the same force on the fingers.

and mixing. In this section, we review three groups of reactions. These groups are organized under the headings of sterilization, time–temperature superposition, and microstructure stability. As we will see, these reactions are rarely chemically well defined, but they are treated intellectually with concepts like those for well-defined chemistries.

9.3–1 STERILIZATION

Sterilization uses heat to kill microbes. In almost all cases, the microbes are undesirable and degrade product quality, potentially compromising the health of the customer. Killing microbes is a good idea: never sicken a customer.

Heat kills microbes by a first-order rate process. We want to estimate the time required for death. Specifically, if the number of microbes per volume is N, then:

$$\frac{dN}{dt} = -kN \tag{9.3–1}$$

where k is a first-order rate constant with dimensions of reciprocal time. If the temperature is constant, then the rate constant k also has one value. In this isothermal case, if the number at time zero is N_0,

$$\frac{N}{N_0} = e^{-kt} \tag{9.3–2}$$

However, sterilization is seldom reported in these terms. Instead of a rate constant, the rate is described by a decimal reduction time D, defined using Equation 9.3–2:

$$0.1 = e^{-kD} \tag{9.3–3}$$

Thus

$$D = \frac{2.303}{k} \tag{9.3–4}$$

Note that decimal reduction time D, which has units of time, is completely unrelated to the diffusion coefficient D, for which this symbol is used in all other sections of this.

We will normally want to sterilize a product long enough to reduce the concentration of microbes N by five orders of magnitude. In other words, we will sterilize for $5D$. The complication comes from the variation of D with temperature. Like the rate constant k, D shows an Arrhenius temperature dependence:

$$D = D_0 e^{E/RT} \tag{9.3–5}$$

where D_0 is a constant and E is an activation energy. Thus, as the temperature increases, D gets smaller. Because E is often large, D gets smaller quickly.

Thus, calculating how fast we sterilize involves finding how the microbial concentration N varies with temperature and time. To make this more specific, imagine we want to use steam to sterilize a can full of liquid product. We begin with an unsteady-state energy balance on the can

$$[\text{energy accumulation}] = [\text{energy added from steam}]$$

$$V\rho \hat{C}_p \frac{dT}{dt} = qA = UA(T(\text{steam}) - T) \tag{9.3–6}$$

where V and A are the volume and surface area for the can, respectively; ρ, $\hat{C}_p$, and T are the density, specific heat capacity, and temperature of the contents of the can, respectively; U is the overall heat transfer coefficient; and $T(\text{steam})$ is the temperature of the surrounding steam. If the initial temperature of the can's contents is T(initial), then Equation 9.3–6 can be integrated to find

$$\frac{T(\text{steam}) - T}{T(\text{steam}) - T(\text{initial})} = e^{-\frac{UAt}{\rho \hat{C}_p V}} \tag{9.3–7}$$

Note that the quantity $(\rho \hat{C}_p V / UA)$ is the characteristic time to heat the can.

We can now estimate the time required for sterilization. We insert the temperature from Equation 9.3–7 into Equation 9.3–5, combine the result with Equation 9.3–4 to find k as a function of time, and use this function in Equation 9.3–1 to integrate numerically to find the microbial concentration as a function of time. This is accurate. For a preliminary estimate, we may prefer heating for three times $(\rho \hat{C}_p V / UA)$ and then heating for five times the decimal reduction time D at T(steam). This will give a good first guess.

Estimating sterilization in this way requires calculating the overall heat transfer coefficient U. The reciprocal of U, often called the overall resistance to heat transfer, is given by the sum of the resistances to heat transfer:

$$\frac{1}{U} = \frac{1}{h_{\text{out}}} + \frac{d}{k_{\text{wall}}} + \frac{1}{h_{\text{in}}} \tag{9.3–8}$$

where h_{out} and h_{in} are the individual heat transfer coefficients inside and outside of the can, respectively; and d and k_{wall} are the thickness and thermal conductivity of the wall itself. As those who have studied heat transfer will remember, the resistance from condensing steam is small, i.e. h_{out} is large. Similarly, the resistance of the wall is small, i.e. d is small and k_{wall} is large.

The key is the estimation of h_{in}. There are two cases. First, if the contents of the can are a low-viscosity liquid, then h_{in} will normally be greater than 1 kW/m^2 K. Second, if the contents are close to solid, h_{in} will be roughly the ratio of the thermal conductivity of water to some characteristic dimension of the can, which we can conservatively take to be the radius. This gives a value of about 20 W/m^2 K as typical for cans with solid contents. As a result, cans of solids will have a smaller value of U, and take longer to sterilize.

9.3–2 TIME–TEMPERATURE SUPERPOSITION

Microstructured products often decay with time. Hand creams phase separate; ice cream gets less creamy; paint will no longer spread smoothly; and chocolate gets covered with a white layer, called "bloom." A carton of orange juice tastes a lot better the day it is opened than it does three days later.

This deterioration in product quality reflects the instability – or at least the metastability – of many microstructured products. Hand creams phase separate because the oil droplets in the emulsion grow with time, a surface-energy-driven phenomenon know as Ostwald ripening. Crystals in ice creams get coarser for similar reasons. Paint is often stabilized with ionic detergents whose effectiveness can be altered by changes in temperature or humidity. Orange-juice flavors degrade once they are exposed to oxygen.

The instability of these products means that many manufacturers put dates on their products. These dates give estimates of the products' shelf lives and implicitly admit product quality cannot be sustained indefinitely. For established products, these shelf lives are known from experience. However, for new products, they must be estimated before extensive experience is available.

Estimating shelf life for new products most often uses a method called time–temperature superposition. Instead of waiting months or years, we measure a new product's properties over days or weeks but at higher temperatures than we expect the product will ever encounter. We then extrapolate to lower temperatures to get the desired shelf life. In this extrapolation, we use an analysis very similar to that used for sterilization above.

To make these ideas more quantitative, we must first choose a key product property. For example, for cold cream, we might choose apparent viscosity at a particular shear. We then assume

$$\frac{d}{dt}[\text{property}] = -k[\text{property}] \tag{9.3–9}$$

where k is a rate constant for property deterioration. Integrating,

$$\frac{[\text{property}]}{[\text{property}]_{t=0}} = e^{-kt} \tag{9.3–10}$$

We now measure the rate constant k at several high temperatures, like 50 °C and 90 °C. We hope that this rate constant shows an Arrhenius temperature variation

$$\ln k = \ln k_0 - \frac{E}{RT} \tag{9.3–11}$$

where E is again an activation energy. From our experiments, we can find k_0 and E, and so we can estimate k at other temperatures, like 25 °C. We can then calculate how fast our product properties will decay and use this to estimate shelf life.

Time–temperature superposition is valuable, well worth trying. In some cases, other variables, like different solvents, can be used instead of different temperatures. This type of procedure is risky because it uses one activation energy for what may be several different steps. For example, imagine the deterioration of properties involves two sequential steps. One, due to mass transfer and described by $k_D a$, varies weakly with temperature; the other, a chemical reaction characterized by k_R, varies strongly with temperature. The overall rate constant k is

$$\frac{1}{k} = \frac{1}{k_D a} + \frac{1}{k_R} \tag{9.3–12}$$

The behavior of k with temperature shown in Figure 9.3–1 illustrates the problems involved. Measurements at high temperature will cause the extrapolation to be seriously in error, especially when we remember that the ordinate is logarithmic. Nonetheless, time–temperature superposition is used regularly. Perhaps

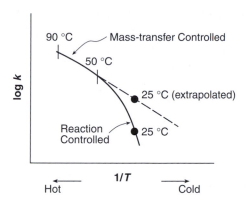

Figure 9.3–1 *Time–temperature superposition for a complex product* In this case, product stability is underestimated.

this is desperation; what else is possible? Perhaps it is conservative; after all, the actual shelf life will be longer than that obtained by extrapolation.

9.3–3 EMULSION STABILITY

The third topic of this section is the stability of emulsions, which are the form of many microstructured products. To cover this topic, we first review the subject of phase equilibria, then discuss emulsion stability, and finally describe what causes emulsion instability.

Phase Diagrams

Phase equilibria can be reviewed by reference to the examples shown in Figure 9.3–2. The gas–liquid phase diagram for one component, shown in (a), is the most easily remembered. The key in this diagram is the dome in the center, which separates liquid on the left from vapor on the right. If the pressure on a fluid is released at a specific, constant temperature, then the volume follows the dashed line. To the left of the dome, the volume is of liquid; crossing the dome at constant pressure, the volume is of a liquid–vapor mixture; to the right of the dome, the volume is of vapor. The maximum of the dome is the critical point. The dotted line within the dome is the limit of stability; fluids falling between the solid line and the dotted line are super-heated liquids or sub-cooled vapors. These are metastable and stay as they are until some upset causes the phases to change. This metastable region is usually and sensibly ignored in introductory discussions.

The second phase diagram, in Figure 9.3–2(b), is for the liquid–liquid mixture of water and triethylamine (TEA). Because there are now two components, there is an additional degree of freedom, accounted for by showing the phase diagram at a single pressure and not displaying the volume. The behavior shown is best reviewed by considering the vertical line at constant composition. This line begins at high temperatures, where there is one liquid phase. At lower temperatures, two phases may form; at still lower temperatures, the system again becomes one phase. The egg-shaped region thus represents where there are two liquid phases. The maximum and minimum of the egg are critical points; the maximum is an upper critical solution temperature (UCST); the minimum is a lower

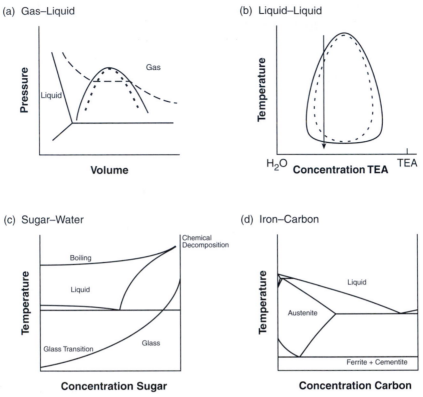

(a) Gas–Liquid

(b) Liquid–Liquid

(c) Sugar–Water

(d) Iron–Carbon

Figure 9.3–2 *A review of phase diagrams* The simple diagrams taught in elementary classes must be extended to describe microstructured products.

critical solution temperature (LCST). As before, the stability limit is shown by the dotted line. Solutions falling between the solid binodal curve and the dotted spinodal line are metastable.

The third phase diagram in Figure 9.3–2(c) illustrates the approximations and omissions which are frequently made in these diagrams. The plot is for sugar in water. Above the boiling line, the vapor phase is thermodynamically stable. The plot is at constant pressure: if we increase the pressure the boiling line moves up. The other lines do not move much if the pressure is changed because the molar volumes of solid and liquid phases are small compared to that of a vapor phase. Immediately below the boiling line, we see a region in which only solutions exist. At lower temperatures are regions in which solid water coexists with solution (on the left), and a larger region in which solid sugar coexists with solution (on the right). Below the eutectic point, the temperature and composition at which water and sugar simultaneously freeze, only solid phases are stable. We can observe the increasing solubility of sugar in water as temperature is raised. There is also a large area of metastability in which super-saturated solutions can form. It shows that if we heat sugar solutions, the sugar may form glass – another form of metastability. This glass region is used commercially to make hard candy, including "sugar-on-snow," that strange confection made in Vermont from tree sap and eaten with dill pickles. Like many metastable materials, sugar glasses can persist for very long periods of time (if they are not eaten). Finally, Figure 9.3–2(d) shows the phase diagram for carbon and iron, probably the most important of all phase diagrams. This diagram, developed empirically by blacksmiths 7000 years ago, shows a variety of solid phases formed by different processes. While these phases are microstructured products, they fall outside the scope of this book, and are described in detail in books on materials science. Their analogs made of liquids are the focus here.

Emulsion Properties

We now move to emulsions, mixtures of one material dispersed in a second, continuous fluid. Such emulsions make up a large fraction of microstructured products. Sometimes the dispersed phase is particles of glasses or solids, dispersed in a continuous liquid; butter, cream, latex paint, and liquid detergents are good examples.

Emulsions have three common properties. First, their viscosity is dominated by that of the continuous phase, μ_0. If the dispersed phase is solid spheres, the overall viscosity μ is given by

$$\mu = \mu_0(1 + 2.5\phi + \cdots) \tag{9.3–13}$$

where ϕ is the volume fraction of the dispersed phase. While this assumes the dispersed phase is solid, liquid–liquid emulsions often behave in the same way. Note that the effect is small – if ϕ is 0.1, the change in viscosity is only 25% – and that it is independent of the size of the dispersed phase. For emulsions, the continuous phase is key.

A second characteristic of emulsions is that they are often opaque and appear cloudy. This is because their structure has a characteristic size, like the average distance between dispersed particles or the diameter of droplets, which is of the same order as the wavelength of visible light. The refractive index difference between the continuous and dispersed phases scatters light. In some cases, this type of effect can be exploited, as in liquid crystal devices (LCDs). Sometimes, it is carefully removed by matching the refractive indices of the two phases.

A third characteristic of emulsions is that their stability can be dramatically enhanced with soaps, detergents, or proteins. Examples of these emulsifiers are shown in Figure 9.3–3. The soap shown is like one described by Laura Ingalls Wilder in her children's books on life on the American frontier. Detergents, which include anionic, cationic, and non-ionic examples, will be found on the ingredients lists of many prepared foods and personal-care products. These species normally retard phase separation in microstructured products. They do so by inhibiting flocculation, when dispersed particles aggregate; coalescence, when they fuse; and coagulation, when the phases separate macroscopically.

The reason emulsifiers inhibit phase separation can be understood in thermodynamic terms. When two immiscible phases are mixed together, their Gibbs free

A Soap:

   Dodecanoic Acid, Potassium Salt

   $CH_3(CH_2)_{10}COO^-\ K^+$

An Anionic Detergent:

   Dodecylsulfate, Sodium Salt (SDS)

   $CH_3(CH_2)_{10}SO_3^-\ Na^+$

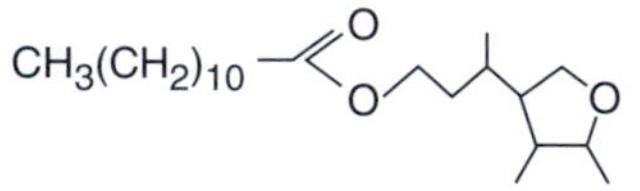

Figure 9.3–3 *Chemical structures of some common emulsifiers* These species increase emulsion lifetimes.

A Cationic Detergent:

   Benzalkonium Chloride

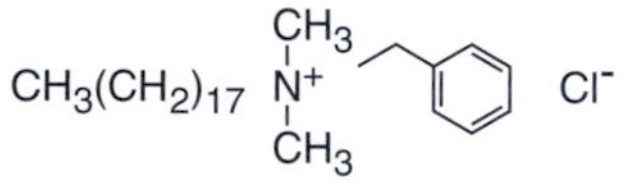

A Non-Ionic Detergent (Tween):

energy is increased because of the interface between them. This change in free energy ΔG, which equals the minimum work of mixing W, is given by

$$\Delta G = W = \gamma \Delta A \qquad (9.3\text{–}14)$$

where γ is the surface tension and ΔA is the change in interfacial area caused by the mixing. For example, imagine we mix $10\ cm^3$ ($10^{-5}\ m^3$) of olive oil in water, as part of making salad dressing. The free energy change for making 2 μm drops is

$$\Delta G = \gamma\,[\text{number of drops}]\left[\frac{\text{area}}{\text{drop}}\right]$$

$$= \left[\frac{23 \times 10^{-3}\ kg}{sec^2}\right]\left[\frac{10^{-5}\ m^3}{\frac{4}{3}\pi(10^{-6}\ m)^3}\right][4\pi(10^{-6}\ cm)^2]$$

$$= 0.69\ J \qquad (9.3\text{–}15)$$

If we add an emulsifier that drops the surface tension to $2 \times 10^{-3}\ kg/sec^2$, we get

$$\Delta G = 0.06\ J \qquad (9.3\text{–}16)$$

Because the free energy of the emulsion is much closer to that of the pure phases, the tendency to phase separate is greatly reduced.

However, this does not explain why some emulsions are more stable than others. While scientists bluster, explanations are elusive. In some ways, medieval recipes are still followed: "Make emulsions with a wooden paddle, stirring always in one direction and only with the right hand." More modern arguments suggest

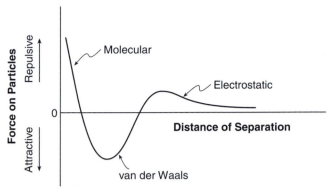

Figure 9.3–4 *Forces between emulsion particles* Long-range electrostatic repulsion can be offset by shorter-range van der Waals attraction.

rules with a stronger scientific basis. First, choose an emulsifying agent that concentrates in the continuous phase. Second, carefully consider a packing parameter, defined as

$$[\text{packing parameter}] = \frac{[\text{volume of emulsifier tail}]}{[\text{area of emulsifier head}][\text{length of tail}]} \qquad (9.3\text{–}17)$$

When this parameter is less than one third, the system forms small droplets of oil in water. When it is about one, the system forms lamellae. When it is much greater than one, it forms droplets of water in an oil continuum.

Instability

Emulsions can be destablilized by changes in concentration, temperature, or electrostatic forces. Changes in concentration are exemplified by the drying of latex paints and glues. These aqueous systems are surfactant-stabilized emulsions of small droplets or particles, often of polystyrene. As water evaporates, the droplets are forced together until they fuse. At this point, the surfactant diffuses into the polymer, and the polymer chains entangle. This fusion is effectively irreversible, which is why latex paint is so hard to remove the day after painting. When the particles are still separated in water, cleaning the spilled paint is much easier.

Emulsion instability caused by freezing involves a similar mechanism. Ice crystals remove water, concentrating the polymer particles until they fuse. Rises in temperature can destabilize emulsions containing non-ionic surfactants because these species often show an LCST. Thus, the detergent phase separates, and the emulsion droplets fuse.

Altered electrostatic interactions are the commonest cause of emulsion instability when stabilization depends on ionic surfactants. These surfactants adsorb to the surface of the discontinuous phase, causing a fixed charge on the particles. This charge causes a repulsive force which is stronger than the attractive van der Waals force, as shown schematically in Figure 9.3–4. The repulsive part of this force between the particles extends over a distance $1/\kappa$:

$$\frac{1}{\kappa} = \left[\frac{E_r \varepsilon_0 k_B T}{\sum_{i=1}^{n} (z_i e)^2 c_i} \right]^{\frac{1}{2}} \tag{9.3--18}$$

where E_r is the dielectric constant, ε_0 is the permittivity of free space, k_B is Boltzmann's constant, e is the charge on an electron, and z_i is the valence of the ionic species whose concentration is c_i. The quantity $1/\kappa$ is called the Debye length.

The quantity $\sum_{i=1}^{n} (z_i e)^2 c_i$, called the ionic strength, is key. If it is small, caused by an ionic concentration of around 10^{-3} M, then $(1/\kappa)$ is about 10 mm, and the emulsion tends to be stable. If it is large, the result of say 1 M ions, then $(1/\kappa)$ is 0.3 nm, and the emulsion is unstable. This is why rivers form deltas. Particles of soil, washed into the river by erosion and stabilized with the anions of natural organic acids, are swept to the sea without much flocculation. There, the salt in the water reduces the electrostatic repulsion between particles, the particles flocculate, the flocs precipitate, and the soil deposits in the delta. We explore this topic and the others in this section in the examples that follow.

EXAMPLE 9.3–1   STERILIZATION OF GELS

The analysis of times for sterilization given above shows that the characteristic time for heating a container filled with solid or liquid is described by a time $(V \rho \hat{C}_p / U A)$. The analysis asserts sterilization for liquid contents should proceed for much longer than this characteristic time. An alternative to this analysis for a solid or gel filled can is to estimate the time τ required for heat to diffuse into the center of the can:

$$\tau = \frac{(d/2)^2}{4\alpha}$$

where d is the distance the heat must travel to be effective, and α is the thermal diffusivity of the gel. This time is much longer than that required to heat up a can of liquid, because in the liquid case heat transfer is dominated by convection, beyond a thin skin next to the can wall, whereas for a solid or gel filled can the only heat transfer mechanism is conduction.

To illustrate the difference in these two cases, we want to compare the characteristic times for heating a can filled with a low-viscosity liquid and with a gel. To make the comparison more specific, assume a metal can 8 cm in diameter d and 12 cm long.

SOLUTION

We first consider a can filled with low-viscosity liquid. The can's surface area A is

$$A = \pi dl + 2 \left(\frac{\pi}{4} d^2 \right)$$
$$= \pi (8 \, \text{cm})(12 \, \text{cm}) + 2 \left(\frac{\pi}{4} \right) (8 \, \text{cm})^2$$
$$= 400 \, \text{cm}^2$$

Its volume is

$$V = \left(\frac{\pi}{4}\right) d^2 l = \left(\frac{\pi}{4}\right)(8\,\text{cm})^2(12\,\text{cm})$$
$$= 600\,\text{cm}^3$$

Thus, the time τ is

$$\tau = \frac{V\rho\hat{C}_P}{UA} = \frac{\left(\dfrac{10^3\,\text{kg}}{\text{m}^3}\right)\left(\dfrac{4.2 \times 10^3\,\text{J}}{\text{kg K}}\right)(600 \times 10^{-6}\,\text{m}^3)}{\left(\dfrac{1000\,\text{J}}{\text{sec m}^2\,\text{K}}\right)(400 \times 10^{-4}\,\text{m}^2)}$$
$$= 63\,\text{sec} \approx 1\,\text{min}$$

This is a characteristic time to bring the can up to sterilization temperature. To ensure that all microorganisms are killed we will keep heating for several times longer than this, perhaps around 10 minutes; this is typical of the time used for canning foods at home.

The can filled with a gel is completely different:

$$\tau = \frac{(d/2)^2}{4\alpha} = \frac{(0.04\,\text{m})^2}{(4 \times 10^{-7}\,\text{m}^2/\text{sec})}$$
$$= 4000\,\text{sec} \approx 1\,\text{hr}$$

Again, this is a characteristic time for heating the can up to sterilization temperature. The actual time required will be considerably longer.

EXAMPLE 9.3–2   CRISPNESS OF A SNACK FOOD

We have been developing a new granola bar, which has wonderful customer acceptance but a questionable shelf life. Customers say the bar loses freshness. We need a way quickly to evaluate this loss. To do so, we make experiments at 50 °C and 40 °C at 100% relative humidity and find the bar stays fresh for 3 days and 24 days, respectively. We suspect this is due less to the higher temperature and more to the humidity, which is 95 g and 50 g H_2O vapor per kg dry air at these two temperatures. Estimate how long the bar will stay fresh at 25 °C, where the humidity is 20 g H_2O vapor per kg dry air.

SOLUTION

While this is like a problem in time – temperature superposition, our experience makes us think humidity, not temperature, is a key. We believe that this is why our competition packages their products in effective but expensive foil wrappers. We also suspect the decay of "freshness" is non-linear, getting faster and faster as the humidity rises. Thus, we guess the shelf life τ is given by

$$\tau = ae^{-bH}$$

where a and b are constants, and H is the humidity. Using the values given

$$\tau = 240e^{-0.046H}$$

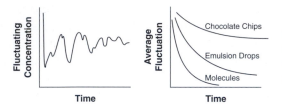

Figure 9.4–1 *Concentration fluctuations during mixing* For microstructures, fluctuations may remain significant, even after long times.

At a humidity of 20 g H_2O per kg air, this gives a shelf life of three months. Before we do any reformulation, we need experiments to confirm the humidity and temperature variations assumed by these equations.

9.4   The Microstructure Toolbox: Unit Operations

Like the other types of products described in this book, microstructured products are designed with a wide spectrum of tools. Their design includes struggles with specifications described in Section 9.2; and estimates of shelf life, in Section 9.3. These design efforts can also include separation processes and transport phenomena, like the diffusion of water and oxygen across packaging materials. Many of these operations are close parallels to similar challenges for other types of products.

Microstructured products also involve significantly different operations in fluid mechanics, especially in terms of mixing and diffusion. These are the operations described in this section. This material is organized around three topics. First, we will explore concentration fluctuations. After all, microstructures explicitly have structure, which implies that the concentration of different components must change at least over small volumes. We must estimate these changes.

Second, we are interested in mixing, both for turbulent and laminar flow. Such mixing is normally aimed at minimizing concentration fluctuations and producing a product which is nominally homogeneous. We will want to estimate how homogeneous this product will be, how long the mixing will take, and how stable the resulting mixture is. Third, we want to review how diffusion occurs, both across a thin film and into a semi-infinite slab. These two cases limit observed behavior. After discussing estimates of diffusion coefficients, we apply these ideas to packaging. We turn to these questions in the paragraphs that follow.

9.4–1 CONCENTRATION FLUCTUATIONS

Imagine we measure fluctuations during mixing as a function of mixing time. Initially, the fluctuations are large as the ingredients begin to blend. Later, these get smaller, oscillating around an average value of concentration, as shown in Figure 9.4–1. Eventually, the average size of the fluctuations reaches a steady-state limit, which can be described by a probability distribution. When we are blending two miscible species, like a solvent and a fully soluble solute, this average size of fluctuations goes to near zero. This is the case we normally consider for products other than microstructured ones.

However, if the species being blended are not of molecular dimensions, the average size of the fluctuations will not approach zero, even after a lot of mixing. For example, the concentration of chocolate chips in chocolate chip cookies will always vary, even after long mixing. Similarly, we anticipate microstructured products will also retain fluctuations in concentration. These fluctuations will be over smaller distances than those for cookies but larger than those for molecular mixtures.

Because these fluctuations can affect product quality, we want to estimate their size. To do so, we define N_i as the number of "particles" in a particular volume V. This number may be of molecules or emulsion droplets or chocolate chips. The average number in this volume $\bar{N}_i$ is the product of the average concentration n_i and the volume itself. We now seek the average variation of this number:

$$[\overline{(N_i - \bar{N}_i)^2}]^{1/2}$$

This is the fluctuation $(N_i - \bar{N}_i)$ (positive or negative), squared (always positive), averaged over all volumes (the overbar), and in the same units as the number itself (the square root).

Often, fluctuations in a fluid at steady state after mixing follow a Poisson probability distribution, defined as

$$p(N_i) = \frac{(n_i V)^{N_i} e^{-n_i V}}{N_i!} \tag{9.4-1}$$

where $p(N_i)$ is the probability that the volume V contains exactly N_i particles. (There is no simple theory for the distribution of fluctuations during mixing before a steady state has been reached.) If this probability is plotted vs. N_i, it begins at $N_i = 0$ with a small but finite value, rises to a maximum, and then decays to zero as N_i becomes large. We can use this distribution to calculate many of the properties of our system. For example, the average number $\bar{N}_i$ is

$$\bar{N}_i = \sum_0^\infty N_i p(N_i) = n_i V \tag{9.4-2}$$

We can also calculate the average size of the fluctuation:

$$\overline{(N_i - \bar{N}_i)^2} = \sum_{N_i=0}^\infty (N_i - \bar{N}_i)^2 p(N_i) = \sum_{N_i=0}^\infty (N_i - \bar{N}_i)^2 \frac{(n_i V)^{N_i} e^{-n_i V}}{N_i!}$$

$$= e^{-n_i V} \left[\frac{(0 - n_i V)^2 (n_i V)^0}{0!} + \frac{(1 - n_i V)^2 (n_i V)^1}{1!} \right.$$

$$\left. + \frac{(2 - n_i V)^2 (n_i V)^2}{2!} + \cdots \right]$$

$$= e^{-n_i V} (n_i V) \left[1 + \frac{n_i V}{1!} + \frac{(n_i V)^2}{2!} + \cdots \right]$$

$$= e^{-n_i V} (n_i V)[e^{n_i V}]$$

$$= n_i V \tag{9.4-3}$$

We really care about the size of the fluctuation relative to the average number of particles σ:

$$\sigma = \frac{[\overline{(N_i - \bar{N}_i)}]^{1/2}}{\bar{N}_i} = \frac{\sqrt{n_i V}}{n_i V} = \frac{1}{\sqrt{n_i V}} \tag{9.4-4}$$

The fractional size of the fluctuation equals the inverse square root of the product of concentration and volume over which the fluctuations are being considered. We will explore this result later in an example at the end of this section. At present, we note that the prediction is reasonable: the bigger we make the concentration or the volume over which we are considering the fluctuations, the smaller becomes the fractional fluctuation.

9.4–2 MIXING

The analysis above describes the fluctuations that come from having a microstructure of a certain concentration. This concentration is frequently determined by the size of the microstructure (the smaller the microstructure, the higher will be its concentration for a given volume fraction). This size is often the result of mixing. Over the next few paragraphs, we review how mixing produces these sizes.

For turbulent mixing, the size is found from correlations of experimental results. One such correlation is

$$\frac{\text{size}}{\text{impeller}} \propto \left(\frac{\text{surface force/area}}{\text{kinetic energy/volume}} \right)^{0.6}$$

$$\frac{1/a}{d'} = 0.04 \left(\frac{\gamma/d'}{\frac{1}{2}\rho(d'\omega)^2} \right)^{0.6} \tag{9.4-5}$$

where a is the drop area per volume, $(6/d)$ for spherical drops of diameter d; d' and ω are the size and angular velocity of the impeller; ρ is the density; and γ is the surface tension. Typical values could be

$$\frac{d/6}{10\,\text{cm}} = 0.04 \left(\frac{[20\,\text{g/sec}]/[10\,\text{cm}]}{\frac{1}{2}(1\,\text{g/cm}^3)\left([10\,\text{cm}]\left(\frac{2\pi 50}{\text{sec}}\right)\right)^2} \right)^{0.6}$$

$$d = 3.5\,\mu\text{m} \tag{9.4-6}$$

Such sizes may grow or "coarsen" with time, after mixing ceases. In its use of a force per area divided by the kinetic energy per volume, a correlation like this is an analog of the friction factor.

In addition to the size generated by turbulent mixing, we must also estimate the time required τ. Again, we must refer to correlations of experimental results, which often are successfully organized by plotting a dimensionless time vs. a Reynolds number

$$\tau\omega = f\left[\frac{(d')^2\omega}{\nu} \right] \tag{9.4-7}$$

Figure 9.4–2 *Mixing in a static mixer* Because of high non-Newtonian viscosity, the mixing is less like conventional stirring and more like shuffling cards.

where d' and ω are again the size and angular velocity of the impeller; and ν is the kinematic viscosity of the fluid being mixed. These results often show a constant value of $(\tau\omega)$ at the high Reynolds numbers required for turbulence. Thus, we just assume $(\tau\omega)$ is a constant, and that for good mixing in a tank of volume V, we need to move ten tank volumes:

$$\tau\omega = 10 \times \frac{\text{tank volume}}{\text{volume swept out by one impeller rotation}} \approx 10\frac{V}{(d')^3} \qquad (9.4\text{–}8)$$

This result works well for turbulent flow.

For laminar mixing, the results are very different. Most microstructured systems are highly non-Newtonian. In many cases, the materials being mixed are highly shear-thinning and so show the equivalent of a high yield stress. Mixing is now less like stirring cream into coffee than shuffling a deck of playing cards. We don't reach a steady state so much as make thinner and thinner cards.

Two methods used for this operation are static mixers and twin-screw extruders. Static mixers are like airplane propellers fixed within a tube. The size produced l, called a "cut size," is given by:

$$l = \frac{d}{2^\nu} \qquad (9.4\text{–}9)$$

where d is the tube diameter and ν is the number of static mixer elements. The results of such a mixer are shown in Figure 9.4–2. The results for a twin-screw extruder are similar:

$$l = \frac{v}{\omega} \qquad (9.4\text{–}10)$$

where v is the pressure-driven velocity in the tube and ω is the angular velocity of the extruder's screw. For some food products, v is around 1 cm/sec, and ω is 10/sec, so the cut size l is about 0.1 cm. Thus, for both laminar and turbulent mixing estimates for design are based on past experience. Product manufacture will usually require additional experiments to confirm and refine these estimates.

9.4–3 DIFFUSION

The last topic we review in the microstructure toolbox is diffusion and mass transfer. While this topic frequently appears in many parts of the chemical sciences, the applications important here are not those important elsewhere. We want to describe simple diffusion models, values of diffusion coefficients, and common problems.

Diffusion is normally described in terms of two important limiting cases. The first, more important case, occurs for steady-state diffusion of dilute solute "1" from a fixed reservoir of constant molar concentration c_{10} across a thin film into a second reservoir at molar concentration c_1. The solute flux j_1, in moles per area per time, is given by

$$j_1 = \frac{D}{l}(c_{10} - c_1) = k(c_{10} - c_1) \tag{9.4–11}$$

where D is the diffusion coefficient, l is the film thickness, and k is the mass transfer coefficient. The diffusion coefficient D, with dimensions of length squared per time, is the language of science; the mass transfer coefficient k, with dimensions of velocity, is the dialect of engineering; but they are the same basic idea. If the thin film is made of a different material than the solutions, then D should be replaced by the permeability DH, where H is the equilibrium partition coefficient between the film and the adjacent solution; also, k is replaced by the permeance DH/l. This steady-state case is central to perhaps 80% of diffusion problems.

The second important case is diffusion from a stirred reservoir at a concentration c_{10} into a large unstirred volume that initially contains the concentration c_1. While this case is less important than the previous thin-film one, useful in perhaps 10% of diffusion problems, it is mathematically harder. The interfacial flux j_1 calculated in this case is

$$j_1 = \sqrt{\frac{D}{\pi t}}(c_{10} - c_1) = k(c_{10} - c_1) \tag{9.4–12}$$

where t is the time that the solution and reservoir have been in contact. This unsteady-state result is a contrast with the thin-film, steady-state result described by Equation 9.4–11.

These two results form the basis for many analyses which allow design estimates to be easily made. We consider three examples. First, consider diffusion across an interface of a solute undergoing a first-order irreversible chemical reaction described by a rate constant k_R. The flux j_1 in this case is

$$j_1 = \sqrt{Dk_R}(Hc_{10} - 0) \tag{9.4–13}$$

In many cases, the new mass transfer coefficient $\sqrt{Dk_R}$ is much larger than k, the value without reaction. This effect is a key in any effort to capture and sequester carbon dioxide. As a second example, imagine a rapid reversible reaction producing an immobile product. The flux j_1 is now

$$j_1 = \sqrt{\frac{D(1 + K)}{\pi t}}(Hc_{10} - 0) \tag{9.4–14}$$

where K is the equilibrium constant for the reversible reaction. This effect is basic to the capture of fluorine-containing species in teeth and to the dyeing of wool and human hair.

The third example is more general, meriting more careful development. Imagine we have a package of fixed volume V and surface area A. The package originally contains no water vapor. Starting at time zero, it is stored in air at constant humidity c_{10}. We want to find the humidity c_1 inside the package as a function of time, and so understand how this humidity can be controlled.

To find this, we write a water balance on the humidity in the package

$$\text{water accumulation} = \text{water diffusing in} \tag{9.4–15}$$

$$V\frac{dc_1}{dt} = Aj_1 = A\left(\frac{DH}{l}\right)(c_{10} - c_1)$$

This is subject to the initial condition.

$$t = 0, \qquad c_1 = 0 \tag{9.4–16}$$

Integrating, we find the desired result

$$c_1 = c_{10}\left(1 - e^{-\frac{A}{V}\left(\frac{DH}{l}\right)t}\right) \tag{9.4–17}$$

The concentration c_1 rises from zero at time zero to c_{10} at large time.

The key to this rise is the time (Vl/ADH). If we seek to keep the contents of the package dry, we can do so by making this time large. We can achieve this in a variety of ways. We can make the package's wall thicker, i.e. we can make l larger. We can use as a packaging material a metal foil, for which D is very small. We can switch from cellophane to polyethylene, thus making H smaller. This simple analysis suggests many routes to an effective design.

This example leads us to a discussion of the values of diffusion coefficients. For fluids, these coefficients don't vary much but cluster around 10^{-5} m²/sec for gases and 10^{-9} m²/sec for liquids. For polymers, good first guesses are 10^{-12} to 10^{-15} m²/sec. For solids, diffusion coefficients at room temperature are much smaller, 10^{-20} m²/sec or less, though these can be strong functions of temperature.

For microstructured materials, the diffusion is dominated by the continuous phase. To illustrate this, imagine we have a continuum containing periodically positioned spheres. If diffusion in the spheres is infinitely rapid, then the diffusion coefficient D in the composite is

$$\frac{D}{D_0} = \frac{1 + 2\phi}{1 - \phi} \tag{9.4–18}$$

where D_0 is the diffusion coefficient of the continuum and ϕ is the volume fraction of the spheres. Thus, if we have 10% spheres, ϕ is 0.1 and D equals $1.3D_0$. Alternatively, if the diffusion in the spheres is zero, then

$$\frac{D}{D_0} = \frac{1 - \phi}{1 + \phi/2} \tag{9.4–19}$$

If we have 10% of these spheres, then ϕ is again 0.1 and D is $0.86D_0$. Thus, changing diffusion in the spheres from zero to infinity changes the diffusion in the composite only by 50%. The continuous phase is dominant.

These results also show that the size of the microstructure does not affect the diffusion coefficient. These results are an analog of those for viscosity, given in Equation 9.3–13, which says that the viscosity of a suspension depends on the volume fraction of the particles, and not on their size. There also, the continuous phase was key, the volume fraction was important and the microstructure size was not. The effect of the microstructure's shape on diffusion is also small. For example, for impermeable cylinders aligned parallel to the surface of a thin film,

$$\frac{D}{D_0} = \frac{1-\phi}{1+\phi} \tag{9.4–20}$$

If ϕ is again 0.1, D equals $0.82D_0$, within 5% of the value of spheres. If the cylinders cross a thin film and are perpendicular to the film's surfaces, then D is $0.9D_0$ when ϕ equals 0.1. The continuous phase dominates the diffusion in microstructures. We reinforce these and the other ideas presented in this section with a variety of examples.

EXAMPLE 9.4–1   FLUCTUATIONS IN CONCENTRATION

If you drink a glass of 2% fat milk, you ingest $10\,cm^3$ swallows with the fat as $6\,\mu m$ droplets. When you eat a chocolate chip cookie, you consume 0.6 cm chocolate chips present at a concentration of 4% with $5\,cm^3$ bites. How much do the fat droplet and chocolate chip concentrations vary?

SOLUTION

For both cases, the concentration in particles per volume is

$$n_i = \frac{\phi}{\frac{4}{3}\pi R^3}$$

where ϕ is the volume fraction and R is the particle radius. Thus, from Equation 9.4–4, for the milk

$$\sigma = \frac{1}{\sqrt{n_i V}} = \frac{1}{\left[\dfrac{0.02}{\frac{4}{3}\pi(3\times 10^{-4}\,cm)^3}(10\,cm^3)\right]^{1/2}}$$

$$= 0.00002$$

There is almost no variation between gulps. For the cookie,

$$\sigma = \frac{1}{\sqrt{n_i V}} = \frac{1}{\left[\dfrac{0.04}{\frac{4}{3}\pi(0.3\,cm)^3}(5\,cm^3)\right]^{1/2}}$$

$$= 0.75$$

There is very substantial variation in the amount of chocolate between mouthfuls. For a problem of greater technical interest, imagine you measure microbial concentrations in $1\,cm^3$ samples of pond water containing, on average, 10^5 microbes per m^3. Then

$$\sigma = \frac{1}{\sqrt{n_i V}} = \frac{1}{\left[\dfrac{10^5}{\text{m}^3}(10^{-6}\,\text{m}^3)\right]^{1/2}}$$
$$= 3.2$$

Even if your measurements are perfect, your data will show huge scatter.

EXAMPLE 9.4–2   MIXER SCALE-UP

We have been making a skin cleaner in 230 cm^3 batches in the lab with very positive results. We want to prepare 5 L batches for larger test panels. However, we have tended to overmix these larger batches so that the larger amounts phase separate over a weekend. The small batches are stable for over a month.

At present, our small mixer has a height of 15 cm, a diameter of 5 cm, and a 1 cm impeller turning at 42 rpm. The mixer runs for only 150 sec. How should the larger mixer be operated?

SOLUTION

We are scaling up the volume by 5000/230, or a factor of about 20. This means the mixer height, diameter, and impeller dimensions should increase by $20^{1/3}$ or about 2.8. We want to use the same Reynolds number so

$$\left[\frac{(d')^2\omega}{\nu}\right]_{\text{large}} = \left[\frac{(d')^2\omega}{\nu}\right]_{\text{small}}$$

where d' is a characteristic size, like the reactor diameter; ω is the angular velocity; and ν is the kinematic viscosity. Thus the new stirring speed should drop by a factor of $(2.8)^2$, to 5.4 rpm. But the time for mixing τ is inversely proportional to this speed, so

$$\tau = \frac{(42\,\text{rpm})(150\,\text{sec})}{(5.4\,\text{rpm})}$$
$$= 1200\,\text{sec}$$

We must mix more slowly for a longer time.

EXAMPLE 9.4–3   PERFUME IN TOILET SOAP

Perfume is a large fraction of the cost of toilet soap, but much of the perfume evaporates before the soap ever reaches the consumer. We want to reduce this effect by putting the perfume in small, solid bubbles that dissolve when the soap is being used. We imagine these bubbles could be made of something like sucrose, using techniques for making candy.

How long will the bubbles take to dissolve and release the volatile perfume?

SOLUTION

The bubbles' walls will dissolve according to the equation

$$j_1 = \frac{D}{R}(c_1(\text{sat}) - 0)$$

where $c_1(\text{sat})$ is the walls' solubility in water, and the bubble radius R replaces the film thickness l in Equation 9.4–11. If the walls are thin, R is approximately constant during dissolution. Thus

$$(\text{wall loss}) = (\text{flux away from wall})$$

$$c_{10}A\frac{\mathrm{d}l}{\mathrm{d}t} = -A\frac{D}{R}c_1(\text{sat})$$

where c_{10} is the concentration of solid wall material. When the dissolution starts, the wall thickness is l_0, so integrating this mass balance gives

$$l = l_0 - \left[\frac{D}{R}\frac{c_1(\text{sat})}{c_{10}}\right]t$$

The bubble breaks open when the thickness l is zero.

9.5   Using the Microstructure Toolbox

In the earlier sections of this chapter, we reviewed the properties of microstructured products and the various tools for their description. These tools include the choice of ingredients, the estimation of shelf life, and the equipment used for mixing highly viscous materials. The tools are drawn from a variety of disciplines, including materials science, statistics, colloid chemistry, and chemical engineering.

These properties and tools are part of the basis for the design of microstructured products. The design is more complicated than the design of the other types of product because it is usually limited by two steps in the design template of needs, ideas, selection, and manufacture. Microstructured product design is often limited by the needs step, because customer needs are frequently difficult to rephrase as product specifications. The design of microstructured products is often also limited by the selection step, because the choice of the best-working alternative requires balancing technical and subjective factors.

We give examples of such designs in this last section. The designs are idealized, but less than those examples given in the earlier sections of this chapter. There, the simple examples aimed to illustrate a single technical point. Now, the examples seek to show the opportunities for invention of microstructured products.

EXAMPLE 9.5–1   SENSORY EXPERIENCE OF DRINK IN THE THROAT

We seek a contract from a multi-million-dollar beverage manufacturer, who has decided on a significant investment in connecting "the sensory experience of drinking" with "the physical properties of the drink." The manufacturer requests

Table 9.5–1   *Three stages of studying throat-feel* The three stages seek a rough survey, details of consumer response, and a model to predict new product attributes.

	Fluids	Descriptors	Physical/Chemical
Stage 1	Three beers, three colas, two milks	"Thick," "smooth," "like," "satisfy"	Average viscosity, temperature in mouth, in throat
Stage 2	One beer (three samples, each with different carbonation), one milk shake, one cola (with additives for yield stress)	"Thick" "Smooth," "creamy"	Viscosity vs. shear, thermal conductivity and diffusivity, coefficient of friction
Stage 3	One fluid with different physical properties		Model fluid mechanics and phase transitions

proposals which explain why we feel "satisfied" and "exhilarated" when swallowing. They think that this understanding of throat-feel will take three people about a year to achieve.

We want to write such a proposal.

SOLUTION

This proposal centers on the translation of needs into physical specifications, and as such exemplifies the difficult "needs" step of microstructured product design. We plan three separate stages for this project, summarized in Table 9.5–1. These stages reflect the conclusion stated by the client, "that seeking [this] mechanism . . . will be difficult." In the first stage, lasting a month, we will guess liquids and descriptors, and use these to define benchmarks. In the second stage, lasting perhaps six months, we will identify a vocabulary of descriptors, and determine how these are related to physical properties. If the results are promising, we will attempt a third stage of experiments which develop a theory for "throat-feel."

Each stage involves three sets of choices. First, we must choose the drinks. Second, we must decide how drinking will be described. Third, we must choose the physical measurements which will be made. Once these decisions are complete, we must decide on the scale used to measure the descriptors, make the measurements of the descriptors, and correlate these descriptors with the physical properties. These choices and correlations will be different for each of the stages, as described below.

Stage 1 Determining the benchmarks. This preliminary stage will use three common, commercially available drinks: beer, soda, and milk. Two beers should be

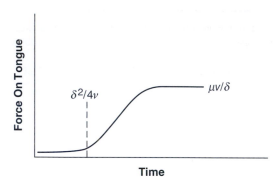

Figure 9.5–1 *Force in the mouth while swallowing beer* The force $\mu v/\delta$ is felt only after a time $\delta^2/4v$. Such a force contributes to beer "creaminess."

chosen which are as different as possible. The sodas should be a cola, a lemon-flavored product, and tonic water. Milk should contain 1% fat or 4% fat from the same supplier.

These drinks will be evaluated by 15–20 untrained persons on the basis of three hedonics and three descriptors. The hedonics are to find what each person likes, doesn't like, and finds most satisfying. (The descriptor "exhilarating" suggested by the client seems hyperbole, at least at this point.) Those being tested will also be asked to rank the samples in terms of three other descriptors: thickness, smoothness, and sweetness. The rankings will be simple comparisons: for example, which drink is sweeter? These data will be compared with evaluations from the literature. After these efforts, we will have defined the region of psychological and physical properties on which we want to focus.

Stage 2 Initial quantification. In the second stage we will extend these ideas with more quantitative measurements. We will replace the milks with a milkshake. We will choose one beer, but carbonate this with CO_2, N_2, or Ne, seeking to understand "smoothness," said to be affected by the gas used for carbonation. We will choose one cola, but use additives which make its viscosity shear thinning. Tentatively, we will drop "like" and "satisfy" and add other descriptors like "creamy" and "crisp." As an aid in this effort, we will use the extensive descriptions of beverages in the literature as a guide. Ideally, we seek independent descriptors, not synonyms or antonyms. Unfortunately, we have little basis for these choices beyond our own judgment. In this second stage, we expect to replace simple comparison scaling with category scaling or ranking. In other words, instead of saying, "beer A is smoother than beer B," we are seeking lists such as, "the smoothness of the liquids is in the following order: beer A > cola C > beer B = milkshake > cola A."

We also expect that the physical and chemical measurements will be more complex. In particular, we will measure apparent viscosity as a function of time and temperature. We expect variations, both because of non-Newtonian behavior of the fluid itself, and because of phase changes in the systems. For example, if we measure the force versus time caused by suddenly applying a fixed velocity to a sample of beer, we may get results like those in Figure 9.5–1: the force remains small for a lag time, and then rises to reach a steady value. This force may be

> **Table 9.5–2 *Parameters used in model of throat-feel*** These are
> used in the third stage of this project.
>
> Pharynx: Cylindrical tube 2 cm in diameter and 10 cm long
> Flow: Velocity 10 cm/sec
> Viscosities: 2×10^{-5} kg/m sec (2×10^{-5} m²/sec) for gases
> 10^{-3} kg/m sec (10^{-6} m²/sec) for liquids
>
> Thermal conductivity and diffusivity:
>   0.024 W/mK (2×10^{-5} m²/sec) for gases
>   0.6 W/mK (2×10^{-7} m²/sec) for liquids

altered by any phase changes due to release of gas from carbonation or melting, either of ice or of fat. This implies measuring forces of a cold fluid – the beverage – in contact with a warm viscometer – the throat. We do not suggest that this particular analysis is exactly what we will want to use; we do assert that this type of thinking is what is needed.

Stage 3 Fluid mechanics of drinking. Finally we try to understand what is being perceived so that we can duplicate or improve the perception by adjusting physical properties. We begin by assuming that what is perceived is force and temperature. To illustrate arguments like those we expect to use, we consider force and temperature in the pharynx, that portion of the throat sitting directly behind the mouth. We assume that flow in the pharynx is caused by gravity, by pressing the tongue towards the roof of the mouth, and by peristaltic motion in the throat. We assume also the geometric and physical properties shown in Table 9.5–2.

We now estimate the flow, force, and temperature in the pharynx for a simple liquid and for a carbonated beverage. The simple liquid is homogeneous. In contrast, the carbonated beverage foams when it hits the warm pharynx, producing a central liquid bolus lubricated by a layer of approximately 0.03 cm bubbles on the pharynx wall.

We can compare these two liquids on three bases. First, we use a Reynolds number Re to characterize the type of flow. For the simple liquid,

$$\text{Re} = \frac{dv}{\nu} = \frac{(0.02\,\text{m})(0.1\,\text{m/sec})}{(10^{-6}\,\text{m}^2/\text{sec})} = 2000$$

where d is the pharynx diameter, v is the liquid velocity, and ν is the liquid's kinematic viscosity. This value is close to the laminar–turbulent transition, where the effort needed to swallow will get much greater.

In contrast, for the carbonated case,

$$\text{Re} = \frac{dv}{\nu} = \frac{(3 \times 10^{-4}\,\text{m})(0.1\,\text{m/sec})}{(2 \times 10^{-5}\,\text{m}^2/\text{sec})} = 1.5$$

where d is the thickness of the gas layer, v is again the liquid velocity, and ν is the kinematic viscosity, but now of the bubbles. This suggests laminar flow, smooth and without upsets.

These different flows cause different stresses τ in the throat. For the simple liquid,

$$\tau = \frac{8\mu v}{d} = \frac{8 \times (10^{-3}\,\text{kg/m sec})(0.1\,\text{m/sec})}{(0.02\,\text{m})} = 0.04\,\text{kg/m sec}^2$$

where again d is the pharynx diameter, v is the liquid velocity, and μ is the liquid viscosity. For the carbonated beverage,

$$\tau = \frac{\mu v}{d} = \frac{(2 \times 10^{-5}\,\text{kg/m sec})(0.1\,\text{m/sec})}{(3 \times 10^{-4}\,\text{m})} = 0.007\,\text{kg/m sec}^2$$

where d is the thickness of the gas layer, v is the liquid velocity, and μ is now the viscosity of the bubbles. There is six times less stress with carbonation.

Thermal effects involve a heat transfer coefficient h, which for the simple liquid is:

$$\frac{hd}{k} = 1.62 \left[\frac{d^2 v}{L\alpha}\right]^{1/3}$$

where k and α are the liquid's thermal conductivity and diffusivity, and L is the pharynx length, about 10 cm. Thus,

$$h = 1.62 \left(\frac{0.6\,\text{W/m K}}{0.02\,\text{m}}\right) \left[\frac{(0.02\,\text{m})^2(0.1\,\text{m/sec})}{(0.1\,\text{m})(2 \times 10^{-7}\,\text{m}^2/\text{sec})}\right]^{1/3}$$

$$= 610\,\text{W/m}^2\,\text{K}$$

The carbonated case is very different:

$$h = \frac{k}{d} = \frac{0.024\,\text{W/m K}}{3 \times 10^{-4}\,\text{m}}$$

$$= 80\,\text{W/m}^2\,\text{K}$$

The simple liquid will feel cooler in the throat than the carbonated drink.

These calculations are not exact; they are estimates. However, we feel that physical models like these can be used to analyze the measurements made in this work, and to suggest new ideas for improved product design.

EXAMPLE 9.5–2   "FRUITY" BREAKFAST CEREAL

Consumer surveys repeatedly stress the public's desire for healthy, natural breakfast cereal. One descriptor which is valued especially highly is "fruity." As a result, our company is interested in developing "fruity flakes." The product will be sold as a lower-priced "house brand," competing with products of Kellogg's and General Mills. The company's marketing group has obtained test data for two doughs based on corn flour and on oat flour. These two doughs have been paired with two flavor packages. Data from these four combinations are shown in Table 9.5–3. These customer scores were obtained with an ordinal scale from 1 to 9, where 1 is the worst and 9 is the best. Our marketing group believes that these data are reliable.

Table 9.5–3 *Product test data with alternative ingredients* The unflavored oats gave a score of 5.3. The flour and flavor are believed to contribute 60% and 40% of the evaluations, respectively.

Customer	Corn flour and flavor #1	Corn flour and flavor #2	Oat flour and flavor #1	Oat flour and flour #2
1	7	9	7	6
2	6	9	8	6
3	5	7	5	8
4	9	7	6	8
5	7	8	8	6
6	4	6	3	4
7	5	6	4	2
8	8	4	3	4
9	4	6	2	4
10	6	5	4	5

We want to outline a possible product and its manufacture. To do so, we must choose a particular dough and flavor combination. We then need to develop a possible process diagram, including a protocol for mixing the ingredients and the flavor. Finally, we want to assess any technical risks, and explore how these can be mitigated.

SOLUTION

In this example, we must both define the need and select a process to meet this need. From previous studies, we expect that:

$$\begin{pmatrix} \text{customer} \\ \text{response} \end{pmatrix} = 0.6 \begin{pmatrix} \text{flour} \\ \text{value} \end{pmatrix} + 0.4 \begin{pmatrix} \text{flavor} \\ \text{value} \end{pmatrix}$$

From the data given, the corn flour scores 7.1, more than the oat flour score of 5.3. Flavor package 1 has a value of 4.6, less than flavor package 2, valued at 5.4. Thus, our best product should be made of corn flour and flavor package 2.

We can find excellent instructions on how to make flaked cereals in the literature from the manufacturers of equipment for making these products. The basic procedures are familiar to anyone who has ever made cookies. First, we blend the dry ingredients, including any butter or other fat. Second, we mix the liquid into the solids, making a paste. Third, we bake the cookies, immediately for drop cookies and after chilling for refrigerator cookies.

Flaked cereal is similar, but with more complicated cooking. First, we mix by weight corn grits with other ingredients, including sugar, salt, and malt. We add water. The resulting dough is cooked in pots 4 feet in diameter and 8 feet long, rotating at 2 rpm. The rotation, necessary to break up lumps, continues for two hours. The cooked dough is dried to 12% water and then "tempered" to make sure its water concentration is uniform. Then the dough is flaked between rollers 20 inches in diameter and 30 inches long. Finally the flakes are toasted at 575 °F

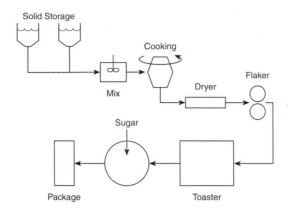

Figure 9.5–2 *A Process for making flaked breakfast cereal* This is inferred from equipment said to be made for this purpose.

in a rotating drum 4 feet in diameter and 15 feet long. Our only real process innovation is to add the flavor and sugar in this last step. We want as much flavor as possible on the flakes' surface, so that we get the most flavor with the least cost and calories. The danger is that, when covered with milk, the first bite of the cereal will taste much better than later ones. This is often a problem with this type of product.

A possible process flow chart for our cereal is shown in Figure 9.5–2. We can estimate the mixing times and flake-forming physics using correlations like those in Section 9.4. However, we will certainly need experiments, best performed in collaboration with the equipment manufacturers. These people will be in ethically ambiguous situations, because some of them will have installed equipment for our competitors. We must be sensitive to this.

Developing this product has only minor technical risk. Our main concerns are the amount of steam needed in the cooking, and the sugar and flavor addition in the flaker. We can mitigate the first risk by careful conversations with the equipment manufacturer. The latter risk will require experiments, probably on a small-scale pilot plant.

EXAMPLE 9.5–3 BROCCOLI PACKAGING

Fruits and vegetables continue metabolism after they are picked. This metabolism produces gases which can accelerate or retard further ripening. Typically, O_2 is consumed and CO_2 is produced at a constant ratio, called the respiratory coefficient, which is equal to 1.2 for most vegetables. For broccoli, the CO_2 production is about 35 mg/hr/kg; the O_2 consumption is 50 mg/hr/kg (Wareham and Persaud, 1999). Other gases can also be important. This is why some suggest enhancing the ripening of winter tomatoes by placing them under a bunch of bananas: the ethylene released during banana ripening accelerates the ripening of tomatoes (and avocados).

We want to develop a 1 L, controlled-atmosphere package for 0.75 kg broccoli. This package should adjust gases within the package to 2% O_2 and 7% CO_2, which is optimal for storage. In meeting this goal, we decide to make a package with two different polymer films. One polymer may be porous. Since this product has needs defined in scientific terms, it is less ambiguous than the previous two examples, and centers on the selection step of the template.

Design the package.

SOLUTION

The need is clearly phrased in scientific terms: a 1 L package for 0.75 kg broccoli with a concentration of 2% O_2 and 7% CO_2. We want to develop a basis for selecting the polymers to be used in this package. To do so, we begin with mass balances on each gas in the package and then see what flexibility we have in choosing polymers. For each gas, the mass balance is

$$V\frac{dc_1}{dt} = r + kA(c_{10} - c_1)$$

where c_1 and c_{10} are the concentrations of species "1" inside and outside of the package; V and A are the package volume and surface area; k is the permeance of the wall of the package; r is the metabolic rate; and t is the time. If the package has 800 cm^2 of area and is in steady state, the balance for oxygen is

$$0 = -0.75 \, \text{kg} \left(\frac{50 \times 10^{-3} \, \text{g}}{\text{kg} \, 3600 \, \text{sec}} \right) \frac{1 \, \text{mol}}{32 \, \text{kg}} + k_{O_2}(800 \, \text{cm}^2)(0.21 - 0.02)$$

$$\times \frac{\text{mol}}{22.4 \times 10^3 \, \text{cm}^3}$$

$$k_{O_2} = 4.8 \times 10^{-5} \, \text{cm/sec}$$

The corresponding result for carbon dioxide is

$$0 = -0.75 \, \text{kg} \left(\frac{35 \times 10^{-3} \, \text{g}}{\text{kg} \, 3600 \, \text{sec}} \right) \frac{1 \, \text{mol}}{44 \, \text{kg}} + k_{CO_2}(800 \, \text{cm}^2)(0 - 0.07)$$

$$\times \frac{\text{mol}}{22.4 \times 10^3 \, \text{cm}^3}$$

$$k_{CO_2} = 6.6 \times 10^{-5} \, \text{cm/sec}$$

Thus, we need a polymer film which has the two permeabilities given. Such a film is hard to find.

As a result, our core team has already decided to make a package with two different polymer films. This will provide much more flexibility than a single film, although the package cost will also probably increase. The core team has decided one polymer film A should be non-selective, and the second film B must have a selectivity greater than (6.6/4.8 = 1.4), the selectivity ratio needed.

We need a criterion to decide on the relative areas of the two polymers. For the non-selective polymer A, we choose microstructured non-woven films of polyethylene, which typically have a pore size of 8 μm, a void fraction ε of 0.3, and a thickness of 300 μm. The permeance of this material is about

$$k_A = \frac{\varepsilon D}{l} = \frac{0.3(0.1 \, \text{cm}^2/\text{sec})}{300 \times 10^{-4} \, \text{cm}}$$

$$= 1 \, \text{cm/sec}$$

where D is a diffusion coefficient typical of bulk gases. In this estimate, we have used the bulk gas value because the pores in the film are so large. In fact, the value of k_A will probably be about three times smaller because of the resistances of adjacent boundary layers.

We next note that the permeance k_1 is the result of the parallel resistances of the two polymers:

$$k_1 = \phi k_A + (1 - \phi)\frac{P_B(1)}{l_B}$$

where ϕ is the fraction of the package made from the porous film; l_B is the thickness of the non-porous film; and $P_B(1)$ is the permeability of the non-porous film to diffusing gas "1". We have one such equation for O_2 and one for CO_2. If we choose a polymer and so specify $P_B(O_2)$ and $P_B(CO_2)$, we can then estimate the fraction of porous polymer ϕ and the thickness of non-porous polymer l_B. This gives us an initial design. We must next decide how to manufacture such a composite material and fabricate it into an appropriate bag.

EXAMPLE 9.5–4   CHEWING-GUM FLAVOR

> Does your chewing gum lose its flavor on the bedpost overnight?
> If your mother says don't chew it, do you swallow it in spite?
> Can you catch it on your tonsils, can you heave it left and right?
> Does your chewing gum lose its flavor on the bedpost overnight?
> M. Bloom, E. Breur, and B. Rose, 1958

Chewing gum is a confection made of chicle, a tree sap, and synthetic rubber, especially butyl rubber. This is blended with about 30 percent sugar in corn syrup and flavoring. While these products are popular, they share a major problem: the flavor in the gum is gone after five minutes. Gum manufacturers seek a new formulation which gives a second flavor burst after ten minutes and which hopefully lasts as long as twenty minutes.

The gum manufacturers are unusually specific about their aims. In addition to the flavor burst, they want the new ingredients to be at least 20 percent flavor. They want any particles added to be less than 100 µm in diameter, and to maintain "desirable mouth-feel." They want the product to be stable for two years. Finally, they want to use only ingredients "Generally Recognized As Safe" (GRAS) by the Food and Drug Administration (FDA).

The obvious answer to this problem is microencapsulation, in which flavor particles are coated with a solid wall. The wall may be wax, fat, sugar, or zein. The flavor may diffuse slowly across the wall, or the wall may suddenly dissolve to release the flavor. However, while microencapsulation is widely used in the food and pharmaceutical industries, gum manufacturers say that it is a failure for chewing gum. No reason why is given; perhaps the repetitive chewing fractures the microcapsules.

Our core team intends to solve this problem. As a benchmark flavor, we choose methyl salicylate. Methyl salicylate, which melts at −9 °C, is the flavor in wintergreen and teaberry. We seek a flavor burst with this molecule after ten

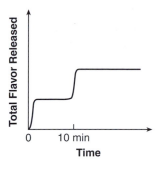

Figure 9.5–3 *Desired flavor release in chewing gum* We seek a second burst of flavor after 10 minutes.

minutes of chewing. To achieve this end, we seek ideas to give this burst. Next, we develop a model for this burst, and identify what the parameters in the model mean. We will select the best ideas which can give this behavior by matrix scaling. Finally, we summarize how we can develop the product further.

SOLUTION

Like the previous problem, this one has relatively well-defined needs. In saying this, we are implicitly assuming that flavor perception is proportional to flavor concentration. While this is reasonable, we also know that the perception of taste tends to saturate, and that changes in flavor concentration may be more easily perceived than a steady flavor concentration. We may be best off with a series of pulses, even though we are seeking a single-flavor jolt.

Thus the tentative goal of this work is to extend flavor release in chewing gum by flavor burst after about ten minutes, as suggested by the sketch in Figure 9.5–3. After significant effort, the core team has reduced the ideas down to four groups, as suggested in Figure 9.5–4. These are permeable spherical micro-capsules ("spheres"), soluble coatings ("coating"), osmotic pumps ("pump"), and hydrolysis reactions ("hydrolysis"). These alternatives will be described and compared in more detail in the paragraphs below. To make this discussion more specific, we use as a model flavor methyl salicylate, as suggested above. Our ability to estimate unknown physical properties is aided by the chemical similarity of this compound to aspirin, whose properties are widely studied. We now turn to the four ideas.

Spheres. Our first idea, spherical microcapsules, consists of droplets of flavor coated with a flexible, permeable polymer coating. The release of flavor from such a sphere depends on the rate of diffusion of the flavor through the polymer coating. For a loading ϕ of 5% of 100 μm capsules, with a diffusion coefficient D equal to 10^{-7} cm^2/sec, a solubility H of 10, and a 30 μm wall thickness δ, we expect a characteristic time for release of

$$\tau = \frac{\delta}{DHa} = \frac{\delta}{DH\left(\frac{\phi}{\frac{\pi}{6}d^3}\right)\pi d^2} = \frac{\delta d}{6DH\phi}$$

$$= \frac{(30 \times 10^{-4}\,\text{cm})(100 \times 10^{-4}\,\text{cm})}{6 \times (10^{-7}\,\text{cm}^2/\text{sec}) \times 10 \times 0.15} = 100\,\text{sec}$$

1. Coated Microcapsule ("Sphere")

Flavor release
by diffusion

2. Soluble, Impermeable Coating ("Coating")

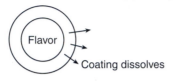

Coating dissolves

Figure 9.5–4 *Ideas for flavor bursts in chewing gum* Coated flavor particles are judged to have the greatest merit.

3. Osmotic Pump ("Pump")

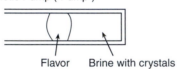

Flavor   Brine with crystals

4. Hydrolysis Reactions ("Hydrolysis")

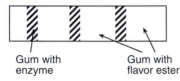

Gum with   Gum with
enzyme   flavor ester

This is faster than we wish. While we could increase this time by choosing a different polymer or a thicker wall, we still would not get the burst after ten minutes which we seek. This idea is a failure and may be why the manufacturers regard microencapsulation negatively.

Coating. The second idea in Figure 9.5–4 is a sparingly soluble 20 μm coating around a 100 μm drop of flavor. The coating is not hygroscopic but is sparingly soluble in water so that it slowly dissolves in the surrounding solution. Because the coating is glassy or crystalline, it is essentially an impermeable barrier for the flavor, which is released in a burst only once the coating is dissolved.

The time required to dissolve the coating t_{lag} is given by

$$c_2(\pi d^2)\delta = k(\pi d^2)c_2(\text{sat})t_{lag}$$
$$t_{lag} = \frac{\delta c_2}{kc_2(\text{sat})}$$

where δ is the coating thickness, 20 μm; k is the mass transfer coefficient from the capsule into the surrounding saliva, about 10^{-3} cm/sec; and c_2 and $c_2(\text{sat})$ are the concentration of the solid coating, and its saturation concentration in saliva. For example, we might choose a coating with physical properties like those of benzoic acid. This has a solubility in water of about 3.2 g/L at ambient conditions and a solid density of 1.3 g/cm. The key time is then

$$t_{lag} = \frac{(20 \times 10^{-4}\,cm)(1.3\,g/cm^3)}{(10^{-3}\,cm/sec)(3.2 \times 10^{-3}\,g/cm^3)} = 813\,sec \approx 14\,min$$

This method gives a burst of flavor near the time desired; moreover, a 20 μm coating on a 100 μm particle represents a flavor loading of over 50% by volume, well above the target of 20%. While benzoic acid is used as a food preservative, it is too toxic for this purpose. We will need an alternative coating. Nonetheless the coating idea has promise.

Pump. The third idea, an osmotic pump, is the most complex, the most expensive, the most generally applicable, and the most difficult to manufacture. It is based on a short length, perhaps 200 μm, of a 200 μm diameter cellulose hollow fiber which is sealed at one end and partly filled with a slurry of salt in salt water. Such fibers are cheap, used for kidney dialysis. Next to the salt water is a drop of flavor; next to the flavor is an air-filled volume.

When this device is wet with saliva, water diffuses across the fiber's wall into the saline. The increased volume pushes the flavor along the hollow fiber. After a while, the flavor is ejected as if from a microscopically sized syringe. The rate of progress of material up the micro "syringe" is approximately given by the following mass balance:

$$\rho \frac{\pi}{4} d^2 v = j_2[\pi\,dl] = \frac{DHc_1(sat)}{\delta}[\pi\,dl]$$

where ρ is the density of the brine; d is the device diameter; v is the velocity of the flavor drop; l is the average length of the brine region; j_2 is the water flux into the brine; (DH/δ) is the water permeance of the fiber wall; and $c_1(sat)$ is the solubility of salt in water. Thus the lag time t_{lag} will be approximately:

$$t_{lag} = \frac{z}{v} = \left(\frac{dz}{4l}\right)\left[\frac{\rho}{\left(\frac{DH}{\delta}\right)c_1(sat)}\right]$$

where z is the length of empty space in the fiber. The permeance of salt across the fiber wall (DH/δ) could be about 10^{-5} cm/sec. Thus,

$$t_{lag} = \frac{(200 \times 10^{-4}\,cm)(100 \times 10^{-4}\,cm)}{4 \times (100 \times 10^{-4}\,cm)}\left[\frac{1\,g/cm^3}{(10^{-5}\,cm/sec)(0.36\,g/cm^3)}\right]$$

$$= 1390\,sec \approx 23\,min$$

This again is in the range which we seek: a pulse of flavor released after a lag.

The pump has the major advantage that the lag does not depend on the flavor used but only on the geometry and materials of the pump itself. This advantage does not seem enough to justify the major disadvantages of complexity and

Table 9.5–4 *Relative benefits of the four ideas for a flavor burst* The most successful is the coating.

	Weight	Sphere	Coating	Pump	Hydrolysis
Gives pulse	0.4	1	8	10	4
Inexpensive	0.2	6	7	2	4
Can work for any flavor	0.2	4	9	10	2
Easy to manufacture	0.2	5	6	1	7
		3.4	7.6	6.6	4.2

expensive manufacturing. While we will continue to consider this alternative, we believe that the idea probably has little value here.

Hydrolysis. Finally, we note that the flavor chosen as a benchmark occurs naturally with a phenolic group. If we can esterify this group, we might make a tasteless compound. If we then add an enzyme or other catalyst to the gum, we could trigger the release of the original flavor. This idea is the antithesis of the pump discussed above. It has the significant disadvantage of being specific to a particular flavor chemical, but it has the advantage of being inexpensive to manufacture. To seem promising relative to the other ideas, it will require much better chemical definition.

Selecting Among the Ideas. These four ideas can be compared by matrix scaling to decide which, if any, merits further development. We choose four criteria for this comparison. The first and most important is that the new product produces a pulse of flavor like that suggested in Figure 9.5–3. While we understand that this pulse will supplement the conventionally released flavor, we feel that this criterion is key, worth at least twice that of any other.

Three other criteria for this new product are cost, flexibility, and ease of manufacture. The new flavor release mechanism should not add more than 10% to the cost of the product. The new system should work for any flavor and not depend on the chemistry of a particular flavor. It should be easy to manufacture, if possible, by slightly modifying existing equipment.

The four flavor delivery ideas discussed here are compared in Table 9.5–4. The scores shown suggest that the coating idea has the greatest merit. Its nearest rival, the pump, is too complex, but its high score suggests thinking about similar chemistry in a different geometry. The coating process should be explored in greater detail.

9.6   Conclusions for Microstructured Products

This chapter describes the design of microstructured products, the most complicated type in our classification. These products often sell for much more than the cost of the ingredients used to make them. That makes this type of product the principal target area for growth of the chemical enterprise. That is why the chapter has special value.

The added value of these products comes for their microstructure, that is, their organization over a size range between 100 nanometers and 100 micrometers. This structure, obtained by specific forms of processing, is frequently not an equilibrium state, but can be either metastable or just kinetically trapped. The science involved is incomplete and only partially known. This makes the design of these products more risky than the design of other types.

In this chapter, we have argued that microstructure design is also hard because the design template has two difficult steps. One of these steps is again the selection between several good alternatives, just as selection was difficult for device design, or molecular product design. This selection involves unfamiliar tools, like time–temperature superposition, and blending shear-thinning fluids with static mixers. The list of tools given is incomplete, a sampling of concepts developed by food science, colloid chemistry, and mechanical engineering. While it is a start, it has big gaps. For example, we have not explained how a set of fluids can be blended to give a particular non-Newtonian viscosity. How to do this blending is not in general known.

The second step of the design template which is difficult is needs, and, in particular, converting needs to specifications. This is hard because many consumer goods are evaluated in terms of qualitative responses, described in popular words and not in scientific variables. The connection between these words and the scientific quantities is often empirical. We have described these empirical connections and suggested how we can go beyond them. Thus we have tried to summarize briefly the entire area of microstructured product design and to provide a start on this type of design.

Problems

1. *Furniture polish*. What is furniture polish? How does it achieve its function?

2. *Bread mix*. Liquid cake mixes are effective and sell widely. Discuss the issues involved in extending this technology to liquid bread mixes.

3. *Eye drops*. The active ingredient in a pharmaceutical ointment for treatment of eye infections is gentamicin sulfate. Gentamicin sulfate is solid but is unstable if stored in contact with water. You want to formulate a product for the efficient delivery of this active ingredient to an infected eye. Write specifications for the product and suggest an appropriate formulation and a possible manufacturing route.

   [*AIChE Journal* (2001) **47**, 2746–2767]

4. *Non-dairy whipped cream*. We want to manufacture a non-dairy whipping cream, with all the physical properties of dairy whipping cream but with health benefits associated with not using animal fat. Write needs for

such a product, including converting consumer preferences into scientific specifications.

5. *Artificial blood.* Some fluorocarbons have very high oxygen solubilities, as much as 10% by weight. They could be used in hospitals and especially on the battlefield, where whole-blood storage is a problem. However, because fluorocarbons are not water soluble, they probably should be used as an emulsion. What should the emulsion structure be? How could it be stabilized?

6. *Ice-cream bars.* Many brands of ice cream are made by coating vanilla ice cream with chocolate. This is popular because of the desirable flavor combination and the satisfying crunch achieved on biting into the ice cream. In southern Africa, the ice cream is manufactured in Cape Town and then distributed throughout the region, including to Johannesburg. The company is perplexed to find that most ice creams sold in Johannesburg have broken chocolate coatings. Explain this observation and suggest solutions.

7. *Surface cleaner in Russia.* A surface cleaner consists of minute abrasive particles suspended in a fluid consisting of water, detergents, and coloring. The product is effective and popular throughout the world. However, one manufacturing company receives frequent complaints from Russia that the abrasive particles sediment, and the product must be vigorously shaken before use. Explain this problem and suggest solutions.

8. *Sustainable salmon farming.* Salmon farming is big business in Scotland, Norway, and Chile. The fish are kept in pens and fed by pellets spread over the surface of the salmon cages with a water cannon. The current pellets are a mixture of food solids and vegetable oil, extruded into 10 mm diameter cylindrical pellets. The oil tends to leak out of the pellets before they are consumed, which wastes nutrition and fouls the water surface. Another problem identified by salmon farmers is that the current pellets "tend to favor the survival of the fattest, to the detriment of the slim." This is because the pellets have a fairly uniform settling rate. The larger, stronger fish are able to reach the surface more quickly and then follow the pellets as they sink. The big, fat fish scare away the slim fish.

   Suggest ideas for improved feeding of salmon.

9. *Low-fat mayonnaise.* Major food companies put a great deal of research effort into producing low-fat versions of their popular but fat-rich products. One such product is mayonnaise. What is mayonnaise? How can it be made with reduced fat?

10. *Non-soap soap.* Since the Dove "beauty bar" was manufactured, the market for non-soap cleansing has grown to over 10^5 tons/year, worth over a billion dollars/year. The motivation behind this product was to make a "non-scumming soap bar." To achieve this, sodium cocoyl isethionate was

chosen as the primary surfactant, with sodium dodecyl benzene sulfonate as co-surfactant. However, the key to the product's success was not only its being non-scumming, but also its being at least as good as soap in all other aspects. The Dove bar had equal "firmness," "lather," "rate of wear," "slipperiness," "mildness," and "fragrance."

Suggest the physical properties which must be controlled in producing the ideal bar of a non-soap cleaner. If prototype non-soap cleansing bars are made in 15 kg batches, suggest how these should be tested to determine which is the best.

11. *Hand-cream manufacture.* A personal-care products company wishes to manufacture a new brand of hand cream which moisturizes the skin to its fully hydrated condition. Market research based on a consumer survey provides the following list of needs:
   (1) The product must be easily applicable and rub in within 30 seconds.
   (2) The product must be "smooth."
   (3) The product must not feel oily or greasy.
   (4) The product must flow when poured from its container, but should not run on the skin.
   Suggest a formulation and manufacturing process for the hand cream.

   [*AIChE Journal* (2001) **47**, 2746–2767]

12. *Improving margarine.* Margarine development has gone through three stages. Before 1960, the focus was on cost reduction. From 1960–1980, the effort was to mimic butter. After 1980, the target was a healthier product. Over this entire history, margarine has remained a water-in-oil emulsion. Solid, crystalline fat stabilizes the water droplets. Liquid fat gives the continuous phase its physical properties. Important consumer properties include "spreading," "appearance," "salt release," "coolness," and "thinness." Problems perceived by consumers are "oily mess," "graininess," "surface moisture," and "brittleness."

   Your company has found that improved health benefits have tended to cause a reduction in product quality, as expressed in its consumer properties. You believe the next generation of margarines needs to combine all the health benefits of current spreads with improved consumer perceptions. What physical properties do you think are important in giving margarine its desirable properties? How could you measure and control these?

13. *Alcopops.* A major drinks manufacturer wishes to enhance the attractiveness of its alcopop drinks, aimed primarily at young women. The alcopops are mixtures of water, food-grade alcohol, sugar, and flavorings. The drinks are sweet to mask the alcohol taste, which many young women do not enjoy, although they do want the effects of alcohol. The drinks are best served as cold as possible, since this masks their cloying sweetness. Your core team believes the best potential product is to cool the bottled drinks to below the

Table 9.14A *Properties of existing mosquito nets.*

	Olyset	Permanet
Manufacturer	Sumitomo	Vestergaard Framdsen
Polymer	Polyethylene	Polyester
Geometry	75 Denier	75–100 Denier
	4 mm × 3 mm mesh	25 holes/cm^2
Active ingredient	20 g/kg Permethrin	3 g/kg Decamethrin

Manufacture	Co-extrude; weave	Extrude; weave; coat
Washes	30	20
Lifetime	5 years	3 years

Notes: A unit of fiber size, a denier is 1 g polymer for 9 km of yarn; one strand of silk is one denier.

freezing point but to keep them in a meta-stable state so the drink remains entirely liquid. With significant agitation, nucleation is initiated and a slush of ice crystals is formed, "as if by magic."

You can try this at home by super-cooling a bottle of water in the freezer and then trying to pour it into another vessel. This works best if the water is pure and the bottle smooth. After a little research, your core team find that some of the flavorings inhibit the nucleation of ice crystals, thus making it easier to store the product as a liquid below its freezing point.

Write final specifications for a prototype to be manufactured on a small scale for consumer testing. What testing would you recommend for the prototype prior to deciding whether to launch it as a commercial product?

14. *Better mosquito nets.* Malaria is endemic in many parts of the world. Malaria infects about 400 million people per year, killing 2 million. One of the best and most cost-effective defenses is a mosquito net draped over the bed. Two such nets, called Olyset and Permanet, dominate the market. Some of their properties are shown in Table 9.14A. One key property is the number of times the net can be washed; another is the resistance of mosquitoes to the insecticide. The Permanet product also contains piperonyl butoxide (PBO), a pesticide synergist that enhances the potency of the insecticide. A major difference is the amount of insecticide and its location. For Olyset, the insecticide is within the fiber; for Permanet, it is on the surface and can be renewed if needed.

As an employee in a development laboratory working for the World Health Organization, you are interested in improving these nets. Prepare a table with perhaps six ideas for potential improvements. Choose two for further development, and select the best one. In this selection, include an analysis of risk and technical calculations.

15. *Medicated patches.* Many over-the-counter medications are available in medicated patches. These often consist of a hydrogel containing the medication and attached to adhesive tape, like a large Band-Aid. These are stuck on the arm, the knee, or the back so that the medication can diffuse across the skin. Those who follow tennis will have noticed Novak Djokovic wearing such a patch on his neck, to help him with breathing problems. Such patches provide relief from pain for many customers.

The medicated patch is often covered with a porous film, called a "non-occlusive backing." This backing allows water from the skin to diffuse across the gel and the backing and hence to evaporate. This prevents the skin from becoming wrinkled. While many manufacturers use this non-occlusive backing, it is patented, and the company which holds the patent has been paid few royalties. This company has now sued to collect these.

Your company makes patches with a non-occlusive backing and has paid no royalties. You must now decide whether to do so or instead to invent an alternative device. So far, you have invented two possible alternatives. Your first "desiccant" invention is to use a non-porous occlusive backing, but to add solid salts like $NaHSO_4$ to the gel. These absorb the water crossing the skin. Many other possible salts are known. Your second "pore" invention is just to punch holes of radius R in the patch.

We want to select which of these inventions is better than the existing benchmark. This benchmark is a porous web, 0.2 cm thick but with 0.1 cm of hydrogel. The water flux through this benchmark is 1000 g/m^2 day for a driving force of 100% relative humidity to an atmosphere of 50% relative humidity. This is 10 times greater than the flux through skin under the same conditions. For the purposes of these estimates, we can assume skin is 0.04 cm thick but the resistance is in the stratum corneum, which is only 20 μm thick. In evaluating these two inventions, include the following:
(1) Calculate the amount of added salt needed to support the "desiccant" invention.
(2) Calculate the hole size needed for the "pore" invention.
(3) Compare the inventions with the benchmark using matrix scaling.
(4) Identify and describe management of the major risks.
(5) Conclude by recommending whether your company should pay royalties or pursue a better invention.

REFERENCES AND FURTHER READING

Arbuckle, W. S. and **Marshall, R. T.** (2000) *Ice Cream*, 5th Edition. Chapman and Hall, New York, NY.

Bourne, M. (2002) *Food Texture and Viscosity*, 2nd Edition. Academic Press, Englewood Cliffs, NJ.

Evans, D. F. and **Wennerstrom, H.** (1999) *The Colloidal Domain*. John Wiley, New York, NY.

Fast, R. B. and **Caldwell, E. F.** (2000) *Breakfast Cereals and How They Are Made*, 2nd Edition. American Association of Cereal Chemists, St. Paul, MN.

Hiemenz, **P. C.** and **Rajagopalan**, **R.** (1997) *Principles of Colloid and Surface Chemistry*, 3rd Edition. CRC Press, Boca Raton, FL.

Kingdom, **A. A.** (2009) *Psychophysics: A Practical Introduction*. Elsevier, Amsterdam.

Schueller, **R.** and **Romanowski**, **P.** (2009) *Beginning Cosmetic Chemistry*, 3rd Edition. Allured Books, Carol Stream, IL.

Stevens, **S. S.** (1985) *Psychophysics: Introduction to its Perceptual, Neural and Social Prospects*. Transaction Books, New Brunswick, NJ.

Wareham, **P. D.** and **Persaud**, **K. C.** (1999) On-line analysis of sample atmospheres using membrane inlet mass spectrometry as a method of monitoring vegetable respiration rate. *Analytica Chimica Acta* **394**, 43–54.

Wesselingh, **J. A.**, **Kiil**, **S.**, and **Vigild**, **M. E.** (2007) *Design and Development of Biological, Chemical, Food and Pharmaceutical Products*. Wiley, New York, NY.

10

A Plan for the Future

This book provides a strategy for designing chemical products. The strategy has two parts. First, it suggests beginning the design with a four-step template of needs, ideas, selection, and manufacture. The needs, usually investigated by marketing, must be converted to concrete product specifications using tools of science and engineering. The ideas, normally around a hundred in total, must be screened to four or so candidates for careful evaluation. Selecting the best of these four requires making approximations of properties. Manufacturing, the final step in the design template, is really a synonym for process design, and it has been made more accurate by digital computation.

The development of this template makes a huge assumption: design has some unknown rules and empiricisms that underlie the creativity involved. The implicit assumption that such rules exist comes from parallels with the development of science. Science makes this assumption in its effort to organize nature, an attitude echoed in the traditional description of scientists as "natural philosophers." The assumption that natural laws exist underlies what is sometimes called "the scientific method." To apply this method, we assume a "law," test its consequences by experiment, and then, if needed, modify the law.

A classic example of the scientific method is Galileo's test of the speed of two falling cannonballs. Today, we understand Galileo's experiment using Newton's second law of motion:

$$m\frac{dv}{dz} = F = mg \qquad (10.0\text{--}1)$$

where m is the cannonball mass; dv/dz is the acceleration, the change in the ball's velocity v with position z; F is the force; and g is the acceleration due to gravity. Because the mass appears on both sides of this equation, it cancels out. This means the velocity of the cannonball does not depend on the mass: a little ball should fall at the same speed as a big one. Galileo is said to have tested this prediction by dropping the two cannonballs from the top of the Leaning Tower of Pisa.

Like science, engineering postulates "laws" and tests these by experiment. However, engineers have traditionally been more willing to modify their laws with empirical corrections, grafted on following experiments. To illustrate this, we return to Galileo's example but with a feather and a coin, traditionally a gold English guinea. The feather and coin are placed in a tube, and the tube evacuated. When the evacuated tube is inverted, both feather and guinea fall at the same rate, just like the cannonballs. However, when the tube is filled with air at atmospheric pressure, the feather floats down much more slowly, the result of friction. Engineers may worry more about friction than scientists do, because friction costs money. At the same time, engineers use the same strategy as scientists: they postulate laws, check the laws by experiment, and improve the laws.

Those in business seem to us to think differently. They understand and respect the laws of accounting, but they seem to have few other rules. They rarely generalize their experiences, trying to describe Galileo's cannonballs, the feather, and the guinea within the same intellectual framework. Instead, they treat each situation as unique, with its own special features. As a result, business schools focus not on natural laws, but on "case studies." Such case studies will normally give a more exact description of the specific case. They will normally not be described as a special case of more general principles, like those of mechanics or thermodynamics.

Chemical product design seems to us to fall between the general but idealized approach of physics and the specific details of a business case study. Like engineering, product design contains a lot of empiricism. Like engineering, this empiricism may speculate beyond current knowledge, as in risk analysis or sensory perception. In writing this book, we have been forcing a way of scientific thinking onto a subject that is only partially scientific. As we have done so, we have repeatedly found our framework didn't work, and we have constantly modified it. After over a decade's work, the framework is better, but it is not perfect. We hope those reading the book will continue to improve the framework, at least for the product area of their own interest. Thus, this book should be regarded as a work in progress.

At the same time, our experience has led us to a set of heuristics that we find guide our development of new products. Sometimes, these heuristics have a basis in science, as in the second law of thermodynamics. Sometimes, they are based in engineering, like learning to concentrate a molecular product before it is extensively purified. Because both authors enjoy bicycle touring, we think of these heuristics as the small toolkit we carry with us when we ride. These tools cannot guide us, but they can make our journey easier. We describe these heuristics, garnered from experience, in the sections of this chapter.

10.1   Using the Design Template

We begin our discussion with the four-step design template itself. This template of needs, ideas, selection, and manufacture is described in Chapters 2–5. Here, we want to abstract this description, to say what our experience has taught us that

Table 10.1–1 *Ten commandments for initial designs* These deal with the four-step design template itself.

1. The four-step design template works well, but it can be expanded if necessary.
2. In using the template, recycle freely – decide explicitly in advance how much you want to return to earlier design steps.
3. The periodic table restricts the properties of chemical compounds, especially of inorganic compounds.
4. Synthetic organic chemistry requires knowledge of at least 1000 reactions.
5. Both mass and mole balances are important, so be sure to make both.
6. Any product should sell for more than the cost of its ingredients.
7. Thermodynamics is much more important to initial design than rate processes.
8. Start any project by reviewing these four phase diagrams:

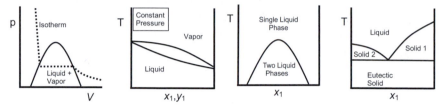

9. Almost all solutions are ideal.
10. Non-ideal solutions include solutes which ionize and mixtures near their consolute or critical points. Gases to watch are CO_2 and NH_3.

is more general even than the generalities of these chapters. We want to identify the parts that we find most important to remember. We want to suggest how, faced with demands for a new product, we can get started.

We have summarized our experience as the ten commandments shown in Table 10.1–1. By expressing our experience in this way, we do not mean to profane the biblical commandments of the Judeo–Christian tradition; rather, we wish to acknowledge the power of this format. While ten is a large number to remember, it seems a minimum size of a list for planning a strategy for design.

The commandments merit discussion. The first two deal with the template itself. The first reminds us that just four steps is arbitrary. For a particular product area, we may need to elaborate on these steps. These elaborations often subdivide the selection step, or they make the manufacturing step much more detailed. Management consultants frequently sell these more elaborate templates and can provide effective guidance for those expert in the detailed science. These tutorials can work well, but they are expensive and can be disruptive.

The second commandment involves deciding beforehand how much we will allow going back to earlier steps within our template. In other words, we should discuss whether if we get well into the selection stage and then get another great idea, we will allow this idea to be included. We have ourselves operated successfully under different rules for this recycle. Once, for a consumer products company, we recycled freely within the template. We really spent an entire week doing all four steps of the template at once, ricocheting between topics wherever the discussion took us. At the end of the week, we had selected five new products that looked commercially attractive. Another time, for a large military contractor, we were strictly forbidden any going back. We spent five months with extensive technical support to define a single potential product.

Each of these strategies can be reasonable. The consumer-products company needed a lot of new potential products and was willing to take substantial risks. The military contractor, who was very risk averse, wanted to ensure the core team did make one product recommendation before a fixed deadline. They wanted to make sure the core team did not get lost in an endless discussion. The second commandment urges that the core team decide how much recycle is acceptable before starting the design.

Commandments 3 and 4 deal with chemistry and the constraints it places on the design. The third commandment just says we cannot have everything we want; we are restricted in practice to perhaps 50 elements. To take a pedestrian parallel, imagine the elements are different Lego blocks. Because we have only a limited number of sizes and shapes, we cannot make any size and any shape of structure. In chemical terms, if we want to change the band gap of silicon by an arbitrary percentage, we cannot do so. We can hope to approximate what we want with gallium arsenide or cadmium telluride, but we cannot have anything we want. Organic compounds allow a somewhat more gentle adjustment of physical properties; but they are still discrete, not continuous. Chemical compounds are limited by the elements in the periodic table.

The fourth commandment deals with organic chemistry, that is, with the chemistry of carbon and of life itself. Organic molecules can be large and complex, so they are more likely to approach anything we want. Synthesizing these target molecules from simple, commercially available reagents is the core of making many molecular products. Doing this synthesis is like speaking a new language whose words are reactions. To be literate in organic synthesis requires knowing about 1000 reactions, just as being understood in a foreign language requires about 1000 words. To be fluent in organic synthesis requires perhaps 10,000 reactions, about the same number of words needed for literary fluency. Thus, this fourth commandment is a warning: don't try organic synthesis yourself unless you are fluent or have accomplished collaborators.

The fifth commandment is a caution to remember the difference between mass and moles. Mass, as weight, is the scale by which we measure our daily lives. You don't go to the market to buy 1.46 moles of sugar; you go to buy 500 g. However, in many cases, especially involving chemical reactions, quantities expressed in moles make more sense. For example, if we burn coke, we could approximate the reaction as:

$$12C + 32O_2 \rightarrow 44\,CO_2 \tag{10.1–1}$$

That is, 12 kg of carbon can be burned with 32 kg of oxygen to make 44 kg of carbon dioxide. While we might understand this equation, we would feel more comfortable with the molar alternative:

$$C + O_2 \rightarrow CO_2 \tag{10.1–2}$$

While this example may be obvious, harder ones frequently go unrecognized in product design. Remember, this fifth commandment: honor both mass and moles.

The sixth commandment, that the ingredients cost less than the selling price of the product, is a check before the core team gets lost in detail. Except for commodity products, we expect the price minus cost will be big. For microstructures, we expect the ingredients will normally cost less than 20% of the selling price. For some molecular products, especially for generic drugs, the margin for profit may be even higher. Making this easy check on how much we can maneuver makes sense.

The remaining four commandments deal with thermodynamics. The seventh says rate processes are less important than energy and mass balances. This commandment, more true for non-commodity products, is hard to accept for many engineers. Engineers can feel that if material is easy, it is not important; and if material is hard, it must be more important. Stoichiometry is easy, and rate processes are harder. Also, some products, especially devices, do depend on rates. Still, mass balances and thermodynamics are the keystone.

Thermodynamic equilibrium, especially of mixtures, is important for product design. How mixtures behave can be reviewed as the four tiny phase diagrams given in the eighth commandment. The first of these, shown on the left, is that of a pure fluid. The volume V per mass or mole is give on the abscissa, and the pressure p on the ordinate. The dome represents the vapor–liquid mixtures. Pure liquid is to the left of the dome; pure vapor is on the right; and the maximum is the gas–liquid critical point. One isotherm is shown by the dotted line. This line begins as a supercritical fluid at high pressure and small volume. If, at constant temperature, the pressure is slightly released, the fluid expands, so the volume increases. When this expansion reaches the dome, the fluid is a liquid at its bubble point: it starts to boil at constant temperature. The dotted line crossing the dome begins as all liquid and evaporates to all vapor. The isotherm to the right of the dome represents the behavior of a gas.

The other three diagrams in the eighth commandment may be less familiar but are more important for chemical products. The second figure from the left is the plot of temperature T vs. composition of the more volatile species in the liquid x and the vapor y. It is the basis of the familiar x–y diagram basic to distillation. A good review of the ideas involved is to consider which curve corresponds to the vapor, which to the liquid, and how the drawing would change if the system had an azeotrope. The next phase diagram, second from the right, is for a liquid–liquid equilibrium at constant temperature but variable temperature T and composition x. The dashed line now represents the splitting of a homogeneous liquid mixture into two phases on cooling. The final phase diagram, on the right, is typical of a mixture of liquid and solid. The mixture's composition is x. Here, we should locate the freezing point of the liquid, the melting point of the solid, and the eutectic. The eighth commandment is really a fast review of phase equilibria.

The ninth and tenth commandments complete this review but for equilibria within phases. Specifically, these commandments suggest considering all solutions as ideal; i.e.

$$\mu_i = \mu_i^0 + RT \ln x_i \qquad\qquad (10.1\text{–}3)$$

where μ_i is the chemical potential of species "i", μ_i^0 is that potential in a reference state, R is the gas constant, T is the absolute temperature, and x_i is the mole fraction. Most solutions are close to ideal. Common exceptions, in the tenth commandment, are ionic solutes and occasionally the gases carbon dioxide (CO_2) and ammonia (NH_3).

The commandments in Table 10.1–1 are the unstated base for starting any product design. We should expect any chemical professional should know them. We do not think they should be formally reviewed before starting the design. However, in our experience, some of the core team will not remember these completely. We urge every professional to review them, at least privately, once the design process is seriously started.

10.2   Specific Types of Products

While we believe that chemical professionals will design effective products, we suspect they may occasionally have trouble getting started. To help, we suggest using more commandments for specific types of chemical products.

COMMODITY PRODUCTS

We begin, in Table 10.2–1, with ten rules for commodity products. These rules are familiar, long used to design processes for commodities. The rules are clearly and accurately detailed in established texts, so only the briefest synopsis is given here.

As the first four commandments in Table 10.2–1 suggest, commodity products are normally made in dedicated equipment, operated continuously, and described by a process flow diagram, where the lines between process units represent pipes. To detail these processes, we normally need thermodynamics and rates for the chemical transformations involved. When the chemicals are organic with molecular weights less than about 200, we can use widely available, easily operated computer programs to estimate thermodynamic properties. The success of these programs was one of the chemical triumphs of the 1980s.

Reaction rate processes are much harder to estimate and normally require experiments. If we have no appropriate experiments, we can use the estimates suggested by the fifth and sixth commandments. These assume the chemistry is so rapid that reagents cannot coexist. They assume any apparent rate is due to the reagents getting together by diffusion. Thus, Commandments 5 and 6 give the maximum rates possible. If the process still looks interesting, the actual rates must be determined by experiment. The rates of heat and mass transfer can usually be estimated from the plethora of correlations developed in chemical and mechanical engineering during the twentieth century. The seventh commandment provides a rule for early estimates.

The last three commandments give details about separations that often involve these rate processes. The eighth commandment urges using distillation whenever possible. The ninth commandment says to use absorption or extraction only if distillation is difficult or impossible, and to remove any solid adsorbent or absorbing

Table 10.2–1 *Commandments for chemical commodities* These are the common basis of chemical process design.

1. The key factor for commodities is cost, so the key design step is manufacture.
2. Operate equipment continuously only if you are making over one ton per day.
3. Use dedicated equipment only if you are making over ten tons per day.
4. Describe your process with a process diagram, which is a flow sheet in space.
5. Homogeneous chemical kinetics is greatly clarified by experiments. If desperate, estimate the maximum possible rate constant k_R as

$$k_R = \left[\frac{8D}{l^2} \right]$$

   where D is the diffusion coefficient and l is the eddy size in the stirred fluid, around 30 μm in liquids.
6. Heterogeneous chemical reaction rate constants are described by

$$\frac{1}{k} = \frac{1}{k_D a} + \frac{1}{k_R}$$

   where k_D is a mass transfer coefficient, k_R is a chemical reaction rate per volume, and a is the area per volume. The maximum rate occurs when k_R is large; the optimum rate is often when the two terms on the right are about equal.
7. Mass transfer coefficients k_D, which are about 1 cm/sec in gases and 10^{-3} cm/sec in liquids, are related to heat transfer coefficients h by

$$k_D = \left(\frac{h}{\rho \hat{C}_p} \right) \left(\frac{D}{\alpha} \right)^{2/3}$$

   where ρ is the density, $\hat{C}_p$ is the specific heat capacity, D is the diffusion coefficient, and α is the thermal diffusivity.
8. Separate by distillation whenever possible, but don't distill if the relative volatility is less than 1.05.
9. Avoid separations that require adding another species, like absorption, extraction, and adsorption. If another species is added, remove it early.
10. Avoid extremes in temperature and pressure. If unavoidable, aim high, not low.

liquid as soon as possible. The tenth commandment says high-temperature processing is better than low-temperature processing. All these commandments have exceptions, but all are supported by extensive experience in the chemical process industries. Commodity chemical process design is well understood.

CHEMICAL DEVICES

The commandments for chemical devices, shown in Table 10.2–2, are based on the same principles as those for commodity chemicals, but they reflect a broadening of objective beyond being the low-cost producer. As a result, as Commandments 1 and 2 say, the key step is not manufacture to minimize cost but selection among alternatives of similar costs but different convenience. These devices are dedicated to one reaction or separation but operate at a much smaller scale than the equipment used for commodities (Commandment 3). They may run in steady state, but they will be turned on and off a lot (Commandment 4).

Table 10.2–2 *Commandments for chemical devices* These products use similar tools as those for commodities but have goals in addition to low cost.

1. The key step in device design is selecting between good alternatives.
2. Criteria for selection are not all technical and include factors other than cost.
3. Chemical devices are dedicated equipment, producing less than 10 kg/day when operating.
4. Operation is normally occasional but may reach steady state.
5. Chemical devices are represented by flow sheets showing flows in space or time.
6. Those devices which synthesize chemicals usually use batch reactors.
7. Recycles are avoided; heat integration is ignored.
8. Devices that effect separations often operate at steady state.
9. The key separation is adsorption; distillation is uncommon.
10. Chemical devices are developed as full-scale prototypes.

Device design is different to commodity design in ways other than scale. As Commandment 5 suggests, the flow sheets for chemical devices may be in space, with lines representing pipes or tubes, like those for a commodity chemical process. The flow sheets may also be in time, with lines representing the next step. Reactions tend to be batch (Commandment 6). Separations tend to be run at steady state (Commandment 8), and these now center on adsorption (Commandment 9). Finally, product details will be refined by building full-size prototypes, which we can rarely afford to do for commodities (Commandment 10). Still, the ideas for chemical devices are basically similar to those for commodities. The biggest change is contained in the first two commandments.

MOLECULAR PRODUCTS

The parallels with commodities are much weaker with molecular products and microstructures. The commandments for molecular products, shown in Table 10.2–3, begin by recognizing the key step is discovery of the target molecule. This key step is the province of medicinal chemistry, pharmacy, and microbiology, subjects beyond the scope of this book (Commandment 1). Most engineers don't have the skills or perspective required in this area, so are unlikely to be effective (Commandment 2). However, once a specific molecule is selected, many engineers and chemists can effectively design the process for its manufacture. Then, as Commandment 3 says, the challenge is choosing among the alternatives that a research chemist will have identified as feasible.

The process designed will be very different to any suggested for commodities. It will use batch reactors (Commandment 6) and run for selectivity for less than eight hours (Commandment 7). Recycles will be rare, almost accidental. Any products should be concentrated before they are purified (Commandment 8). If the product is made with microorganisms, that is, by fermentation, the production fermentor will be much larger than the research fermentor. Scale-up of the fermentor will be at constant power per volume (Commandment 9). To recover

Table 10.2–3 *Commandments for molecular products* Most of the engineering occurs after the target molecule is identified.

1. The key design step in design is discovery, which falls outside our design template.
2. If you are an engineer, don't try to do drug discovery. Drug release or dosage, maybe.
3. Once a target molecule is identified, the key step in molecular or product design is selecting between good alternatives.
4. Because speed is needed, screen alternatives quickly, without detailed thought about cost or efficiency.
5. Expect to use generic equipment shared with other products.
6. Use batch reactors for quality control. Run for selective reaction, not for rapid reaction.
7. The only important reaction rate constant is [eight hours]$^{-1}$, because this means the reactions can be finished in one shift.
8. Concentrate before you purify.
9. Fermentors are often scaled up at constant power per volume. Their stirring is one third of the cost of drug manufacture, product separation is another third, and everything else is the final third.
10. Separate fermentation products by sequential filtration, extraction, adsorption, and crystallization. However, design the separation in the reverse order.

Table 10.2–4 *Commandments for microstructures* These products may be hard to design when their specifications are not based on physical or chemical measurements.

1. These products often have two difficult design steps: needs and selection.
2. In the needs step, try to convert consumer assessments into instrumental measurements.
3. In the selection step, choose between good alternatives.
4. The result of the design will be less a flow sheet than a recipe.
5. Microstructures are often non-equilibrium mixtures whose properties are dominated by the continuous phase.
6. These mixtures have unusual thermodynamics. For example, solubilization of organics in water is a strong, non-linear function of detergent concentration, and solubility of non-ionics in water can decrease with temperature.
7. Check product stability using time–temperature superposition.
8. Consider diffusion as across a thin film or into a semi-infinite slab. Any other model is too complex.
9. To make a mixture of fluids, add the discontinuous phase to the continuous phase while stirring. To make a mixture of a fluid and a solid powder, slowly add liquid to powder.
10. Most microstructured products can be destroyed by shear, so always think about laminar flow.

the product from the fermentation broth, start by considering a sequence of filtration, extraction, adsorption, and crystallization (Commandment 10).

MICROSTRUCTURED PRODUCTS

The design of microstructured products is different from all other chemical products, as summarized by the commandments in Table 10.2–4. First, the design template has two difficult steps, needs and selection (Commandment 1). The needs step is often expressed as consumer goals which have no clear meaning in physical or chemical terms (Commandment 2). Finding these meanings seems to us an area with limited science. The second hard step, selection between alternatives, is not that different to the same hard selection step for devices or molecular products (Commandment 3). Any flow sheet of a process will be a recipe, a sequence of physical and chemical steps (Commandment 4).

The other commandments describe how microstructure chemistry differs from the normal generalizations that we learn for gases, liquids, and solids. The microstructures are complex, often metastable, and involve frequent exceptions to heuristics like "solubility increases with temperature" (Commandments 5 and 6). These metastabilities should be checked by time-temperature superposition (Commandment 7). Diffusion should be modelled as simply as possible, remembering the continuous phase dominates the process. Properties of microstructures depend strongly on the order in which ingredients are combined (Commandments 8–10). These rules for the different product types will normally help get the design process underway.

10.3   Conclusions

This chapter summarizes the arguments that make up the book. It suggests using a design template of needs, ideas, selection, and manufacture as a basis for the design of chemical products. This template is an approximation, a strategy paralleling that used throughout science and engineering. This strategy is different from that used in business, where a focus on case studies allows details of a particular situation to be considered carefully. The template used here attempts generality, not detail.

This design template is similar to, but simpler than, many others described in the literature. Because your job will have specific objectives, you may find it helpful to add more steps to the four used here. Most often, these additional steps will be decisions required in the later design steps. This often makes sense but should be done cautiously. Remember, many more product failures are cancelled too late than too early. This suggests the design template may need more divisions in early steps, like needs, than in the later steps, like manufacture.

In this chapter, we have also tried to extract heuristics, called "commandments," that we have ourselves found help structure our own thinking. We hope you will change these, developing rules of your own that work better for you; please add your own with pride. Our only advice is limit yourself to a small number. Moses might have had less impact had he come down from Mt. Horeb with 63 commandments instead of 10.

Finally, we should re-state the reason for this book. Over the last 50 years, the chemical industry has gone from a growth business to a commodity business. The products of this commodity business, like ethylene, polypropylene, and nitric acid, have social value and will continue to be made cheaply. Newer chemical products can offer new social value as well as greater chances for profit. These include new devices, new active molecules, and new microstructures. After all, a cow is really a chemical reactor for converting vegetable carbohydrates and proteins into microstructured animal protein. The cow is 6% efficient. What other large-scale reactor is so wasteful? Why can't we make good cheap meat without a cow?

This book tries to provide tools to let you start to answer questions like this.

FURTHER READING

Deresky, **H.** (2007) *International Management: Managing Across Borders and Cultures*, 6th Edition. Prentice-Hall, Englewood Cliffs, NJ.

Dul, **J.** and **Hak**, **T.** (2007) *Case Study Methodology in Business Research*. Butterworth-Heinemann, Oxford.

Peters, **M.**, **Timmerhaus**, **K.**, and **West**, **R.** (2002) *Plant Design and Economics for Chemical Engineers*, 5th Edition. McGraw-Hill, New York, NY.

Seider, **W. D.**, **J. D. Seader**, **J. D.**, **Lewin**, **D. R.**, and **Widagdo**, **S.** (2008) *Product and Process Design Principles: Synthesis, Analylsis, and Design*, 3rd Edition. Wiley, New York, NY.

Turton, **R.**, **Bailie**, **R. C.**, **Whiting**, **W. B.**, and **Shaeiwitz**, **J. A.** (2008) *Analysis, Synthesis, and Design of Chemical Processes*, 3rd Edition. Prentice-Hall, Upper Saddle River, NJ.

Wesselingh, **J. A.**, **Kiil**, **S.**, and **Vigild**, **M. E.** (2007) *Design and Development of Biological, Chemical, Food and Pharmaceutical Products*. Wiley, New York, NY.

Yin, **R. K.** (2008) *Case Study Research: Design and Methods*, 4th Edition. Sage Publications, Los Angeles, CA.

Product Index

Subject Index